Protean Behaviour

The butterfly, a cabbage-white,
(His honest idiocy of flight)
Will never now, it is too late,
Master the art of flying straight,
Yet has — who knows so well as I? —
A just sense of how not to fly:
He lurches here and here by guess
And God and hope and hopelessness.
Even the aerobatic swift
Has not his flying-crooked gift.

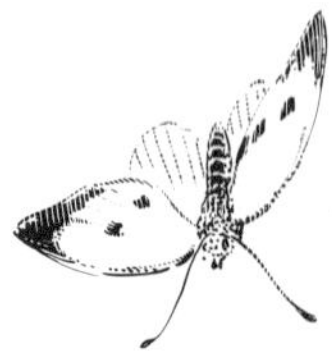

Robert Graves, (1965). *Flying-crooked*

Though this be madness, yet there be method in't.

Shakespeare, *Hamlet* II, ii

Protean Behaviour

The Biology of Unpredictability

P.M. Driver

and

D.A. Humphries

CLARENDON PRESS · OXFORD
1988

Oxford University Press, Walton Street, Oxford OX2 6DP
Oxford New York Toronto
Delhi Bombay Calcutta Madras Karachi
Petaling Jaya Singapore Hong Kong Tokyo
Nairobi Dar es Salaam Cape Town
Melbourne Auckland
and associated companies in
Berlin Ibadan

Oxford is a trade mark of Oxford University Press

Published in the United States
by Oxford University Press, New York

British Library Cataloguing in Publication Data
Driver, P.M.
Protean behaviour.
1. Animals. Behaviour. Predictability
I. Title II. Humphries, D.A.
591.51
ISBN 0–19–857170–4

Library of Congress Cataloging in Publication Data
Driver, Peter M.
Protean behaviour: the biology of anarchization/P.M. Driver and D.A. Humphries.
p. cm.
Bibliography: p.
Includes indexes.
1. Animal behavior. 2. Behavior evolution.
I. Humphries, David A. (David Andrew) II. Title.
QL 751.D725 1988 591.5'1–dc 19 88–15094
ISBN 0–19–857170–4

Set by Graphicraft Typesetters Ltd., Hong Kong
Printed in Great-Britain
at the University Printing House, Oxford
by David Stanford
Printer to the University

Preface

In his splendid television series, *The Ascent of Man*, Jacob Bronowski said 'ask an impertinent question and you are on the way to a pertinent answer'. In 1961 we asked a question which was highly impertinent, for it implied that we, by making observations and taking thought, had recognized a universally-applicable biological phenomenon which had been overlooked by everyone else. This universal phenomenon we now see as the principle of anarchization, although our question all those years ago was simply 'can it be that organisms have often evolved means of behaving unpredictably in order to overcome opponents?'

We began our answer by cataloguing our knowledge in a classification which developed into Chapter 4. We also found that Chance and Russell (1959) had published a short paper in which they coined the term 'protean displays' for behaviours which are unpredictable in certain details. We thus had a precedent and a label for our endeavours. But we little thought that this would develop into a subject so large and with so many ramifications. It is now clear that the protean principle — the principle of anarchization — has implications far beyond the predator–prey situation. It has been our privilege to explore some of these implications for the first time.

In 1971 J.Z. Young noted 'As members of a communicating species what we seek is to make statements that shall interest our fellow-men, and with which they will find they can agree'. In these times of ever-increasing complexity and proliferation of the sciences, with effective communication between disciplines becoming more and more difficult, we have deliberately written this book in the universal English which any literate person can understand, though it would have been easier, in some chapters, to use technical jargon. A knowledge of living systems is helpful to the reader of the more technical sections of the book, but we include no statistics — at least, not in the sense of highly sophisticated mathematical techniques for revealing otherwise hidden relevance. Indeed, our thesis is at present beyond such statistics. They are not needed for the general theory nor suitable for particular instances. We do however look forward to the development of techniques for testing protean theory quantitatively.

In order to give some idea of the types of phenomena covered by our investigations, we begin our chapter sequence with descriptions in everyday language of a variety of protean displays. An unusual form of order exists in this seemingly random assortment of behaviours, and in the rest of the book we explore this paradoxical situation. Chapter 2 explains our approach, and Chapter 3 puts protean phenomena into context with other forms of conflict and competition. In Chapter 4 we classify protean behaviours, with some account of their effects, and in Chapter 5 we describe the unitary components of protean displays. In Chapter 6 we consider how these operate, and in Chapter 7 we discuss their possible evolution. Chapter 8 extends the protean principle to more static phenomena: the distributions of individuals in space and their variations in form. The final chapter summarizes the foregoing, illustrates the relevance of the protean principle to other areas of biological study, and begins a dialogue between protean theory and the study of human behaviour and evolution.

We have tried to make each chapter intelligible as a separate unit. At the same time the chapters follow a logical sequence and each of Chapters 3 through 8 has a summary. The book may thus be tackled piecemeal, skimmed for its essence, or perused more thoroughly in formal sequence. However, as our main theory concerning the universal application of anarchization is somewhat unorthodox, we recommend the reader to follow the chapters as they are presented, for they form a natural progression of information and argument. The meat may be easier to swallow following the aperitif, hors d'oeuvre, and fish. But those of sound digestion may skip the first three chapters, except for the definition on p. 36, and proceed directly to the weightier matters in Chapter 4 and after. We hope they will be with us for our special dessert in Chapter 9.

As another unorthodox scientist, Hans Selye (1956), wrote in his important work *The stress of life*, 'To discover does not mean to see, but to uncover sufficiently that many can see and continue to see forever'. We could not have uncovered the principle of anarchization without the work of so many others who, like us, have amassed volumes of painstaking observations of the living world. Without all this other work to compare with our own we could not be so sure of our thesis. Naturally, we hope that others will largely agree with our statements although, in Gilbert White's (1788) words 'we do not pretend to say that they are perfectly void of mistake, or that a more nice observer might not make many additions, since subjects of this kind are inexhaustible.' We would therefore be grateful for the identification of errors or omissions. For when we are correct we have the privilege of teaching others. But if we can be corrected we have the continuing privilege of being taught. Whatever the results, we trust that we will stimulate exploration into the fascinating realm of functional unpredictability.

PMD

1988 *DAH*

Acknowledgements

During the years over which this book has been developed we have, of course, received help from a large number of people. We are grateful for much encouragement, assistance, criticism, and advice, even if the last has not always been taken. We cannot mention everyone who has been involved — we trust that those not named will take our gratitude as read — but the following have contributed significantly (we omit titles for reasons of space and, in this context, irrelevance): B. Bailey, N.C. Blurton Jones, D. Bowring, P. Boyle, L.P. Brower, L. Butcher, D. Bygott, A.J. Cain, N. Cairns, M.R.A. Chance, A.H. Clarke Jr., M.J. Coe, A. Cole, M.I. Collins, W.C. Corning, J.H. Crook, M.J. Cullen, G.Cunliffe, J.E. Davies, A.J. Driver, D. Driver, M.J. Dunbar, M. Evans, J.C. Fentress, D. Goodwin, E.C. Grant, T. Halliday, J. Hanby, W. Hansen, A. Hardy, C.J.O. Harrison, T. de Havas, W.G. Hayes, D.A. Henshilwood, C.C. Hemmings, A.L. Hochbaum, M.E.H. Howe, D. Ilsley, P. Jarman, P. Hope Jones, H. Kegnasty, M. Kenchington, B. Kilmer, H. Kruuk, H.T. Lees, K. Lorenz, J.V. McConnell, H.J. Midwinter, W.G. Monteith, D. Morris, S.R. St. J. Neill, M.M. Nice, C.W. Nicol, G. Parker, O.S. Pettingill Jr., J.C. Pernetta, J.R. Platt, W. Pollock, J.W.S. Pringle, A. Rapoport, G. Robinson, J.S. Roote, W.M.S. Russell, J. Schaeffer, P.J. Sheldrick, P.G. Shinkman, L.B. Slobodkin, J.N.M. Smith, J. Stanley, N. Tinbergen, B.N. Walker, B. Webster, M. Wheeler, C. Whitehead, R.S. Wilcox, and B. Wolf.

In particular, we would like to thank M.E.H. Howe, C.W. Nicol, and B. Wolf for invaluable research and general assistance and support; D. Morris, H.J. Midwinter, and W.M.S. Russell for stimulating discussion and criticism; N. Tinbergen and J.R. Platt for support and facilities; and T. Halliday, B. Bailey, C. Whitehead, and R. Gillmor for illustrations. Significant parts of the MS, at various stages, have been read by D. Bygott, J. Hanby, W.H. Hansen, A.J. Cain, G. Parker, and N. Tinbergen, and we are grateful to all for their advice. We would also like to thank the staff at Oxford University Press for their assistance and support. Most of all we are very pleased to thank our wives Kay Driver and Dorothy Humphries for their long years of help and encouragement.

Some of this work was done while D.A.H. was working in the Department of Zoology, University of Newcastle, at the Ethology Laboratory, Uffculme Clinic, Birmingham, and the Department of Biological Sciences, the University of Aston in Birmingham, and while P.M.D. was working in the Department of Zoology, McGill University, the Department of Extra-Mural Studies, Bristol University, the Mental Health Research Institute, University of Michigan, Ann Arbor, and the Animal Behaviour Research Group, Oxford University. We are grateful for all the facilities provided. Additionally, P.M.D. is pleased to acknowledge support from a Carnegie Arctic Research Scholarship and the Arctic institute of North America while at McGill, a Post-Doctoral Fellowship and Research Training Grant while at Ann Arbor, and the NERC while at Oxford.

Notwithstanding the assistance acknowledged above, the authors have supported most of this work themselves; the responsibility for any errors therefore resides largely with them.

Contents

List of Figures xiii

List of Tables xv

1 Introduction and variations 1

Some well-known examples of behaviour which illustrate the protean principle

Variations on a theme 2
The single erratic display
The multiple erratic display
Mobbing behaviour
Scatter
Injury-feigning
Convulsoid behaviour
The protean principle in human situations 9
Conclusion 9

2 Context and attitude 11

The authors' approach: evolutionary and multidisciplinary

Systematic unpredictability 11
The authors' interests and attitude 14
Conclusion 17

3 Strategies and tactics 20

Animal conflict and opportunities for anarchization. Protean behaviour defined

Problems of focus and definition 21
A survey of relevant animal interactions 24
Predator–prey interactions
Co-ordination and control systems: the targets of protean weaponry 33
Protean behaviour: a definition 35
Summary 40

4 Catalogue and classification 40

The forms of protean behaviour

Primary and secondary protean displays 41
The single erratic display 44

Derived erratic displays 72
The multiple erratic display
The united erratic display
Scatter
Distraction
Immobility
Convulsoid behaviour
General comments 115
Summary 116

5 **Elements of unpredictability** **119**
The unitary foundations of protean displays
Change in speed 120
Change in direction of movement 122
Changes in form, colour, sound, timing, and type of movement 124
Variation in the distance between points of change 132
Change in number, behavioural role, or relative position within a group 132
Differences between the displays presented to the predator on different occasions 135
General comments 137
Summary 138

6 **Conflict and confusion** **140**
The mechanisms whereby opponents are confused
The validation of protean theory 140
Protean displays as inducers of conflict 141
Factors influencing conflict intensity 147
Experimental evidence for conflict induction 149
Factors connected with prey density and numbers 153
Protean displays as learning blockers 156
Protean displays as part of the behavioural mosaic 158
Summary 159

7 **Evolution and anarchization** **161**
Avenues for the development of protean displays
Injury-feigning 163
Mobbing 170
Anarchization 172
The united erratic display 174
Single erratic displays 176

Convulsoid behaviour 182
Immobility displays 188
Anarchization and the diversity maturation hypothesis 192
Antiprotean devices 194
 Aerial plankton feeders
 Terrestrial insectivores
 Aquatic predators
 Quadrupedal hunters
Summary 203

8 **Dispersion and diversity** **206**

Irregularity of occurrence and diversity of attributes at population level

Searching image and area-concentrated search 207
Scattered distribution and the confusion of predators 208
Clumped dispersions in aquatic prey 212
Some processes mediating searching image formation 215
Adaptations which counter searching image behaviour 216
 Change of colour and form
 Dispersions
 Cryptic fragility
 Polymorphism
Heterosematism 220
Apostatic polymorphism and reflexive selection 223
Ruffs, anarchization, and evolutionarily stable strategies 228
Some further avian polymorphisms 231
Anomaly and aspect diversity 233
 Aspect diversity in web-building spiders
Anarchization of appearance in insects 236
Behavioural polymorphism 237
Display diversification — a universal strategy? 240
 Display diversification in avian vocal mimicry
Blood parasites and beech nuts 242
General comments 246
Summary 247

9 **The integration of irregularity** **249**

Proteus and human affairs. Summary and suggestions

Proteus and altruism 253
Proteus and deflection 257
Proteus and human activities 261
Proteus and human evolution 269
Conclusion 280

References 281

Author index 311

Species index (English) 318

Species index (Latin) 325

Subject index 330

List of Figures

1.1	The flight path of a rabbit pursued in the open	3
1.2	The multiple erratic display of impala	5
3.1	Diagram showing some of the factors involved in defining an organism's 'niche'	24
4.1	The non-protean irregularity of threat of attack experienced by a 'sky-pointing' gannet	42
4.2	Hypothetical comparison between protean flight pathway and fortuitously erratic pathway	43
4.3	The single erratic display of snipe	49
4.4	The single erratic display of different snipe species	50
4.5	Lapwing song flight	51
4.6	'Whiffling' descent of geese	52
4.7	The single erratic display of stickleback	53
4.8	The erratic display of dancing midges	55
4.9	The single erratic display of mosquito	56
4.10	The single erratic display of psychodid fly	57
4.11	Single erratic reaction of lacewing to bat	58
4.12	Reactions of moths to ultrasound	59
4.13	The single erratic display of silver-fish	60
4.14	The single erratic display of collembolan	60
4.15	The single erratic display of sandhoppers	61
4.16	The 'crazy dance' of weasel hunting vole	64
4.17	'Play-fleeing' of rabbit	65
4.18	Escape behaviours of cuttlefish	68
4.19	Protean scent trail of fleeing fox	73
4.20	Reactions of grey mullet disturbed in rock pool	75
4.21	Lapwing: song flight routes of four territory owners	76
4.22	The united erratic display — general form	77
4.23	Tactics used in encounter between peregrine and startings	78
4.24	Wader united erratic display — detail	79
4.25	Wader wing and tail markings: principal patterns	81
4.26	The scatter display — typical appearance	86
4.27	Covey of partridge scattering from sparrowhawk	87
4.28	'Inflorescence' of flattid bugs	91
4.29	Diagram of shorehopper display	92
4.30	Injury-feigning postures of black-winged stilts	97
4.31	Wing-thrashing injury-feigning of old-squaw duck	98

4.32	Old-squaw injury-feigning—field example	99
4.33	Distraction display of bustards	102
4.34	Mobbing distraction of snake by hornbills	103
4.35	Mobbing-type behaviour of mongoose	104
4.36	Mobbing of fox by gulls	105
5.1	The single erratic display of duck flea	120
5.2	The single erratic flightpath of chironomid fly	123
5.3	The jump of the skipjack beetle	126
5.4	Flightpaths of jewel flies	127
7.1	Model of the organization of distraction display in Anatinae	166
7.2	Model of the organization of fox feeding behaviour, including the 'crazy dance'	185
7.3	Bat countering single erratic display of moth	196
7.4	The bubble net of the humpback whale	197
7.5	Lurcher dog catching rabbit	203
8.1	Some morphs of *Cepaea*	221
8.2	Ruffs and reeve	228
8.3	Some examples of bizarre Hemiptera	238
8.4	The South American alligator bug	239
9.1	Some racquet-tailed birds	259
9.2	Protean deflection devices in flying targets	260

List of Tables

2.1	Information and entropy compared	13
7.1	Combination and succession of postures in fox crazy dance	187
9.1	Summary: protean phenomena in biological armament	250–1

1
Introduction and variations

'Consistency is the last refuge of the unimaginative.'
Oscar Wilde

Conflict between animals is a constant feature of the living world. All animals, from protozoans to people, are a threat in some way to others. They may fight among themselves for territory, food, or mates. They may compete with individuals of ecologically similar species for space or 'niche'. From time to time they may be forced to join battle with others for their lives in predator–prey encounters. And in man — only in man — unnecessary strife is common.

In the animal kingdom as a whole these conflicts and competitions, though of baffling complexity and variability to the biologically naïve observer, display certain definite patterns and repeated categories of adaptation. These patterns in turn reflect the various kinds of biological equipment possessed by the protagonists. The details may vary enormously between species but again and again we encounter already-known tactical principles and recognize familiar weapons in new guises. The principles and materials of conflict can therefore be ordered and classified, revealing parallel developments of weaponry in quite unrelated animals. Even the human arms race, sophisticated though it may be, reflects many tactics and techniques common in the animal world, some of which were possessed in simpler form by primitive man. In a sense the human arms race is simply a special example of a much broader and incredibly more complicated biological arms race that has been experiencing innovation, elaboration, and diversification since the dawn of life by the slow processes of natural selection within innumerable competing species populations.

Our survey of the biological arms race points very strongly to a widely ignored, yet fundamental feature which is common to many categories of conflict and is expressed in diverse ways, from individual appearance and behaviour patterns to group behaviours and population dispersions. Though manifested in many different guises this feature is a basic principle whose widespread importance has so far not been recognized. It is the *protean principle* whereby the effectiveness of weapons or tactics is increased by endowing them with characteristics that cannot be predicted by the opponent, thus conferring a distinct advantage upon the user. A weapon is normally thought of as possessing a definite action: its operation is regular and essentially similar from one occasion to the next. But, however powerful the weapon, this very regularity of operation renders it vulnerable to prediction and therefore to the development of effective countermeasures. This vulnerability is reduced by the protean principle that decreased predictability undermines the effectiveness of countermeasures.

The scope of this principle is far-ranging; our work suggests that animals, including man, have been unconsciously benefiting from it from earliest times. The reader may reasonably ask why — if it is a fundamental and universal feature of the biological arms race — the importance of this principle has not been recognized before. The answer lies in the very nature of the scientific method. Science has largely developed as the study of regularities, and scientific enquiry has therefore been ill-attuned to the detection of irregular information. Protean phenomena are apparently inconsistent; their irregularity and unpredictability render them at first sight unsuitable subjects for scientific study. As Barber (1961) noted, 'No part of the aim of normal science is to call forth new sorts of phenomena; indeed those that will not fit . . . are often not seen at all. Nor do scientists normally aim to invent new theories, and they are often intolerant of those invented by others.' We believe, with Kuhn (1970) that such attitudes may be defective. This point is taken up later. Meanwhile, this chapter is designed to give a broad generalized introduction to the operation of protean phenomena in nature so that the classification and analyses which follow can be more readily understood, however disparate the examples may originally appear.

Variations on a theme

The single erratic display

Rabbits, like many other animals, escape from fast-moving predators, when surprised in the open, by 'jinking' in the *single erratic display* (Fig. 1.1). This looks like dodging, but isn't. Dodging is a technique for a last-second evasion by a sudden switch in course away from the pursuer's path. In contrast, the rabbit's twisting and turning single erratic display is seen in encounters covering a considerable range of predator-prey distances. Sometimes the rabbit will jink at the very last moment, and this does look very like dodging. At other times the jink takes place long before the predator is within striking distance. And repeated jinks may occur, even when the predator does not pursue. More often than not a jinking rabbit escapes its pursuer. This technique is so effective that all groups of animals with fast-fleeing species have incorporated it into their antipredator behaviour.

Fig. 1.1 The flight path of a rabbit pursued in the open; a very common example of *the single erratic display* — the most universal of all escape mechanisms. The major features of the display vary unpredictably from one occurrence to the next.

The multiple erratic display

Many fishes of the shallow seas, like grey mullet, sand eels, and 'whitebait' — a general term covering many species: a multitude of fins — commonly swim inshore in the summertime and may be trapped in small pools when the tide recedes. Under these conditions the fishes' normal anti-predator reaction of bunching together and swimming in one fast-moving zigzagging school may be changed by the narrow confines of their enclosure. When disturbed the school may zigzag with slightly increased speed, but it soon breaks up with the individuals darting about the pool as if in a frenzy, showing no co-ordination at all between individuals. If an attack is not pressed, then they may regroup, only to explode again into wild confusion when threatened.

Those who conduct marine biology courses, and their students, know how exasperating it can be to try to catch fishes with a small hand net. It is not merely their speed that makes the fishes so elusive; it is, in particular, the unpredictability of their individual erratic movements, their criss-crossing paths, and their continually varying visibility that so often make it impossible to fix one's attention on any one fish for long enough to develop an effective attack with the net.

Such behaviour patterns — if these variably structured happenings can be called patterns — occur in many kinds of animals where groups confined to small areas by physical barriers or behavioural constraints are attacked by predators. One of the best eyewitness accounts of this kind of display is that of Sir Percy Fitzpatrick (1907) in his collection of South African hunting stories, *Jock of the bushveld.* When stalking a group of impala he suddenly found himself in their midst; and just as suddenly they burst into action:

> The place was a whirlpool of racing and plunging impala; they came from every side and went in every direction as though caught suddenly in an enclosure and, mad with fear and bewilderment, were trying to find their way out. How many there were it is quite impossible to say: the bush was alive with them; and the dust they kicked up, the noise of their feet, their curious sneezy snorts, and their wild confusion completely bewildered me. Not one stood still. Never for a moment could I see any single animal clearly enough or long enough to fire at it; another would cross it; a bush would cover it as I aimed or it would leap into the air, clearing bushes, bucks, and everything in its way, and disappear again in the moving mass. They seemed to me to whirl like leaves in a wind eddy; my eyes could not follow them and my brain swam as I looked.

This is as good a description of the structure and effects of a multiple erratic display as one could hope to see. We illustrate this behaviour in Fig. 1.2.

Mobbing behaviour

A group of people 'ganging up' on someone is an ugly sight. They jostle and snarl with menacing looks and threatening gestures which, sooner or later, become open aggression unless redirected or put down by superior weaponry. The victim may bluster but generally is cowed, simply trying to fend off as much of the attack as possible. In humans this behaviour is grotesque, but it has elements in common with the mobbing behaviour of other animals. Small perching birds, for example, will gather round an owl if they find one during the day, hurling avian abuse (locatory alarm calls) from all directions and seldom suffering any effective counter-attack. Swallows will mob birds of prey in the air in a similar manner, flying in from all directions and 'buzzing' the victim until it turns away — although it is much superior in individual armament. Gulls are just as effective in driving away people or foxes from their nesting colonies. They swoop and chatter, some-

Fig. 1.2. The multiple erratic display of impala. This is an uncoordinated group of individual single erratic displays, though there may be a common directional tendency.

times striking with foot or wing, achieving *en masse* what would be far beyond them as individuals.

But it is not a simple summation of the physical strength of the attackers that enables them to overcome an otherwise superior opponent. Common to all these examples of mobbing is the difficulty of predicting who is going to attack next, and from which direction. The victim — an accurate term for the central animal in the encounter, though it may be a snake or a wolf — has to divide its attention among all of its attackers and therefore cannot select any individual for a punishing counter-attack. Such a counter-attack would necessitate a concentration upon one individual to the detriment of defence against the others. The best policy is to avoid open conflict.

A simple example of the effectiveness of introducing an unpredictably irregular element into an otherwise equal and regular biological competition is given in one of the authors' class demonstrations. It represents a simple mobbing situation with one central figure and two attackers. The attackers aim blows alternately at the central figure — representing the predator — who blocks the blows in standard karate fashion. If the speed of the attack and defence activity is increased until the regularly alternating attacks are as fast as they can be without overcoming the predator's defence system, then we have a representation of an equal and maximally developed armament situation. This is similar to many examples of 'regular' weaponry in the biological arms race.

The demonstration of successful defence is now stopped and it is explained that the attackers, who have been striking with a regular left–right–left–right sequence, are to go away and decide upon some unpredictable change in their pattern of striking. The predator remains ignorant of when they will institute this change. The predator then re-starts the sequence, but at the pre-arranged point, when the weapons are otherwise equal, the attackers unpredictably change their strike pattern. For example, the sequence may now be left–right–left–right (change) right–left–right–left. When the change occurs, one or more blows land on the predator (much to the enjoyment of the audience). Many predator–prey interactions in the wild can be seen to involve elaborations of this basic principle.

Scatter

Ducklings of all kinds, diving species or not, will 'crash dive' when attacked or startled. Eider ducklings at least, from their very first hour in the water, will submerge in less than half a second if anything untoward suddenly occurs in their immediate environment — if the mother duck, or a human parent-object to whom the ducklings have been imprinted (which is how we know about it) makes a sudden swirl in the water; if a gull or hawk swoops upon them; if a gun explodes in their vicinity; if they are attacked by a bigger duck, parent or otherwise. In all these circumstances, and other similar ones, the ducklings disappear beneath the surface in the traditional 'fraction of a second'. But this defence is only partial; the birds must soon reappear at the surface to breathe. It would seem that all a predator has to do is wait for its intended prey to exhaust its air supply, then strike as it surfaces.

The mere speed of the crash dive however is not the whole story of the ducklings' protection by submergence, for after the ducklings have dived they scatter. Any predator waiting at the point of disappearance for ducklings to reappear is therefore likely to be disappointed. Moreover, a predator cannot learn where the prey is likely to reappear because each individual of the scattering bunch follows a highly erratic course under water, like the jinking of a rabbit. Occasionally in fact a duckling's escape path turns back upon itself and the bird may reappear close to the original point of submergence. But that is all to the good, for the predator has usually moved on. The pattern of underwater scatter of ducklings is therefore a non-pattern. The only time a predator can deliberately overcome this tactic is when the water is calm and clear, for then a bird flying above may be able to see the ducklings beneath the water and grab one as it surfaces. But this happens seldom, for natural water is not often both clear and calm. The underwater scatter of ducklings has therefore transformed the vulnerable clustering of prey into a wide and irregular spacing. This so defeats detailed prediction by predators that the otherwise weaker prey often escapes.

Injury-feigning

One of the more complex and confusing aspects of animal behaviour which has puzzled naturalists for centuries is the injury-feigning display of birds and other animals. This display occurs as an antipredator response in many species during the breeding season, most strongly in birds when they have hatching eggs or recently emerged young. The parent flops away from the brood as if crippled and provides the predator with a clearly attractive stimulus. The predator's attention is drawn away from the brood, even if it has already discovered that prize, and it follows the apparently injured bird further and further away from the vicinity of the nest or brood until its patience is exhausted. By this time the predator is so disoriented that even if the prize has remained where it was when action started, it is much less likely to be found. If the prize was a nest of eggs, the predator may return and search for it, perhaps successfully, but young birds will have had plenty of opportunity to scatter and hide or take to water, and have a good chance of escape.

In order for the predator's attention to be held long enough for all this to be successful the injury-feigning bird must maintain a high degree of both avoidance and attraction. A human equivalent would involve a consciously delicate balance between enticing and fleeing. The naturally erratic behaviour of an injured animal would seem to facilitate the performance of this double role. A bird with an injured wing or leg will naturally flutter or stumble along, occasionally falling over or stopping momentarily for breath. A water bird will be expected to thrash through the water, with occasional half dives and changes of direction. And an extra burst of speed from time to time is quite natural as the apparently doomed bird gathers what appears to be its failing strength. These variations in the behaviour of the prey animal, though somewhat unpredictable in their occurrence, are generally to be expected in an injured animal, and the predator follows in apparent eager anticipation of an easy capture.

However, a predator following an injury-feigning bird usually finds that it is not easy to capture after all. If the predator increases its effort, the prey seems to gather a fresh supply of energy from somewhere.

And somehow the apparently uncoordinated changes of speed and direction of the prey's movements make it very difficult to catch. Several observers have reported instances of an injury-feigning bird escaping its pursuer, but nobody seems to have observed one being caught.

Controversy continues as to the mechanism and evolution of injury-feigning behaviour. Is the bird caught in its own neurological conflict between fleeing and attacking? Is its co-ordination blocked by fear? Has it evolved a complex form of behavioural mimicry? Or is natural selection maintaining a mechanism which produces a delicate balance between confusion and order in the bird's brain? Whatever the answer, injury-feigning displays provide another, rather unexpected example of the value of being unpredictable.

Convulsoid behaviour

Injury-feigning displays are not the only form of bizarre behaviour that has protean characteristics. Moths, lizards, and mice are harmless creatures compared with the fierce predators that hunt them. Yet when attacked they may perform wild, convulsive displays whose sudden onset and unexpectedly changing details effectively deter the predator from further attack for a while, and they may escape. Many of the dull-grey-brown moths that hang on fences, walls, and clothes in cupboards, drop to the ground when touched. There they may flop, skip, and jump as if demented or burnt in a candle flame. Or they may fall and remain motionless as if dead; but even then they will go into their St Vitus' dance routine if poked. The immediate effect is startling. One may experience this display a hundred times and still be so startled that one's attempts to pick up a skipping moth are greatly hindered. One is so unprepared for the convulsive outburst from such a small creature that one is — momentarily at least — taken aback. And that moment may be enough for the creature to take cover or flee. The general pattern of the convulsion is rather like that of a vigorously flapping fish out of water; a type of behaviour notoriously difficult to counteract.

Conventionally, convulsions are usually thought to be some kind of malfunction, symptoms of a horrible disease, yet we shall argue that they have undoubtedly saved many a prey animal from capture, and that convulsive behaviour is widespread in the animal kingdom, not as a malfunction but as a useful adaptation. It is only social custom to consider all convulsions as nasty. This is a human aversion to departures from the behavioural norm. Epileptic or hysterical fits in a member of a family are, for example, often kept secret. But do these convulsive states in man always reflect damage to the machinery of the brain? Or can the study of adaptive convulsions in animals shed any new light on these unfortunate human conditions?

The protean principle in human situations

The protean principle operates also in *specifically* human situations including sport and war. A disturbing contemporary demonstration of the techniques of the mythical Proteus in confusing his opponents is seen in paramilitary subversive activities. In all cases one knows more or less what is going to happen in particular circumstances at one of a variety of times and with varying intensity and effect. But we cannot deal easily with these problems: their operation and occurrence have too great an element of unpredictability for us to be able to plan effective safeguards. One cannot protect continuously all the possible targets against all the many alternative offensive actions which even a weak dissident group might mount: the manpower requirement and cost would be prohibitive. However much one knows about the structure of the various behaviours which may occur, a lack of information about which particular item is to be selected and especially its timing and siting means that much knowledge is wasted because it cannot be applied appropriately, i.e. at the right moment and in the right place. Effective guerrilla warfare is based on the same principles. The only really effective countermeasure is an efficient intelligence system giving a greatly increased knowledge of the behaviour of the protagonists so that predictability of their activities can be improved. As in all other categories of warfare, human or otherwise, the employment of weapons or tactics with unpredictable characteristics is of great advantage to the user, being very difficult to counter.

Conclusion

In the simple examples presented in this chapter we see the protean principle at work, either making it hard for the opponent to decide where the displayer is going to be next, or what it is going to do, or which individual within a displaying group will be next to become temporarily vulnerable or threatening. These examples are not new to the biological literature. Similar instances have been described many times. But attention has nearly always been focused on their *regular* aspects: on the fact that pursued animals flee or dodge, that grouped animals spread out when attacked, that mobbing deters predators or at least advertises their presence, that disturbed moths show odd uncoordinated movements as if nervous function is deranged, that terrorists cause damage and fear.

The significance of the protean nature of these displays has largely been unrecognized. Even where their irregularity has been specifically commented upon, it has been treated as a minor aspect of that particular behaviour, an oddity viewed within a narrow framework, not as an important principle of general applicability to many behaviours of quite different outward form. We shall argue that the protean principle has in fact been of service again and again in the living world, in many different kinds of animals — and even plants, and in conflicts of widely differing form and weaponry. There seems to be an inexorable logic in biological conflict which, through the slow action of natural selection or the swifter accumulation of human cultural experience, throws up similar sets of solutions for similar problems, wherever they occur. Protean phenomena constitute one such set — a highly significant one.

Whilst it is perhaps curious that protean phenomena have been so well described but so poorly appreciated, it is also fortunate for the development of our thesis that some of the theoretical analyses and relevant experimental investigations, which might have taken years to carry out, have been done already in other contexts; for other reasons, certainly, but with scarcely less relevance for that.

What we have attempted in this book is to develop the broader theory, adding new ideas, but above all drawing together the diverse evidences and theories to show their significance in the light of the protean principle.

2
Context and attitude

'The quest for certainty blocks the search for meaning.'
Erich Fromm (1947)

'It is not informative to study variations of behaviour unless we know beforehand the norm from which the variants depart.'
P.B. Medawar (1973)

Systematic unpredictability

This book is about the application of unpredictable phenomena in the living world. But the concept of unpredictability is rather a difficult one to embrace. Unpredictable with respect to what, and to whom? And what is the relationship with randomness, and probability? We cannot — and need not — deal with the mathematics of this, but some clarification is necessary.

The word 'unpredictable' has been in frequent use only since the 1880s (*OED*), and its usage has accelerated over recent decades, presumably because with the growing pace of change in modern society the need to predict events in social, economic, technological, and military spheres has become increasingly important and the failures of prediction correspondingly more significant. Yet unpredictable occurrences have been of major significance in the natural world for a very long time indeed — almost certainly from the beginning of life and probably since the origin of matter. Before life began, unpredictability *sensu stricto* did not exist, because there was nothing living to attempt the act of prediction. Randomness, however, in the sense of fortuitous chance occurrence, operated then, as now, in the patterns of juxtaposition of atoms and molecules. When one — or some — of these meetings resulted in the appearance of living things with their capacity of monitoring the environment as a necessary part of the maintainance of their steady state, predictability came into being. Regular and systematic phenomena, from celestial cycles to the physical composition of the environment, and from the distribution of food to the behaviour of enemies, could all be responded to adaptively. Regular problems could be met by systematic solutions.

Unpredictable phenomena have always been more difficult to deal with, whether in the form of minor irritations such as a freak storm or rock fall or the unusual behaviour of an enemy, or as major upheavals. One of the most significant operative features of early human evolution must have been the development of an interest in the unpredictable and the ability to develop countermeasures. This is still seen in many aspects of human behaviour today. We shall return to this point in the final chapter.

Certain unpredictable phenomena may, unavoidably, be catastrophic, as in major environmental calamities, and in mutations. It might even be said that mutations are the ultimate protean phenomena for, in the natural state, they are always unpredictable and usually of considerable significance to the mutating organism, but they would, in fact, fall outside our definition (p.36).

Now, as an example of the philosophical difficulties which may arise with respect to the significance of ideas of randomness and unpredictability, let us take the concept of entropy and its relevance to the protean principle.

The word 'entropy' was originally introduced as a means of facilitating the calculations and clarifying the results of thermodynamics. In physical terms the entropy of a system is generally regarded as a measure of its degree of disorder — the *OED* definition of the term. In the words of Miller (1965): 'The disorder, disorganisation, lack of patterning, or randomness of a system is known as its *entropy*. It is the amount of progress from improbable to probable states'. It might be thought that, as protean phenomena are apparently disorganized, random, and without pattern, they could be said to be of high entropy. But this is not so.

The probable state referred to by Miller is that which all materials move towards if they receive no energy input. For example, if one opens a container of gas in a closed room of air, the interdiffusion of molecules results in a uniform distribution of all molecules to produce equilibrium throughout the room. That equilibrium is the condition of maximum probability and disorder, and greatest entropy for that system.

But a 'probable' state is not necessarily the same as a 'natural' state. J.Z. Young (1971) notes that although most people regard growth as a natural process it is in fact unnatural 'since it involves incorporation of less ordered materials into the highly ordered living systems, contrary to the general tendency for increase of entropy'. The complex nature of organisms is thus regarded as improbable or unnatural, their breakdown into basic units only being avoided by the organized expenditure of energy.

However, although organisms may be unnatural or improbable in the basic physical sense, they are quite natural to the organisms them-

Table 2.1 Information versus entropy

Information	Uncertainty
Negentropy	Entropy
Signal	Noise
Accuracy	Error
Form	Chaos
Regularity	Randomness
Pattern or form	Lack of pattern or formlessness
Order	Disorder
Organization	Disorganization
Regular complexity	Irregular simplicity
Heterogeneity	Homogeneity
Improbability	Probability
Predictability	Unpredictability

From Miller (1965), p. 201.

selves and to other organisms reacting to them. This natural orderliness is maintained through evolutionary time by the gene-flow, and is manifested by the species' taxonomic characters and species-specific behaviours. It is characterized by information in the technical sense, i.e. 'the degree of freedom that exists in a given situation to choose among signals, symbols, messages or patterns to be transmitted' (Miller 1965). In cybernetics, information is defined precisely, in mathematical terms, as negative entropy.

The relationship between entropy, information, and protean behaviour is not simple however. If Miller's list of the opposing characteristics of information and entropy is considered, it can be seen that the characteristics of protean displays fall partly into one column and partly into the other, (Table 2.1). For example, the displays often involve highly improbable events and increased complexity as well as irregularity and an approach towards apparent randomness and disorganization.

The apparent contradictions arise from the fact that biological systems, including behavioural ones, involve an increase in information and organization and a decrease in entropy. But through evolutionary time predators have come to recognize, and often overcome the hitherto superior regularities of the prey. The various animal groups have then come to counter opposing armaments further by decreasing the relevance to the opponent of information available in their structures and behaviour, using the protean principle, and showing an apparent disorganization of their behaviour structure. But as Burgers (1963) has pointed out, any arrangement represents some form of order or pattern of regularity when viewed mathematically. And as primary protean phenomena, at least, are ordered and patterned by natural selection and have evolved because they decrease the likelihood of the individual's dissolution, they must be following the general evolutionary trend of decreased entropy while at the same time providing less readable information.

We may summarize thus. Protean behaviours, like organisms themselves, are improbable and unnatural in the thermodynamic sense. They are probable and natural to the actors showing them but so improbable to reactors as to be unpredictable. They are highly ordered in their mechanisms and causation, but so organized as to appear disordered. They do not follow any mathematical rules of randomness, but they are sufficiently random to overcome the predictive abilities of reactors. They are systematically unpredictable.

This brief discussion may have relevance to the considerations of 'deterministic chaos' which are proceeding in a number of disciplines at the time of writing (1985/86).

The authors' interests and attitude

There is more to the establishment of the context of our subject than the pinpointing of the phenomena we are considering. The nature of a study area depends to a considerable degree upon the investigator's attitude to it. Faced with identical complex phenomena, different researchers will often tend to notice different points of interest. Their attention is drawn to different aspects of the phenomena, and different lines of theoretical development and experimental investigation will naturally lead, in many cases, to the description of separate — though complementary — facets of the overall reality.

This diversity of approach is common, indeed valuable, in science. To a large extent the points which attract interest, and the direction and methods of the subsequent investigations, depend on the investigator's previous training and past experience. But this individual setting of attitude and interest may hinder communication between different disciplines and lead to needless misunderstanding. Some remarks on our general approach are thus necessary.

Ideally, we believe, the investigation of any major aspect of animal behaviour should be a multidisciplinary operation involving ecologists, ethologists, geneticists, experimental and physiological psychologists, neurologists, biochemists, and others according to the subject (Driver 1968; Hess 1966). Different loadings can be given to the various specialities according to the nature of the phenomenon under investigation, but the basic orientation should be a broad biological one.

Of course, the progress of biology has partly depended on the division and re-division of the study of living systems into categories which have generally become more accessible to human understanding as they have become smaller and more restricted in scope. But certain large categories of living systems' study remain of compelling interest in their own right, however complex and resistant to understanding they may be. This applies particularly to the major behavioural and ecological categories of adaptation of which protean behaviour is but one example. Only an all-embracing synthesis which takes account of the parts in relation to the whole, as well as to each other, can overcome the difficulties inherent in the study of such problems. A vital key to that synthesis is evolutionary theory.

One of the reasons why Charles Darwin's (1859, 1871) concept of evolution by natural selection was such an outstanding contribution to the development of understanding of living systems was that it provided the biological integration which brought the diverse facts of ecology, anatomy, physiology, and behaviour into a deep and subtle interrelationship. Naïve teleological explanation has now largely been replaced by teleonomy — the scientific search for principles of adaptation (Williams 1966; Monod 1972). Reinforced by population genetics, and more recently in behavioural matters by game theory (Maynard Smith 1974; Caryl 1979), this evolutionary approach is a highly fruitful one.

Our own approach is essentially Darwinian. We believe that evolution by natural selection has been operationally responsible for the majority of the principal physical and behavioural characteristics of animals. At the same time we note with approval — as do Gould and Lewontin (1979) — Darwin's (1859) conviction, stated at the end of his Introduction to *The origin of species* (and usually ignored), that though natural selection has been the main means of modification, it has not been the exclusive one.

An important aspect of the explanation of individual items of behaviour and structure is their context within fundamental modes of adaptive relationship between organisms and their environment. These basic modes of adaptive relationship can be discovered, and their nature clarified, by meticulously sifting, within carefully limited and defined ecological contexts, the myriad individual adaptations of a great diversity of species, searching for parallels of form and process, convergence of pattern and principle, within often contrasting biological materials. Such sets of adaptations as aerial escape, Batesian and Müllerian mimicry, camouflage, and ritualization of signals have all evolved independently, and many times, in diverse groups of animals as solutions to particular types of environmental problem. And all these intricate and beautifully balanced sets of adaptation, products of unseeing but effective chance, are perceived from our human standpoint as reflections of the overriding drive to survive — if necessary at the expense of other animals.

Features at every level of the hierarchical structure of life, down to its biochemical basis, are influenced by the evolutionary process, by the capacity of species' populations for balanced dynamic phylogenetic change. And at every level the shaping of features relates to survival, the continuing propagation down the generations, of the relevant genes. The dominant theme of all life, the process of evolution by natural selection, is manifest in every living process. Thus, while slight changes in biomolecular processes may affect the behaviour of the whole individual and thereby the population, so do environmental influences upon the population ultimately affect individual animals and their component parts at all levels. Conventionally, this is recognized in the countless intricate physiological and behavioural mechanisms that promote the survival of the individual organism: homeostatic mechanisms centred upon informational transactions that enable the organism — or, failing that, a sufficient number of its close kin — to maintain identity as a unique arrangement of materials in an environment which constantly threatens this identity (Young 1957, 1971). The organism represents a complex of such transactions at any point in time; it is in a sense a stage in a process of constant exchange of information between its component parts and with other entities. Exchanges occur in various dimensions: between mechanisms at the same level, between different levels of organization — from the molecular to the population level — and in developmental and evolutionary dimensions. It follows that the behaviour of an organism cannot be fully explained in terms of transactions taking place in one dimension only (Driver and Corning 1968). Biochemical mechanisms cannot provide a sufficient explanation of behaviour, any more than the set of principles relating to a white rat's reactions to the lever of a Skinner box. A broad, multidimensional approach is essential for the synthesis of adequate explanations, and this means an approach which is biological in the widest sense. For natural selection 'operates at the macroscopic level, the level of organisms' (Monod 1972).

Among the many disciplines which are concerned with behavioural transactions, that which has the broadest basis in biology is ethology, straddling the gulf between neurophysiology and sociobiology. Simply but comprehensively called by Lorenz (1965) and others 'the biology of behaviour', it is basically concerned with the whole animal, that is, the normal animal in its natural environment — the only unit which exhibits, in a natural integrated fashion, all of a species' principal organic transactions. But we must be clear that ethology is not just a simple matter of careful observation of wild animals. As Medawar (1973) put it:

> Anybody who thinks that ethology consists of a passive imbibition of the information proffered by nature has still much to learn. . . behaviour structures do not declare themselves in an obvious way. Their identification depends on an imaginative conjecture on the part of the observer which further observation may or may not uphold.

An organism's behaviour reflects the state of its organs and tissues, just as they in turn reflect the state of their own constituents. It also ultimately comes to reflect, through the process of natural selection, the state of the group to which the organism belongs, and the ecosystem of which it is a part. An important consequence of this shaping by natural selection is that — witin the limits of contemporary knowledge — the only way in which an organism's features, including its behaviour, can be *fully* understood is by studying its relationship to its natural environment.

Conclusion

We make no apology for the fact that the descriptive parts of this book have more of the style and approach of a nineteenth century natural history, rather than of a contemporary, scientific text. In fact, we believe that this is a strength of the work, for without experience of a significant and representative sample of the enormously complex panorama of animal behaviour in the field, and as a whole, no one can safely deduce general principles. Such experience requires clear and often detailed observation, of which we have jointly amassed several thousand hours.

It is not always necessary to change good observational data by translation into mathematical terms. Notwithstanding the important developments made in biomathematics in the last few decades, mathematical techniques can still only cope with the analysis of relatively simple and rather limited biological systems. Their application to the elucidation of the higher levels of biological transactions has often been inappropriate, requiring such grossly inaccurate simplifying assumptions that the resulting mathematical model has curiosity value only. Nevertheless, mathematics has long enjoyed a high place in the esteem hierarchy of the sciences, and grossly inappropriate models have sometimes attracted undue respect. As one distinguished biologist who is also a mathematician has put it (Slobodkin 1965):

to insist on the theory of biology conforming to aesthetic standards derived from extant mathematics is illegitimate, and is, in fact, an imposition of metaphysical criteria on the empirical world. Empirical sciences must develop their own standards of quality and cannot take refuge from the necessity of thought in the shadow of Newton or Euclid.

We see no significant change since this criticism was written, though the model developed by Holling (1965) dealing with the ecological/evolutionary consequences of the behavioural qualities of predators and their prey is a rare example of biological integrity taking precedence over mathematical elegance.

In the Croonian Lecture to the Royal Society, 1972, Niko Tinbergen began his discussion of 'adaptedness' as follows:

Every time one follows the life of a generation from its start to successful reproduction and sees how few individuals achieve this end, and how many fall by the wayside, one marvels at the intricacies of adaptedness — of the ways in which animals surmount the innumerable obstacles that the environment puts in their way. It is a fascinating occupation to analyse the animal's achievements — to discover in ever-increasing detail the demands imposed by the environment and to confront these requirements with what the animal actually does to meet them. For this one has to study the effects rather than the causes of behaviour, and to compare them with the pressures exerted by the environment. This study has of course to relate the behaviour to the natural habitat — the situation in and for which natural selection has moulded and stabilised each species.

The examples of protean behaviour that we have given in the first chapter illustrate a category of adaptedness which has previously received far less attention than it deserves. Protean displays form an adaptive strategy widely used by animals for surmounting some of the 'innumerable obstacles that the environment puts in their way'. In our work we have, indeed, primarily studied the effects of an animal's behaviour in its natural environment, only later classifying and analysing, suggesting modes of causation and mechanism, and pointing to lines of adaptive modification which animals may have followed in their evolutionary utilization of the protean principle.

At a time when experimentation with sophisticated apparatus is in high fashion — what Snow (1964) called 'the euphoria of gadgets' — we should remember the comment of the great physiologist Claude Bernard (1865) when he pointed out that 'experimental reasoning is absolutely the same, whether in sciences of observation or in experimental sciences' and that

a naturalist observing animals in all conditions necessary to their existence, and deducing from these observations consequences verified and controlled by

> other observations . . . such a naturalist uses the experimental method even though he performs no experiments, properly speaking.

A carefully worded anecdote, based upon sound observation, can be more valuable than a whole battery of irrelevant experiments, however 'sophisticated' or 'elegant' and 'statistically significant' they may be. And anecdote is only a preliminary in ethologial science; it builds into carefully designed sets of observations and, where it is necessary to reveal hidden features of the behavioural systems under scrutiny, it develops into experimental manipulation in the field or laboratory.

A better understanding of the value of the ecological approach to the explanation of behaviour has developed rapidly in the last decade with the rise of sociobiology. We would only add that — as Charles Darwin showed in so much of his work — this approach has been good science for a very long time.

Our attitude is summarized admirably in the words of T.H. Huxley (1860). Notwithstanding his formidable knowledge of biological science he advised a correspondent:

> Sit down before fact as a little child, be prepared to given up every preconceived notion, follow humbly wherever and to whatever abysses nature leads, or you shall learn nothing.

3
Strategies and tactics

> 'The very urgency of this central biological problem of self-preservation is reflected in the variety and specialisation of nature's adaptive experiments in offence and defence, as it is also in the parallel inventions and contrivancies of man.'
>
> *H.B. Cott* (1940)

Our next task is to survey the principal recognized strategies and tactics used by predators and their prey in the unceasing struggle for existence. In nearly all categories we shall see opportunities for the application of the protean principle. Equally frequently we shall see relevance to human conflict too. To biologists much of this may seem unsophisticated and superficial. Indeed it is. But we hope this book will interest scientists from a variety of disciplines, and laymen too. We therefore include the chapter as a means, perhaps, of making clearer to the uninitiated the universality of the protean principle. Field biologists and behavioural ecologists may prefer to turn straight to Chapter 4. But please note that towards the end of the chapter (on p.36) we give our full definition of protean displays.

We can think of no better beginning than to continue this chapter's introductory quotation, taken from the preface to Cott's *Adaptive coloration in animals* (1940). He writes:

> The primeval struggle of the jungle, and the refinements of civilised warfare have here very much the same story to tell. In both realms we see the results of an armaments race and an invention race, which has led to a state of preparedness for offence and defence as complex as it is interesting. In both, methods mainly similar are employed: we have the evolution of speed, on land, in the air, and under water, by pursuer and pursued; the employment of stealth and surprise, of deception and ambush; the display of warning signals and of alluring baits; the elaboration of smoke-screens, traps, nets, parachutes, of electrocution and booby-traps; the adoption of fossorial and nocturnal habits; the development of poison, and of deadly apparatus in the form of fangs or stings or arrows for its injection into the bodies of enemies or prey; the protection afforded by armour-plating, spines and barbed wire; the use of chemical warfare which is practised, for instance, by certain insects; and of poison gas, by creatures like the skunk.

To this list we now add the major category of protean phenomena. It is no reflection on Cott that he did not recognize the protean principle in his grand survey, for the very nature of protean phenomena makes them unusually difficult to describe and classify in an orderly scientific fashion. Furthermore, any challengingly new scientific thesis must necessarily await the appropriate combination of the accumulation of data, the maturation of established ideas, and the appearance of a new generation of scientists with the fresh experience and insights which enable them to utilize new qualities of information.

The struggle for existence or gene perpetuation, whether we see it ecologically, in terms of competition for environmental needs represented in an n-dimensional hyperspace, or ethologically, in terms of the stimulus–reaction chains linking two individual organisms, is an arms race in which the opponents are continually battling, in the evolutionary sense, for superiority of weapons. The human arms race, like many other principal categories of human behaviour, is simply a special production of a play well-known on the grand evolutionary stage. As Cott indicates, the sets, props, costumes, and principal characters and themes are essentially the same in both human and non-human productions. The only major difference would seem to be in the mode of origin of the weaponry. In man, ideas are the basis for outwitting one's opponents and can be modified if unsuccessful. All other animals, and plants, must be viable entities at all stages of the evolution of their defensive structure or strategies. Man, to a degree, thinks ahead, whereas other animals' defences are, as it were, chosen for them by the forces of natural selection.

It is relevant at this point to consider the use of the terms 'strategy' and 'tactics'. The objection is sometimes made (for example, Louw 1979) that the term 'strategy', widespread in the biological literature, is misused, in that it is applied to such unthinking processes as biochemical and physiological adaptations, whereas its original and proper meaning pertains to generalship in war. However, as Murtagh (1979) has noted, unconventional usage of words may be preferable to coining new and unwieldy labels as long as the variations in meaning are clearly explained. We support this view and therefore announce that our usage of the word '*strategy*' infers a general principle such as crypsis or warning colouration, conscious or not; and by '*tactic*' we mean a more particular device such as hissing or bouncing.

Problems of focus and definition

The study of relationships between organisms concerning their environmental and metabolic requirements is extremely complex, but we can see that as food chains and webs have built up through evolutionary time, the utilization by organisms of the available resources has developed along various lines. These we point to in this chapter; understanding the controlling principles is a more difficult matter. Although Darwin put us on the right lines, and Fisher (1930, 1958), Haldane (1932), Williams (1966), Maynard Smith (1974), and others have kept us on them, a general philosophy that encompasses both Darwinian selection and modern biological knowledge still eludes us. Take, for example, Dawkins's belief (1976*a*) that we are 'robot vehicles blindly programmed to preserve the selfish molecules known as genes'. Such a statement can hardly be regarded as 'truth' or involve 'science' as he claims, because it is not testable. The principal criterion of a scientific theory is, as Popper (1968) has convincingly demonstrated, 'its falsifiability, or refutability, or testability'. Until biological science reaches a stage of development where it can satisfy this criterion with respect to such as the 'selfish gene' theory, we cannot agree that 'it is reasonable for us to picture an animal as a machine designed to preserve and propagate the genes which ride inside it' (Dawkins and Krebs 1978). These statements beg far-reaching semantic and epistemological questions. Terms like 'selfish' and 'machine' for example, are human artefacts designed to describe certain types of behaviour and structure as seen by man. They are essentially anthropocentric. Machines are man-made objects which copy, not often very efficiently though sometimes more impressively, evolved human transactions with the environment. 'Selfishness' is a term given by man to certain characteristics which he thinks he understands. If we accept that man has evolved from pre-existing animals, with ever-increasing complexity of structure and behaviour, how can we give an anthropocentric term like 'selfishness' to a lower animal, let alone a collection of nucleic acids?

On the other hand it seems equally naïve to assume that organisms consciously or unconsciously assess the significant environmental factors and behave appropriately for the optimum production of young (Wynne-Edwards 1962). Not even man can do this — indeed, we should not expect him to. Wynne-Edwards's original thesis was that animals have evolved various behavioural and physiological adaptations for the avoidance of over-population. There is much circumstantial evidence in its favour but more recent work, particularly by Maynard Smith (1976), has indicated that it cannot be so. Wynne-Edwards (1977) now accepts that social behaviour has not evolved primarily for reasons of population control.

These extreme points of view — that major control phenomena are dominated by genes on the one hand or group processes on the other — are reminiscent of other divides in biology such as the instinct-learning schism of the 1960s (Lorenz 1966*b*; Schneirla 1966; Driver

1968). We prefer a more integrated view, one which tries to consider the whole as well as the parts — for reasons explained in the previous chapter. It seems eminently reasonable to us to concentrate upon the individual organism because it — the organism — represents the one continuing, though mutable, unit of organization which focuses all the processes of living materials in their constant battle with the environment. As C.L. Prosser (1965) put it: 'evolution can be understood only in terms of the action of natural selection on the whole organism, and study at all levels of complexity is necessary'. This seems to be the underlying assumption of the principal published works on predator-prey interaction (Cott 1940; Edmunds 1974; Curio 1976) and it is that which we follow throughout this book.

The level of the individual organism is the most convenient, as well as the most logical focus for our work. Lest it be thought unreasonable to try to focus the various levels of biological organization in this way, we offer an integrational diagram (Fig. 3.1). This illustrates the interaction of various environmental phenomena which, we suggest, may help to define an organism and its niche. The diagram takes account of predation, food chains, the pyramid of numbers, environmental pressures — i.e. habitat requirements —, and the cycle of some nutrients. Two food chains are represented, one terrestrial, one aquatic, connected at their ends. In life there would also be cross-connections which, with numerous connections to other food chains, would form the classical food web (see Elton 1927; Allee *et al.* 1949).

This diagram is, of course, speculative and imprecise, but it does indicate the possibility of regarding an organism in terms of some of the interacting factors governing its place in nature — its 'niche'. We can suppose that adaptations to environment tend to maintain an organism in its niche, while predation will of course remove certain individuals. Genetic changes may result in the re-adaptation of certain groups, and competition may eliminate others. But in all cases it is the specific combination of operative factors which delineates the '*n*-dimensional hyperspace' that is the organism's niche.

A further demonstration of the possibilities for integrational exercises is the work of Russell (1961) which emphasizes the process of *combinatorial selection*. This would seem to be a general principle of operation in biological systems involving the scanning of the environment to ascertain its pressures, and the selection of the best available operative units to deal with these pressures. It would appear to function at all levels of biological organization including the molecular, cellular, organ, organism, gene pool, and population levels. We have profited greatly from this work.

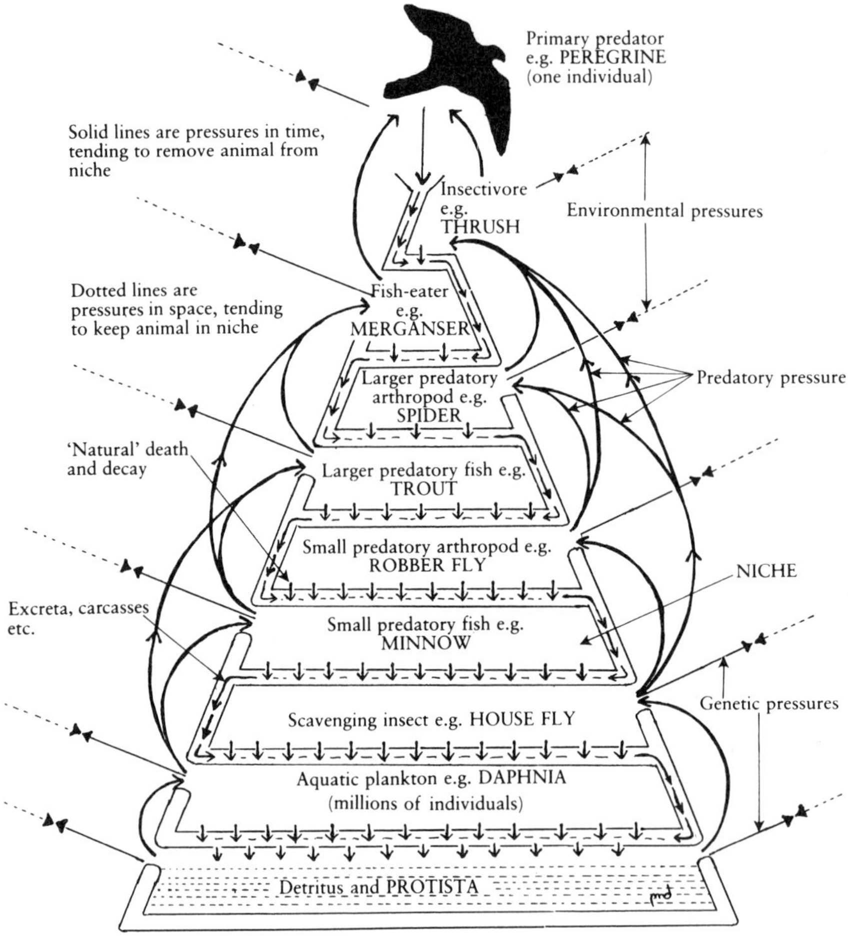

Fig. 3.1. Diagram showing some of the factors involved in defining an organism's 'niche'. Two interconnected food chains are shown, each of six trophic stages. The 'pyramid of numbers' is incorporated, as is the cycle of certain nutrients (see text).

A survey of relevant animal interactions

Like Edmunds (1974) we divide animal interactions into

Competition for food, shelter, mate, territory.
Partnership where both partners (mutualism) or only one (commensalism) benefit.
Predation in which prey is usually killed.
Parasitism in which the host is debilitated but usually not killed.

The first two categories do not enter significantly into our survey, but it will become evident that various aspects of competition between animals *can* be influenced by the protean principle. In particular, territorial disputes may often involve behaviours similar to those seen in predator–prey interactions.

The principal lines of development in the competition for the necessities of life with which we are concerned are those in which there is serious conflict between individuals. These are the two major categories of predator–prey and parasite–host relationship. Each of these categories has in itself a great variety of lines of adaptation. The predator–prey relationship we shall deal with in detail below, but because of the great variety of possible lines of evolutionary development open to the numerous and diverse species of animals in the parasite–host category, we shall see that in some cases protean phenomena have become operative here also.

Predator–prey interactions

In this book we are principally concerned with the predator–prey area of conflict in which one animal is killed in order that another may continue to live. As in other categories of conflict, successful lines of predator–prey association involve a continual struggle for existence in which superior armaments are offset by countermeasures and the balance of nature fluctuates about a norm. As Allee *et al.* (1949) put it

> Interactions between different species of organisms and interactions between organisms and their environment produce selection pressures. Reciprocal genetic patterns evolve by means of such selection and produce interspecies adaptations, interdependence, and integration. Harmful disoperation between species eliminates itself.

As predator–prey relationships have developed through evolutionary time, a variety of strategies and tactics have been evolved by the predators to overcome the prey's reluctance to be captured, and by the prey in order to avoid capture by predators.

The simplest strategy of predation is probably to lie in wait and pounce upon unsuspecting prey. Leopards, octopuses, and certain snakes are prime examples. Their prey's defences are centred upon highly developed senses, stealth, and camouflage. In various phylogenetic lines, diverse strategies and tactics are seen as variations on this basic method, the examples already given being, in fact, highly evolved forms using a long-established and successful technique. In many cases, human counterparts will come to mind. Herons use an essentially similar technique but stand immobile in the open, as the aquatic animals which form the bulk of their prey respond largely to movement (see p.199). Here, the prey's defence is based upon fast reaction times and speed of escape movements. Predators upon invertebrates in this basic category can also wait immobile without hiding, and most of them — many amphibia, and insects such as the mantids, for example — may themselves be the prey of larger predators and therefore tend to be camouflaged (Edmunds 1974; Wickler 1968). The protean principle operates here when a polyphagic predator — of insects say — fails to capture a prey animal which unpredictably flashes deterrent eye spots, after it had captured many which showed no such behaviour (Brower 1971). It may also operate in the unpredictable — not just the unexpected — occurrence of a predator in a place from which it has previously been absent. Schaller (1972) notes how a lion will usually capture prey it comes across unexpectedly.

The actual pursuit of prey is an obvious development from the basic technique considered above, though accounts of predators' abilities in this respect are often exaggerated. Many well-known pursuit predators, such as lions and the broad-winged hawks (Accipitridae), use surprise and pursuit over a short distance as their principal technique. Lions seldom exceed 50 km/h when charging, and that only for short distances. And the accipiters show a distinct preference for hunting in well-wooded country such as parkland where their manouevrability and capacity for short bursts of speed enable them to capture birds taken unawares. Regular prey countermeasures in such instances include highly-developed sensory attunements and, particularly in gregarious species, the use of warning signals. Protean countermeasures centre upon fast, unpredictable movements which frustrate the charge or swoop, and the predator has to set up the situation afresh. The predators are very often unsuccessful, except when they hunt in groups (Schaller 1972; Kruuk 1972; Bertram 1979).

Speed and acceleration are of primary importance to prey animals in the situations described so far. This has resulted in some predators becoming even faster. The cheetah is a classic example (Schaller 1972), sometimes reaching 100 km/h but, as with the lion, only for short distances. Among birds the sharp-winged hawks or falcons (Falconidae) are outstanding examples, different species showing a range of size and speed according to their preferred prey. The fastest is probaby the duck hawk or peregrine, *Falco peregrinus*, which, in a 'stoop' or dive upon its prey, may exceed 240 km/h. Most falcons pursue their prey in open flight; prey defences include dodging, taking cover, or outflying the predator. As we shall see, protean phenomena may also be of vital importance.

Another pursuit variation is the combination of speed and endurance of dogs such as the Cape hunting dog *Lycaon pictus* of Africa (Schaller 1972) and the dhole, *Cuon alpinus*, of Asia. These animals often maintain pursuit until the prey is exhausted and/or overcome by weight of numbers. Any failure is often due to protean displays by the prey — such animals as zebra, wildebeest, and gazelles. Hyenas also hunt in packs but even they may be beaten by protean movements of the prey (Kruuk 1972). Wolves may run down their prey but usually when the animal is somehow already incapacitated (Mech 1970).

Both Schaller (1972) and Kruuk (1972) document the greater success of predators when hunting in groups. Bertram (1979) has listed the advantages of hunting in numbers. But we must remember that two lions need to be more than twice as successful as one lion if they are to benefit from the co-operations. This is by no means always the case — largely through protean displays by the prey. In fact, when a single lion has captured a small animal such as Thomson's gazelle it is likely to pull it away from other lions. This is less common with larger prey (D. Bygott pers. comm. 1979).

Insects are a favoured prey of many animals. The bats (Order: Chiroptera) feed largely upon flying insects, as do many birds. We later show how some insects have developed protean flight responses as a major weapon against the aerial plankton feeders.

It is beyond the scope of this book to survey the evolved ramifications of insect predator–prey relationships. Some insects, for example the dragonflies (Order: Odonata), have even become aerial predators (and have to cope with the protean movements of their prey!). But in the group as a whole, we can see most of the major adaptations in prey behaviour which have been developed in the rest of the animal kingdom and these may therefore be usefully considered here.

As in other groups we see the development of speed as an escape mechanism, many of the true flies (Order: Diptera) being good examples. Equally developed are a variety of protean escape movements. We also see a multitude of examples of camouflage or crypsis, both by structure and colouration, with complementary behaviours, as in the mantids already mentioned and other orthopteroids such as crickets and grasshoppers. We will see later that crypsis may have protean properties when taken in association with related events.

Mimicry is also highly developed in insects, both Batesian mimicry in which an edible mimic copies a distasteful model, and Müllerian mimicry in which both model and mimic are distasteful. Other categories of mimicry have been described, and named after their discoverers, and the total picture is extremely complex. Basically, mimicry is an effective antipredator device because the models and their mimics are warningly coloured (aposematic) and by experiencing the model's taste the predator quickly learns to avoid objects which have that appearance. Behavioural mimicry often accompanies structural mimicry, some mimetic butterflies behaving atypically for their group but very like their models. Some short-horned grasshoppers (Family: Acrididae) mimic butterflies in both colour pattern and behaviour when in flight. The flight path of the butterfly model is itself protean, and an added protean element is involved when a grasshopper takes flight and apparently becomes a butterfly.

Many unpalatable mimetic butterflies are variations within a polymorphic species, other morphs of which are palatable. Such is the case with *Papilio dardanus* in Africa, *P. lysithous* in South America, and *P. polytes* in Ceylon. We show later in our discussion of apostatic polymorphism (p.223) that in palatable populations, individuals may derive a selective advantage when the population is polymorphic (Clarke 1969). Simply stated, the predator concentrates on the more obvious morph, ignoring the others, until it is too rare to find easily. Then it switches its attention to another morph. Thus, all morphs benefit in the long run. It will be interesting to see if the selective advantage of apostates (a protean phenomenon) applies in these tropical papilionids. It may even apply when butterfly or other insect species exist as different morphs in different regions. Such species are not, strictly speaking, polymorphic, but their effect upon an insectivorous bird migrating through the different populations may be the same as if they were.

Many well-known insects are armoured, beetles for example, and we shall return to this point. Others may bluff an attacker into thinking they are more dangerous than they actually are. For example, earwigs (Order: Dermaptera) have formidable-looking pincer-like organs at the tip of the abdomen which, like members of the beetle family Staphylinidae, they curve up in a threatening manner when provoked. This may be regarded as a form of deterrence. Other well-known examples of insect deterrence are seen in the butterflies, and a number of other forms, which flash brightly-coloured eye spots on the hindwings when disturbed. These eye spots are normally concealed under the fore wings when the animal is at rest, and their sudden appearance gives the impression that the animal is much larger and more dangerous than is actually the case. Such activity, as we have indicated, may be protean with respect to a predator's previous experience.

A particularly fine example of behavioural deterrence is given by Brower (1971) in the case of the larva of the sphinx moth *Leucorhampha*. This larva, some 10 cm long, lives upside down on vines at the edge of the forest. It is normally difficult to see but when disturbed it lets go with all but its hindmost legs, swells up its anterior to reveal highly realistic eyes, and sways in a snake-like manner. When touched, these larvae lash their heads like a striking snake.

This example allows us to introduce the point that adaptations of prey animals to their predators may change in ontogeny as well as in phylogeny. Moths of the family Sphingidae, for example, may have mechanisms in the larval stage such as that just described, but may be cryptic, or use eye-spot flashing as adults. The phenomenon of insect metamorphosis allows larval insects a different life-form, sometimes a quite different habitat, from the adults, thus enabling a greater variety of protective mechanisms to be utilized during the life history of the species. This principle applies also to other animals with a distinct metamorphosis in their life history, the tadpoles of amphibia, for example, having, on the whole, different predators from the adults. A protean display in one life stage may thus protect another stage which only has regular protective mechanisms.

A broad consideration of the Insecta enables us to point to another strategic principle employed by a wide variety of prey animals, which is the principle of reducing the number of times a predator will encounter and recognize its prey. This is achieved in a variety of ways including various tactics for crypticity, taking cover when the predator is active — this is often associated with nocturnal habits — and a somewhat random dispersion of the prey individuals in the habitat. Many of these tactics would seem to be connected with the fact that a vertebrate predator, particularly the more advanced forms such as birds and mammals, will quickly learn to seek a particular type of palatable object — often after only one ingestion. Birds in particular appear to develop a specific attention mechanism — often referred to as the 'searching image' phenomenon — whereby they are rewarded for concentrating upon a particular form of food object (pp.207–37). Obviously, if an insect, or any other animal, looks like a stone, a thorn, or a bird-dropping, it is not likely to be ingested by a predator except by accident. A quotation from Brower (1971) illustrates the range of crypticity in the insects:

> In forests, the sunlight shining down through the canopy of leaves and vegetation is broken up to present a marvelous dappling effect of light and dark spots upon the surface of the lower vegetation and the ground. Many insects bear bright iridescent and metallic colours, and although in a museum cabinet they gleam as fine jewelry does, they are virtually invisible in their natural habitat.

He notes that in a Costa Rican rainforest a chrysomelid beetle looked like a sun-spangled raindrop upon a leaf, and that other insects resemble many aspects of the non-living substrate. Some seashore or desert grasshoppers, for example, look like sand, while others are hardly distinguishable from small rocks. In forests the leaf litter, dead bark, and dead leaves still attached to their stems all have counterparts so closely resembling them that only an experienced naturalist will even be aware that the insects are in the area. These resemblances can be unbelievable. C.B. Williams recounted (pers. comm., 1952) how he was searching for the tropical spider which spins a special web on a leaf so that the 'spider sitting in web' looks like a bird-dropping complete with run-off. He thought he saw one, but on close inspection decided it was, after all, the model not the mimic. And as he moved away so did the spider — from the centre of its bird-dropping web!

Brower (1971) also notes how plants too have been copied by numerous species of insects. Membracid bugs, for example, may resemble thorns, berries, leaf buds, or bracts of the plants from which they suck sap. They often arrange themselves around the stems in a way which is appropriate for that part of the plant. Many katydid bugs look like leaves. In Trinidad one of these looks like a leaf that has been nibbled by a beetle or attacked by fungus. Another species in Costa Rica looks like a caterpillar-chewed leaf. Other species blend so well with the moss-covered trees that they can only be found by their moving antennae.

This account of insect crypticity gives only a fraction of the total picture of this strategy as employed by the group, but it does give a good idea of the principle. This principle is employed to equal effect by animals in many other groups, and always, of course, necessitates a combination of behavioural and structural adaptation. Cott (1940), Wickler (1968), and Edmunds (1974) provide the best summaries. Such crypticity is so incredibly well developed in many cases that one would think other protective devices unnecessary, yet, as we shall see (p.233), the protean principle may apply even here.

If — as we believe biologists should — we pay more attention to the *effects* of an animal's behaviour, it may be seen that crypticity is similar to hiding and variations in dispersion in that it reduces the predator's chance of finding the prey and of recognizing it. Of course, many predators have techniques for uncovering or discovering prey in their hiding-places. Many invertebrate-feeding ground birds, for example, employ techniques for removing or probing into, or beneath, the ground litter under which their prey are hiding. Examples include many of the pheasants (Phasianidae), thrushes (Muscicapidae; Turdinae), sparrows (Ploceidae; Passerinae), and plovers (Charadriidae). By these tactics the predators may be successful in both finding and recognizing their prey, and further prey tactics have been developed in response to this, including protean displays as we shall see later.

As far as the prey's dispersion in the habitat is concerned there is evidence that the more sparse the dispersion, the greater the difficulty the predator has of finding an individual prey animal in a given time (Croze 1970; Smith 1971, 1974*b*). In some species as we have seen, this effect may be enhanced by colour polymorphism, the dispersion of any one morph being considerably more sparse than that of the population as a whole (Clarke 1962*a*).

Up to this point we have dealt largely with terrestrial animals, and indeed we must continue to do so, for space does not allow us to consider the additional variations in tactics developed by aquatic animals as a whole. We must note, however, that many of the principles referred to so far may also be observed in aquatic animals. The principal variations in pursuit tactics, for example, are seen among fishes and their predators. The principles of camouflage, deterrence, and mimicry; the reduction in sightings by predators through hiding, crypticity, and apostatic polymorphism (p.223) all apply in the sea as on land. The polychaete annelids, the crustaceans, the molluscs, and the fishes have all developed a remarkable variety of mechanisms in marine or fresh waters which match the strategies and tactics of their terrestrial relatives or descendants. We can only refer to a few of them in this chapter. Others will be used in other chapters to illustrate specific points.

Some types of predator–prey interaction are more readily adopted by aquatic animals. It is easier to produce a 'smoke screen' if you are aquatic. A number of marine organisms, including squids, have taken advantage of this, ejecting clouds of 'ink' which may cover their retreat — usually a protean zigzag — when they are attacked. Alternatively, in deep-sea or nocturnal animals, the cloud may be luminous. Or it may be compact and act as a diversionary lure.

The use of poison as an offensive or defensive weapon is equally common in the sea and on land. Marine animals such as the weever-fishes (Trachinidae), and some sea urchins (Class: Echinoidea) have poisonous spines, equally as effective as the sting of the terrestrial scorpions (Order: Scorpiones). Poison glands are found in all major groups of multicellular animals from coelenterates such as the sea anemones and jellyfishes to the American short-tailed shrew, *Blarina brevicauda*. Many animal poisons are deadly, whether in scorpions and spiders, fishes, or snakes; at least 30 000 humans die every year from snake bites alone. Some animals — for example, ants and certain snakes — can even squirt jets of poison some distance.

It is difficult to counter the use of poison as a weapon, except by the development of natural immunity to the toxin, though a behavioural adaptation is seen in the fish *Balistes* which turns over the sea urchin *Diadema* by aiming a jet of water beneath it, thus avoiding the poisonous spines and gaining access to the vulnerable ventral side (Fricke 1971).

A.J. Cain (pers. comm. 1984) has reminded us of the multiplicity of toxins in snake venom. This is believed to be a predator response to the (protean) variability of immunity in prey species and individuals.

Chemical defence can be effective without being deadly of course. Skunks have developed the stink glands common to the family Mustelidae to a very high degree, and can squirt an extremely offensive secretion accurately for several feet. Their strikingly patterned coats are clearly of aposematic significance.

The use of pitfalls and nets, like the use of deadly poisons, offers relatively little scope for countermeasures. However, the loose wing scales of butterflies and moths enable them to escape from spiders' webs (Eisher *et al.* 1964), the protean principle operating through the unpredictable capture of insects which can escape quickly. In the spider *Argiope*, at least, this is counteracted by a long bite as soon as the insect hits the web, whereas other kinds of insect are first wrapped in silk (Robinson 1969*a*).

Continuing the comparison of terrestrial and aquatic strategies and tactics, we must note the use of armour in both environments: the beetles we have already noted in this respect; other terrestrial examples include the woodlice or sowbugs (Crustacea; Order: Isopoda), some of which can roll up into an armoured ball when disturbed. This habit is shared by some other animals including the pill millipedes (Myriapoda; Sub-class: Diplopoda), the armadillos of the mammalian Order Edentata, and the pangolins or scaly ant-eaters (Mammalia; Order: Pholidota). The spines and quills of hedgehogs and porcupines (Mammalia; Orders: Insectivora and Rodentia) can also be regarded as armour, and hedgehogs such as *Erinaceus europaeus* also roll into a defensive ball. An obvious example of an armoured mammal is the rhinoceros.

In the aquatic environment armour is common, its development assisted by the support of the medium. The Crustacea are particularly well armoured and so are Pisces such as the porcupine fishes (Didontidae) which inflate to erect the spines when attacked, and the boxfishes (Ostraciontidae) which are encased in bony plates. The latter group includes a number of brightly-coloured species which secrete virulent poisons when attacked. The shells of molluscs and the dermal adaptations of reptiles are also for mechanical as well as physiological protection, of course, as are the skins of cartilaginous fishes such as the sharks and rays with their special dermal denticles. In all of these cases, protection by armour is usually so effective that the protean principle need not be evoked by the prey and is of little use to the predator.

We must now consider two other major strategies developed largely as defence mechanisms in prey animals and which utilize the protean principle to a considerable degree. These are the use of convulsive displays and of immobility. We cover these in some detail in later chapters, and here we need only note their context.

Convulsive behaviours occur in a wide variety of situations, of species, and of habitats. Typically, they occur as a response to attack when more normal channels of escape are blocked or when the predator's armaments are overwhelmingly superior in other respects. In the former case, an animal caught in the open far from cover may show an apparently uncoordinated sequence of spasms often referred to in the natural history literature as 'death throes'.

The effect is to delay the predator's final attack, as a result of which the prey may sometimes escape. In the latter case we find many examples among the moths which sit against matching backgrounds. When such insects are disturbed they often fall to the ground and perform a remarkable sequence of erratic bouncing and skipping. The insect's predators find such an animal difficult to grasp.

Immobility of the death-feigning type may occur in similar situations, that is when the prey is actually discovered. Cryptic immobility comes into a different category. Many animals, again of a wide variety of species, may become immobile and apparently lifeless when attacked in the open. If the attacker is not sufficiently hungry to begin its meal immediately, it may relax its attention enabling a proportion of the prey animals, which unexpectedly 'come back to life', to fly, run, or swim away. The protean principle is seen here in the unpredictability of the apparent reviviscence.

Co-ordination and control systems: the targets of protean weaponry

Before going on to consider how protean behaviour transforms many of these strategies and tactics we must bring into focus the co-ordination and control systems of the protagonists. The evolution of physical structures and behavioural mechanisms are, of necessity, closely linked. The central nervous mechanisms and the stimulus features of the protagonists must be systematically related to one another. Since these mechanisms and configurations are the immediate controlling factors of predator-prey interactions they are of fundamental importance in the theory and principles of protean behaviour. First, we must consider their basically systematic nature.

It is generally accepted that hunting behaviour is released by specific stimuli, appropriate to the particular type of prey normally taken by the predator. Not all the detectable features of the prey necessarily affect the predator's releasing mechanism — the predator's sensory attunements (Driver and Corning 1968) are concentrated upon the most appropriate stimulus available. For example, the key stimulus of movement by any small object will release attack in the frog, the hunting wasp, dragonfly nymphs, hunting spiders, newly-hatched cuttlefish, and many others. But hunting behaviour is seldom a single unit; it is built up of a series of integrated movements, each with its own particular releasing stimulus. Baerends (1950) notes that, in the water bug *Notonecta glauca*, approach is released by vibrations of the water surface, the attacking leap by the sight of a moving object, piercing by tactile stimuli, and finally, sucking by chemical stimuli. Tinbergen and his co-workers, over a period of years, revealed a similar situation in the hunting behaviour of the wasp *Philanthus triangulum* (Tinbergen 1958). This wasp takes only honey bees, recognizing potential prey by its approximate size as it flies by. Then, if the scent of the prey is correct, the wasp will grasp it. And finally, if the captive is hairy enough to be a honey bee, it will be stung and taken back to the wasp's burrow in the sand.

Observational and experimental evidence suggests that each stimulus not only releases the appropriate behaviour, but also the possibility of response to further key stimuli. Thus, a successful predator shows a regular chain of dependent responses, each linked through a releasing mechanism with a specific stimulus. Such systematic relationships between key stimuli and elements of behaviour may be acquired either during ontogeny or during phylogeny (Driver 1974; Eibl-Eibesfeldt 1961; Räber 1949; Leyhausen 1965). Baerends (1950) points out that, since every releasing mechanism has its own combination of releasing stimuli and will, during the course of evolution, develop towards a better connection with the appropriate object, this must reciprocally influence the object responsible for these stimuli — assuming that it is another living organism (see also Gilbert and Raven 1975). Therefore, genotypic variations in the stimuli provided by the prey will be evolutionarily selected towards a poorer fit with the predator's releasing mechanisms, or for affecting those mechanisms in such a way as to disrupt the adaptive flow of the hunting sequence.

It is well known (Baerends 1950; Chance and Russell 1959) that the prey may act upon these systematic relationships in three basic ways. Firstly, prey animals may deny the key stimuli for attack by being cryptically coloured or by remaining immobile in the presence of the enemy, or both (Sumner 1935; de Ruiter 1952; Herrebout *et al.* 1963). Secondly, they may provide key stimuli releasing non-hunting activities, as in the warning colours of wasps and other noxious animals, and the eye-spot patterns of insects, particularly Lepidoptera (Blest 1957*a*). And thirdly, the prey may provide modifying stimuli which release attack in such a way that it is directed to a less vulnerable or less important part of the prey, or to a less important site. Examples include the eye-spot patterns on fore and/or hind wings of many Lepidoptera, the apparent reversal of body posture in certain fishes and Lycaenid butterflies, and a wide variety of avian distraction displays (Cott 1940; Edmunds 1974; Armstrong 1947).

It is significant that the systematic, regular features of all these developments make possible both ontogenetic and phylogenetic correction on the part of the predator. For example, de Ruiter (1952) shows that immobility and cryptic form and colour in stick caterpillars (*Ennomos* and *Biston*) initially prevent attack by jays and chaffinches, but, once the bird discovers one caterpillar, it begins to direct its attention to both caterpillars and twigs, and some jays were able to detect and attack caterpillars specifically. Such a learning process is well within the abilities of many birds. Furthermore, Blest (1957*a*) demonstrates that the escape responses of yellow buntings, *Emberiza citrinella*, show a rapid habituation to the eye-spot displays of the peacock butterfly *Nymphalis io*. Bird predators' ontogenetic corrections of response may thus operate in relation to both positive and negative reactions.

Several searching studies have been made of various aspects of the biological arms race: for example, Beukema (1968), Blest (1957*a*), Brower (1969), Cott (1940), Darwin (1871), Holling (1965), Robinson (1969*b*), and de Ruiter (1952). But these have all involved regularly systematic and predictable structure–behaviour complexes, including traditional weapons such as claws and teeth and their usage, structural and behavioural camouflage and mimicry, various locomotory methods of fleeing, and biochemical methods of defence. Even the sphingid moth displays described by Brower (1971), though unexpected and well-nigh incredible, are substantially predictable in form and occurrence once one has learned what to expect.

Much behaviour involves erratic or apparently random movements — as in running over uneven ground (p.43) or dodging irregular attacks — or are unexpected as in the insect displays described by Brower. But these are not sufficiently unpredictable to warrant the term *protean* in the strict sense in which it is most usefully employed. There may be clues as to the causation and occurrence of such irregular responses which can make them predictable. Protean displays are so irregular and, apparently, unsystematic that their onset or nature cannot be predicted in enough detail for the reactor to be able to cope with them.

Protean behaviour: a definition

In 1959 Chance and Russell drew attention to 'displays which are sufficiently unsystematic to be quite unpredictable in detail', and which might serve as a device against predators. They gave two relevant examples: the colour changes of the cuttlefish *Sepia officinalis* (Holmes 1940), and the disoriented running and convulsing of the deermouse *Peromyscus maniculatus* (Chance 1954, 1957). These they named 'Protean displays' after the accounts of the mythical Proteus who 'frustrated would-be captors by constantly changing his shape, so that they had nothing systematic to which to react'. This original suggestion has been subsequently supported by our own work (Driver and Humphries 1966, 1969, 1970; Humphries 1970; Humphries and Driver 1967, 1970).

We believe that the protean concept can be applied to a large number and variety of encounters between animals, particularly interspecific encounters of a predator-prey nature. We define **protean behaviour**, as typified in these encounters, as follows:

Protean behaviour is that behaviour which is sufficiently unsystematic in appearance to prevent a reactor predicting in detail the position or actions of the actor.

We stress that though such behaviour often appears random, it is not random in the sense that the protean acts of the actor are not functionally connected with the behaviour of the reactor. They are connected in a biologically systematic manner, but a manner which is, at the same time, so structured as to appear unsystematic to the reactor. What is important to the actor is that the behaviour should be unpredictable enough to overcome the reactor's armament. For no one knows how appropriate mathematically-defined randomness is to the general predator–prey situation. We hope that work in the future will indicate how close to mathematically defined randomness a specific piece of animal behaviour must be, to be effectively protean to a natural reactor.

In the next chapter we show how widely the protean principle seems to apply in the animal kingdom. Protean unpredictability plays an important part, for example, in such well-known behaviour patterns as injury-feigning, scattering, and mobbing. All such behaviours, though their general pattern is familar to human observers and probably to animal reactors in general, include elements which are unpredictable in occurrence or form, or both.

It is probably significant that this important aspect of the behaviour of so many different kinds of animals, including man, has attracted very little scientific attention. The lack of attention itself may be an illustration of the effectiveness of protean behaviour, for scientific study usually involves a search for regularities in natural phenomena, and protean behaviour embodies unpredictable irregularities as a basic principle, thus confusing observers.

If we now give some illustrations of how protean phenomena have become built into the more regular strategies and tactics already described, our thesis should become clearer. Firstly, let us take the behaviour of a resting small tortoise-shell butterfly, *Aglais urticae*, when disturbed by man. It may employ any of several variations in its basic repertoire of escape behaviour. It may fly up in a rising spiral to a height of over 6 m, descending after several minutes and alighting with its wings spread, showing their colourful upper surfaces; it may fly away low and fast, zigzagging wildly, finally diving into the vegetation with its wings closed, showing only the cryptically coloured lower surfaces; it may settle with wings open after a zigzagging flight; or it may not fly at all but simply drop down in a cryptic attitude and remain motionless for several minutes. If the butterfly is touched during the cryptic phase it either fails to respond or suddenly flies away or skips about convulsively in the grass, flashing its bright wing colours at each skip, finally becoming immobile and cryptic again.

Analysis of this behaviour shows two well-known aspects of escape — fleeing and immobility. Fleeing takes the insect away from the dangerous vicinity of the potential predator, and cryptic immobility makes it difficult for the predator to find it after it has stopped fleeing. In themselves, fleeing and immobility are not protean. If the butterfly always responded to disturbance by a straight flight at constant speed for a constant distance and dropped into the grass without changing its appearance, it would be a fairly simple matter for a pursuer to capture it. Protean effects are introduced by the zigzagging, by the changes in apparent colouration, and by the variety of kind and degree of responses given in similar situations by different individuals, or by the same individual on different occasions. This illustrates two of the main ways in which protean properties may be manifested: by irregularity in the sequence of single acts or systems of acts in one individual, and by behavioural polymorphism between individuals. Chance (1959) has named the latter 'polyethism'.

Chance and Russell (1959) suggest that protean behaviour may affect a predator's orientation mechanism or its mood. Most protean phenomena in the small tortoise-shell butterly have the former effect, preventing accurate orientation by a predator. (Any lepidopterist or other naturalist who has attempted to catch flying butterflies will be only too familiar with what we mean.) Cott (1940) also remarks upon the association between flash coloration and sudden changes of movement and is of the opinion that these function to 'misdirect or confuse' predators. An effect on mood, substituting escape reactions for attack in the predator, at least temporarily, may be brought about by the convulsive, flashing skips of the butterfly when disturbed during the cryptic phase. Chance and Russell (1959) point out that 'Sudden unsystematic events are frightening for many animals, on much the same grounds as eye-spot patterns'. We have ourselves observed this reaction in a wide variety of animals in the field. And, as we have mentioned in the case of skipping moths, the immediate human reaction to the skipping of butterflies is to be taken aback.

As another well-documented example of the incorporation of protean behaviour into a prey animal's corpus of antipredator strategies and tactics, we take a study of the behaviour of impala, *Aepyceros melampus* (Jarman 1974). The impala is one of the most abundant of the African antelopes, living in herds of varying size, according to season, and preferring the grassland–woodland edge. As with most large herbivorous mammals, impala show a wide variety of antipredator behaviour. First, they reduce the chance of contact with predators by habitat selection, in both space and time. For example, they lie down in open areas at night, and choose appropriate times and places for drinking and giving birth. Secondly, they maintain a general alertness most of the time and have appropriate alarm calls and other reactions to co-ordinate attention upon a detected predator. Detection is often followed by movements which result in keeping the predator in sight. Thirdly, active physical tactics may be used by a mother when her calf is attacked by a small predator such as the jackal. These may be successful. Fourthly, there is the flight response to a large diurnal predator such as the lion which attacks by a final swift charge. If the charge is detected at, or near, its onset the response is simple, straight-line running. If the predator is detected at some distance the herd runs away fairly well bunched, but if it is not detected until it is fairly close, the individual impalas' flight paths will diverge and the herd will fan out before closing up again in co-ordinated flight.

When, however, the predator is not detected until it has almost gained its target the herd explodes, as it does when a shot is fired, with animals bounding high in the air, not necessarily covering ground horizontally in the process. They may even collide in mid-air. Then one or two animals will start running and the herd reunites in direct flight. This is the same reaction as that quoted from Fitzpatrick in Chapter 1. This example illustrates the protean principle of explosive scatter (see Chapter 4) in which the predator is presented with a multitude of confusing stimuli so affecting its orientation and reaction time that an individual will be captured only by chance. It is probably significant that this protean display occurs only *in extremis*, when the prey's other defence mechanisms are unable to operate. If the explosion occurs in a somewhat confined space, then another protean phenomenon comes into operation — the multiple erratic display in which each individual flight path may return upon itself, adding confusion to confusion.

Protean movements are sometimes seen on other occasions in impala, as, for example, when individuals in a herd fleeing almost at full speed may perform huge vertical leaps, during which they may twist, and displace sideways slightly. This is commonly seen when a herd streams across a road in front of a car. An essentially similar reaction is commonly seen in domestic sheep released from a confined space or passing through a gateway under pressure. This could well have the same effect as the shuffling seen in the united erratic display, when individuals fleeing in a co-ordinated group change their relative positions within the group thus making it difficult for a predator to fix his attention for any effective length of time upon one individual.

In the following chapter we shall be dealing at length with the variety of ways in which protean phenomena transform more predictable strategies and tactics, but there is one further example we would like to point to here. This concerns the situation when large flying birds such as bustards (Otidae) are attacked on the ground by flying eagles. The bustards are so large and bulky, being adapted for running and a largely terrestrial existence, that their powers of flight are poor compared with those of eagles. Their defence tactics in this situation therefore involve fleeing by running rather than by flying. Even so, the eagle would often catch a bustard if some other bustard did not run across in front of it while in full pursuit, distracting its attention and redirecting its pursuit. If this tactic is continued, the bustards usually win (Fig. 4.33, p.102).

At the beginning of this chapter we mentioned that some parasite–host interactions have developed protean features. We must consider these before passing on to describe protean displays in a more systematic manner. One category of parasite–host relationship where protean phenomena play a definite part is the insect ectoparasite category. The parasite is often an active animal, running, jumping, or even flying actively, moving around upon its host, or between host individuals, with considerable facility. Such parasites are common on birds and mammals, particularly the fleas (Order: Siphonaptera) which are found in great variety on both groups, the wingless flies (Nycterybiidae) found on bats, and the flatflies (Hippoboscidae) found on birds. We consider the movements of these parasites to be highly protean as they move through the fur or feathers of their hosts. Indeed, certain aspects of the behaviour of fleas (Humphries 1970) seem to have developed protean characteristics in response to the host's attempts to dislodge them. The unpredictable manner in which these insects appear, disappear, and reappear in some unexpected part of the host's surface covering when the host is grooming or being handled, strongly suggests a hitherto unrecognized point of behavioural convergence. We later suggest the operation of the protean principle in the parasite–host relationship at the population level also (p.242).

Finally, we note that the employment of protean tactics by prey animals is responsible for the development of antiprotean devices by some predators. We survey these at the end of Chapter 7.

Summary

It can be seen that human strategies and tactics in sport and warfare are largely manifestations of phenomena used more widely in nature. We define our usage of the terms 'strategy' and 'tactics' as applied to this broader stage.

Our focus is on the whole organism. We give our reasons for this and note other examples of integrational exercises where the organism provides a logical focus for studies reflecting activities at many levels of biological organization.

We present a brief outline of the types of interaction between predator and prey and indicate where utilization of the protean principle can affect the outcome.

The target of protean weaponry is the co-ordination and control systems of reactors. We note that systematic weaponry allows both ontogenetic and phylogenetic correction and adaptation by reactors. Protean phenomena block this corrective ability.

* * * * * *

Protean behaviour *is defined as that behaviour which is sufficiently unsystematic in appearance to prevent a reactor predicting in detail the position or actions of the actor.*

* * * * * *

We present summaries of the defence mechanisms of a butterfly (the small tortoise-shell) and an antelope (impala) to illustrate how the protean principle can contribute to the overall armament.

4
Catalogue and classification

> When we no longer look at an organic being as a savage looks at a ship, as something wholly beyond his comprehension; when we regard every production of nature as one which has had a long history; when we contemplate every complex structure and instinct as the summing up of many contrivances, each useful to the possessor, in the same way as any great mechanical invention is the summing up of the labour, the experience, the reason, and even the blunders of numerous workmen; when we thus view each organic being, how far more interesting — I speak from experience — does the study of natural history become!
>
> *Charles Darwin* (1859)

Primary and secondary protean displays

Irregularity may enter into the composition of agonistic behaviour in various ways. For example, the irregularity of aggression experienced by a 'sky-pointing' gannet, *Sula bassana*, signalling its intention to take flight and suffering attacks of varying intensity as it proceeds (Fig. 4.1) is irrelevant to our thesis. Such irregularity is fortuitous and has no special protean significance.

Again, when dogs, wolves, or hyenas attack prey such as antelope or deer, variations in attack behaviour arise in part as by-products of the differing motivational levels and physical constitution of the actors and the conformation of the ground, just as the prey's movements — and indeed survival — will often depend upon its fitness as well as the terrain.

In many interspecific encounters, however, including those already mentioned, irregularity and unpredictability appear as endogenous, intrinsic features of the prey behaviour. It is displays of this type which we regard as being primarily protean in nature, i.e. their protean features would seem to have been selected for as a natural weapon. Figure 4.2 may help to distinguish between protean and non-protean behaviour.

The unpredictable features of *primary protean displays* are the product of natural selection. Other behaviours which may be confusingly unpredictable because of immediate chance interactions and inevitable

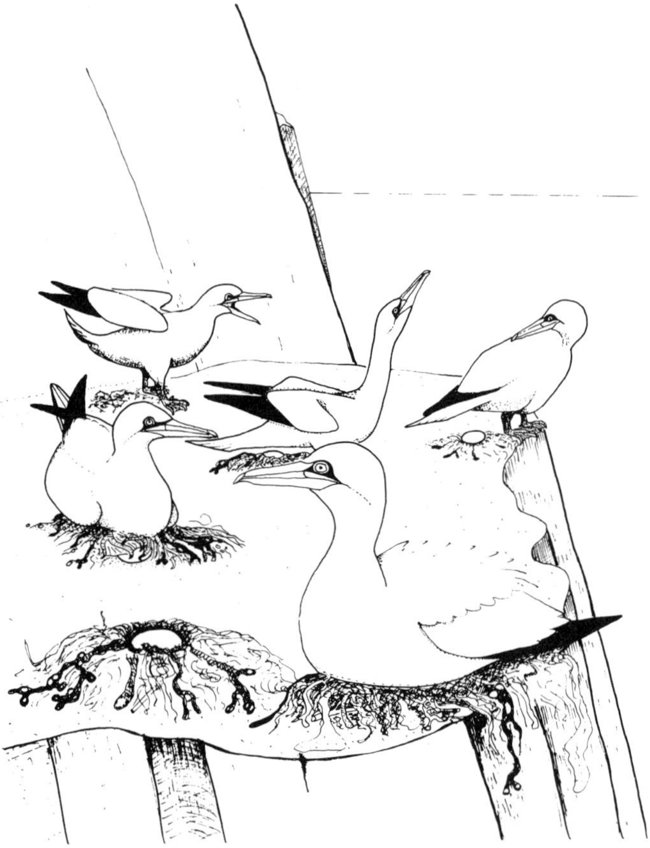

Fig. 4.1. The non-protean irregularity of threat of attack experienced by a 'sky-pointing' gannet. The 'sky-point' is a ritual behaviour signalling 'I wish only to take flight', thereby reducing aggression in a crowded colony.

variations in the biological and physical systems involved can be regarded as *secondary protean displays.* Both primary and secondary protean behaviours may have a confusing effect, but only the former has been selected for because of that effect.

Even so, what we call secondary protean behaviours may well become grist for grinding by natural selection. As Monod (1972) puts it, 'Even today a good many distinguished minds seem unable to accept or even to understand that from a source of noise natural selection could quite unaided have drawn all the music of the biosphere'. Thus, as soon as a gene is selected because it puts a jink in a flight path and hence enhances survival, we have started down the evolutionary road towards another protean display.

Primary protean displays, therefore, are usually those shown by

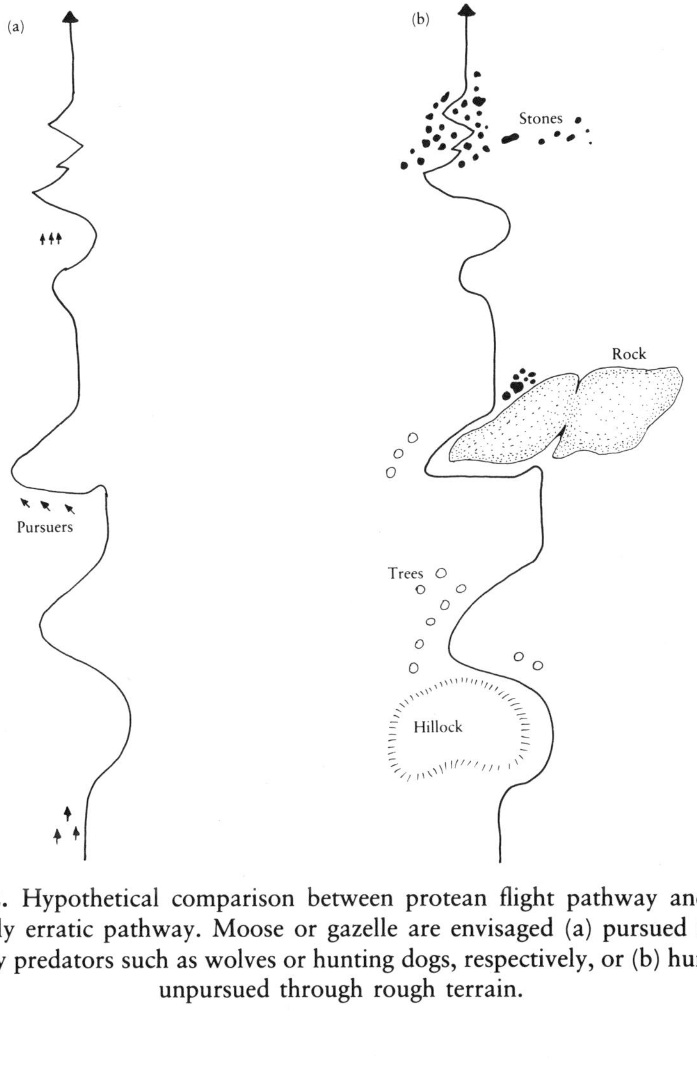

Fig 4.2. Hypothetical comparison between protean flight pathway and fortuitously erratic pathway. Moose or gazelle are envisaged (a) pursued in the open by predators such as wolves or hunting dogs, respectively, or (b) hurrying unpursued through rough terrain.

species populations as a whole, not just by a few individuals. In many cases these displays are common to much larger groups, even Phyla and Classes. This chapter is devoted largely to the classification of these primary protean phenomena.

In surveying the widespread occurrence of primary protean displays in nature — both from our own experience and that of others recorded in the literature or communicated to us personally — we find that certain categories of display are apparent. These categories fall into a classification based partly upon the form and partly upon the function of the display. Although it would be possible to develop a taxonomy based either upon form or upon function, we have decided on a classification which follows the terminology already widespread and generally accepted in the biological literature. A rigorous re-definition of the categories of escape behaviour is certainly overdue, but we shall not attempt this difficult task here.

The single erratic display

The *single erratic display* appears to be the principal form of protean display, being found throughout the animal kingdom (Humphries and Driver 1967). Indeed, it was an example of this display (Chance and Russell 1959) which first alerted us to the phenomenon of protean behaviour. Chance and Russell refer particularly to the reactions of the deermouse *Peromyscus*, and other small mammals, when they are attacked *and* prevented from performing systematic escape movements. They show disoriented running interspersed with 'catalepsy and wild aggressive attacks'. The result is a complex protean display based on erratic flight behaviour.

Long inspection of the literature and observation of living animals has revealed an enormous number of instances of the occurrence of the single erratic display — the unpredictable locomotory behaviour and erratic course of an individual animal. In some instances the records may be of movements which are unpredictable in appearance to man though this unpredictability is not functional against reactors under natural conditions. We have, therefore, been careful to include only those examples in which we consider the behaviour to be functionally protean. Even so, we have an embarrassment of riches in the number of available examples of the single erratic display. It is probably the most widespread and most successful of *all* locomotory antipredator devices.

The most widely-used pattern of single erratic display is the well-known zigzag movement of a fleeing animal. Tinbergen (1951) notes that 'Birds and mammals of steppe and tundra show a very interesting convergence in that they run away from predators in a zig-zag course'. For illustration he uses the European hare, *Lepus europaeus*, and refers to accounts of this behaviour in one of the large flightless birds — the nandu, *Rhea americana* (Portielje 1925), and in the ptarmigan, *Lagopus lagopus* (Kratzig 1940). Spinage (1962) writes: 'Most of the animals of the plains when pursued, whether by man or by beast, run in a zig-zag manner, weaving from side to side'.

Our belief that such single erratic displays are automatic and endogenous, and not a directional response to imminent attack, is supported by the observations of Leakey (1969):

Hares count upon escaping from their enemies — serval cats, jackals, and dogs — by sprinting over a short distance and then suddently 'jinking', turning sharply to the right or left and doubling back on their tracks so that the animal chasing them overruns them. As a boy I found that I could catch them after observing that a fleeing hare always carries its ears at an angle until just before it jinks, when it lays them flat on its neck. By always sidestepping to the right at the moment a hare put his ears down, I had a 50–50 chance that he would come straight into my hands and be caught!

The black-tailed jack-rabbit, *Lepus californicus*, even if undisturbed, 'when running, every so often...leaps much higher than usual' (Anthony 1928). This is an example of protean insurance, a form of display used when a predator has not been detected but when the prey is vulnerable to surprise attack. When jack-rabbits are pursued the leaps are exaggerated and incorporated into an erratic flight path. Another simple example of protean insurance is the occasional side-slip shown by wood-pigeons, *Columba palumbus*, in normal free flight, presumably as a protection against their principal predator the peregrine, *Falco peregrinus*. More complex examples are given later.

In antelopes the bounding aspect has been very highly developed. This is especially striking in the case of the springbok, *Antidorcas marsupialis*. 'The name springbok arose from the habit of these bucks when startled or at play, of leaping high into the air with body curved, legs held stiff and close together, and head down; at the same time the line of long white hair (on the back) is displayed like a fan by the action of certain skin muscles' (Fitzsimmons 1920). This 'pronking' display (similar to 'stotting' in other species) seems to have a number of functions, but the protean element is certainly inherent in it. The animals leap not only to a considerable and variable height — up to 3.5 m — but also in any direction, and when several animals display together the effect is particularly confusing. This is admirably illustrated by Millais (1895). It is likely that the protean nature of this display has been developed, in part, to offset the attention-drawing qualities of the bouncing animal with its dorsal signal line. A predator may have its attention drawn to a particular animal, but it will not be able to predict the details of its movements, especially when several animals are pronking together. Several other antelope species behave in a similar manner. These include the oribi, *Ourebia ourebi*, and the dikdik, *Madoqua* spp. In the gazelles the display is also very highly developed, though it may differ in detail from that of the springbok. The 'stotting' of Thomson's gazelle, *Gazella cuvieri*, is a well-known example.

Our suggestion that protean elements are important in these bouncing displays is supported by the observations of many workers, including Schenkel (1966) who regards the pronking and high leaps of impala, *Antidorcas melampus*, as both intraspecific warning signals and a display serving to confuse predators. This is the interpretation we prefer. We return to this later (p.181) when we consider the evolution of such displays.

There is much more observational evidence in the important work which has been done in the Serengeti ecosystem of East Africa. Here the most abundant small herbivore is Thomson's gazelle. With many predators, including lion, leopard, cheetah, hyena, wild dog, and jackal, the gazelles clearly have a problem. Their only defence is to flee, initially in a straight line or a large circle, 'but with erratic zig-zags when the predator is very close' (Edmunds 1974). Kruuk (1972) followed 43 hunts by hyenas of which 33 per cent were successful. Schaller (1972) indicates that the success of wild dogs is greater (49 per cent for adult gazelles). Schaller however, studying lion, found that only 26 per cent of hunts resulted in capture, most gazelles escaping by virtue of the agility — and, we believe, the unpredictability — of their evasion of the final charge. Sometimes they would escape by jumping completely over the predator. Edmunds (1974) states that the lions' low capture success with gazelle 'is entirely due to the efficiency of the gazelle's defence by erratic jumping'. Clearly, dogs and hyenas, coursing predators hunting in groups, will have an advantage over a single stalking predator such as the lion. Even so, an appreciable number of gazelles avoid capture and this must depend in part at least on their special version of the single erratic display.

Kruuk's observations (1972) describe the effectiveness of the single erratic display of Thomson's gazelle very well:

> They both ran, seemingly as fast as they could, for about 4 minutes, by which time the hyena had almost caught up with the tommy. Then the tommy turned a very sharp corner which the hyena was slow to follow; it took him some time to change course and catch up with his quarry again. When almost within touching distance, the tommy again turned sharply to one side and gained a lead of 3 m or 4 m. This process was repeated 8 times; then the hyena began to slow down and gave up the chase, having gone at full speed for 8 minutes.

In another example Moss (1976) notes how dikdik and gerenuk, *Litocranius walleri*, avoid lion or leopard by keeping them in view or by running through the bush, twisting and turning around obstacles, and hiding.

An African mammal which has developed the single erratic display to an astonishing degree is the springhare (Springhaas), *Pedetes capensis*, a rodent-like animal of uncertain affinities. The hindlegs are kangaroo-like and the tail is well-developed for balance; in fact it may be longer than the rest of the animal, each measuring about 40 cm. It is a distinct aid in turning, the animal being able to swerve and zigzag at speed with great ease. The ability to 'jink' in this species is more highly developed than in any other mammal so far observed, and is combined with changes of speed to make the display a particularly effective one. Its effectiveness is reflected, perhaps, in the unusually low breeding rate of the species, only one, sometimes two, young being produced annually. Twinning, in fact, is found in less than one per cent of pregnancies, at least in Botswana (Butynski 1979). The same worker found that 'the juvenile Springhaas, when it first emerges from the burrow, has feet and ears ninety seven and ninety three per cent of their adult size respectively, and appears to be nearly as capable, at least physically, of coping with predators and other inevitable hazards as the full-grown individual'. This accelerated development of the appendages is even greater than it is in precocial animals such as ducklings and fauns and would seem to be directly connected with the springhare's remarkable jinking ability.

A distinctive and easily-observed single erratic display is that of the eastern grey squirrel, *Sciurus carolinensis*. We have observed this animal in captivity and in the wild both in North America and in Britain where it is feral. When it is caught unawares in the open by a predator at some distance or when playing or engaged in sexual chase, this species shows the typical single erratic display of so many mammals, swerving, zigzagging, and stopping at irregular intervals until it reaches a tree or other object up which it can climb.

Thurber (1953) has eloquently made a most apposite comment concerning a grey squirrel at his country home: 'he frequently runs out towards my automobile when I start down the driveway, and then hesitates, wavers, jumps forward and backward, and occasionally would run right into the car except that he is awfully fast on his feet and that I hurriedly put on the brakes . . .'. This is typical of the grey squirrel which seems to 'lose its head' when surprised away from refuge. We have several times seen the display described by Thurber, the squirrel leaping erratically about in front of an automobile and occasionally running at the vehicle, instead of away from it, sometimes with fatal consequences. It is noteworthy that this single erratic display becomes much more pronounced in frequency and irregularity of turns when the animal is approached closely by a predator or predator-substitute. It is as if the animal's internal control system has broken down. This will have relevance to our later comments on convulsive behaviours.

We have already referred to the single erratic display of the European rabbit, *Oryctolagus cuniculus* (p.2). This is one of the displays which we have observed on many hundreds of occasions. As with squirrels, one is never able to say when, or which way, the animal is going to jink. This display may be seen frequently, not only in land animals, but also in animals of the air and of the water, as in waders — particularly snipe, *Capella gallinago* — when taking flight, and in many species of fishes when disturbed.

Among birds, Witherby *et al.* (1947) record the single erratic pattern of movement for snipe, jack snipe (*Lymnocryptes minimus*), and solitary sandpiper, (*Tringa solitaria*), some of the many species in which we have seen it ourselves. It is normal for the snipe species — and others such as the redshank, *Tringa totanus* — to fly like this on take-off, particularly when flushed. We have seen an especially fine example of the display in the bar-tailed godwit, *Limosa lapponica*. It seems that all wader species use this display as an antipredator device, particularly the Scolopacidae, (sandpipers, snipe, etc.) which are most vulnerable to peregrines.

The single erratic flight display of snipe, *Capella gallinago*, is extolled by Richard Jefferies (1879). Advising those hoping to shoot snipe, he notes 'This singular zig-zag flight so deceives the eye as almost to produce the idea of a spiral movement. No barrel can ever be jerked from side to side swiftly enough, no hair trigger is fine enough, to catch him then. . .'. This passage admirably describes the fast unpredictable changes of direction which make single erratic displays so effective (Fig. 4.3). An additional problem facing predators of snipe is that different snipe species show specific differences in the display (Fig. 4.4) adding further confusion to the natural intraspecific variations between individuals and between successive displays of the same individual.

The avian single erratic display is by no means restricted to waders. Kruuk (1964) for example, remarks on the frequent unpredictable changes in flight direction of isolated black-headed gulls, *Larus ridibundus*, in the presence of a flying peregrine falcon. We have observed the single erratic display in many species of gulls, crows, and ducks when chased in flight by a member of their own or a similar species. And the seed snipe (Thinochoridae) are so-called because of their extremely snipe-like flight on take-off. They are not particularly snipe-like in physical appearance and are not closely related to the true snipe (Scolopacidae), but they do have similar predators.

A more complex version of the avian single erratic display is shown by the lapwing or peewit, *Vanellus vanellus*, particularly when avoiding an aerial attack. The aerial breeding display of the lapwing (Dabelsteen 1978) is a variable sequence of flapping, zigzagging, and tumbling, the general pattern of which is stereotyped though the details of

Fig. 4.3. The single erratic display of snipe, showing characteristic variations in angle and frequency of turn.

the bird's movements during each component are irregular (Fig. 4.5) At the same time the wing beat produces a 'wup-wup-wup' sound during level zigzag flight, adding to the confusing effect. The display has the usual territorial and epigamic functions and has a number of features evolved to emphasize the signal content, including enlarged wings in the male. But the very fact that the bird is advertising itself so successfully in various ways means that it is more likely to be seen by other birds, predators included. The erratic nature of the display therefore probably acts as a form of protean insurance. In the presence of a predator the changes of speed, direction, and angle of turn are noticeably increased in frequency and, apparently, in variability. Other lapwing species have similar flight patterns (Stiefel 1964).

Spencer (1953) also drew attention to the 'crazy flight' of the peewit, and it has been likened by Lind (1961) to a similar display in the black-tailed godwit, *Limosa limosa.* The latter consists of 'a quick ascending flight, sometimes with monotonously repeated short calls, alternating with steep dives inclusive of leaf-scattering and quick zigzag flight low over the ground, often with sudden changes of flying direction'. Here again the display has features which are likely to act as protean insurance and are emphasized in fleeing.

The same principle seems to have been followed in the evolution of the territorial display of the sugarbirds of South Africa — two unusual species (*Promerops cafer* and *P. gurneyi*) of the family of honeyeaters (Meliphagidae). These small birds are highly vulnerable during their

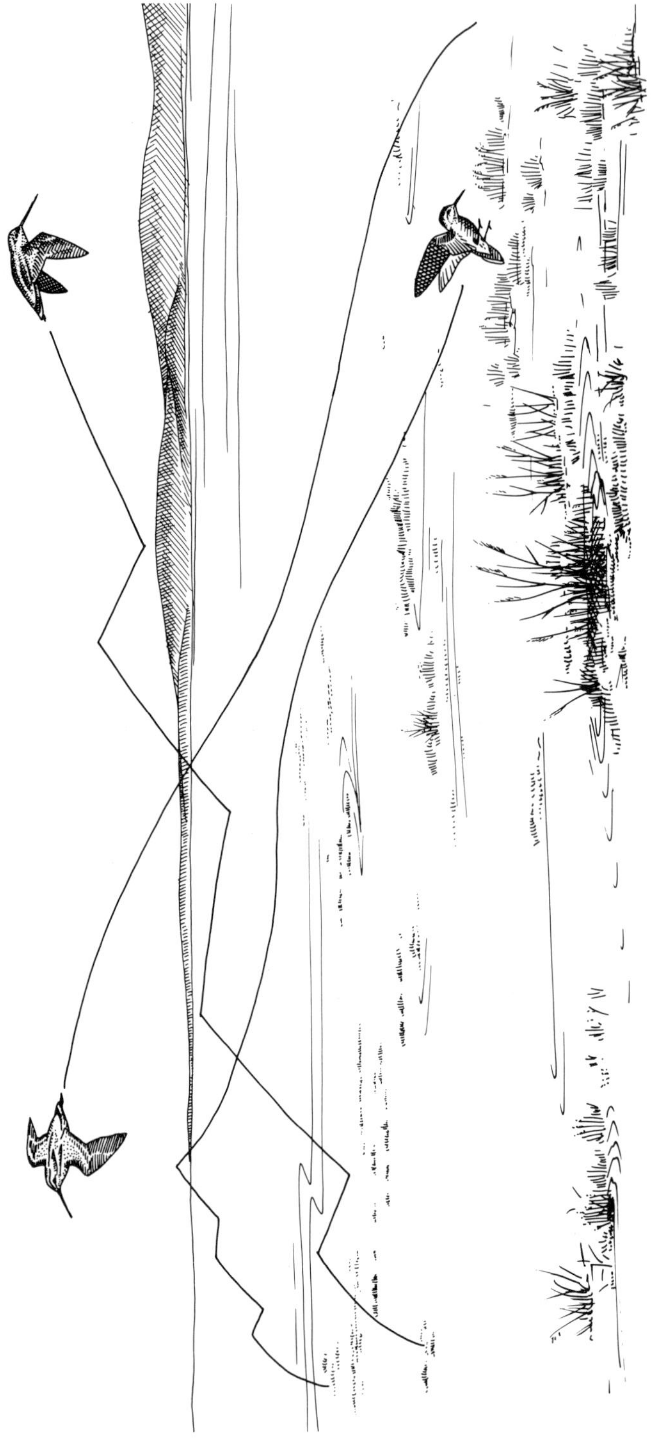

Fig. 4.4. The single erratic display of three snipe species, showing interspecific variations in the generalized form of the flight path (great snipe top left, common snipe top right, and jack snipe).

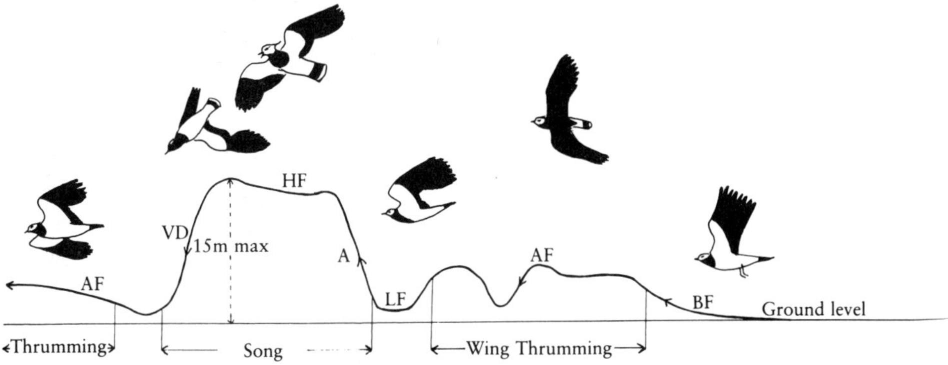

Fig. 4.5. Lapwing, typical song flight. Initials are for Butterfly Flight, Alternating Flight, Low Flight, Ascent, High Flight, Vertical Dive. A thrumming sound is produced by the wings in Alternating Flight. (After Dabelsteen 1978).

displays but fly in a manner that makes them very difficult to catch. They have long tail streamers — much longer than the body in some males — and fly in erratic, jerky arcs around their display sites, the tail streamers flickering behind them. The birds necessarily make themselves conspicuous in the process, but the form of the display acts as protean insurance. Interestingly, the sugarbirds are entirely restricted to areas of *Protea* vegetation — that large group of plants named for its confusing number and variety of forms.

The falling single erratic display of geese when avoiding aerial attack, together with their essentially similar method of landing from a height normally, is so well know among wildfowlers and ornithologists as to have its own term: 'whiffling' (Witherby *et al.* 1944). The general appearance is of a fast-falling leaf — a side-slipping, high-speed, zigzagging fall, the changes of which cannot be predicted with accuracy (Fig. 4.6). As White (1939) described it for the white-fronted goose, *Anser albifrons*: 'They tumbled out of the sky, side-slipping, stunting, even doing spinning nose-dives'. Millais (1913) quotes a similar form of descent in the three species of scoter *Melanitta fusca*, *M. nigra*, and *M. perspicillata*. We have seen this in many other species of Anatidae; in many waders, including the dowitcher, *Limnodromus griseus*, and the lesser yellowlegs, *Tringa flavipes*, as well as those already mentioned; and in the crows *Corvus corax* (raven), *C. corone* (carrion crow and hooded crow), *C. frugilegus* (rook), *C. monedula* (jackdaw), and *P. pyrrhocorax* (chough). Lind (1961) notes that the 'leaf-scattering' in the black-tailed godwit 'is often performed by solitary birds and by birds in flocks when they fly down to the resting or feeding places'. For the lesser yellowlegs, Tucker (in Witherby *et al.*

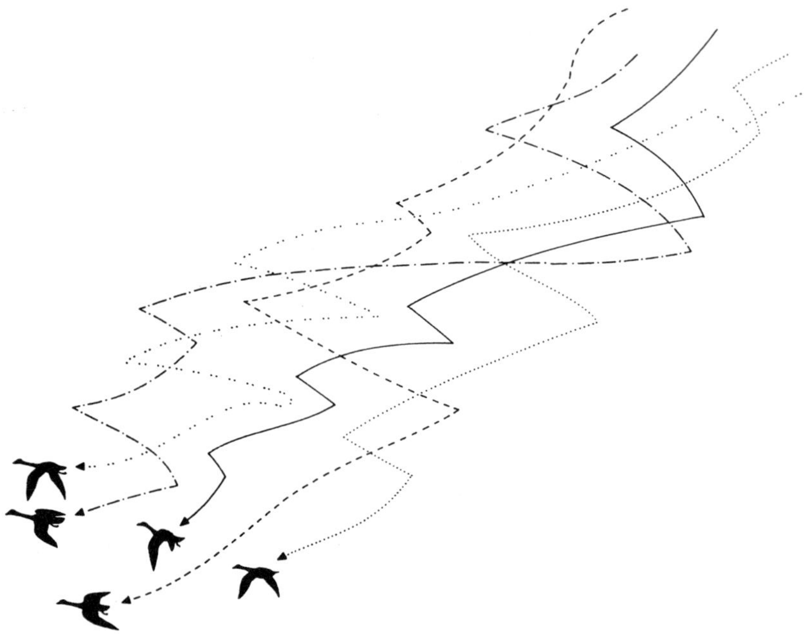

Fig. 4.6. The single erratic pattern as seen in the 'whiffling' descent of geese coming in to land. Extra confusion is brought about by the simultaneous display of several birds.

1944) notes 'flocks alighting often tumble down in zig-zagging fashion at great speed'. It would seem to be common for birds to descend in this manner, particularly waterfowl, waders, and corvids, and even the smaller perching birds do likewise on occasion (Jones 1961).

Clearly, this descending single erratic pattern is an efficient method of losing height in order to land, but the protean insurance principle is probably operating in that, when a bird is necessarily concentrating on the manoeuvres involved in landing, it will be of survival value if the unpredictable form of these movements automatically protects it against interception. We would not however agree with Lind (1961) who suggests that in the black-tailed godwit 'As in other birds, it is presumably a pure escape reaction'. It would seem to function equally well for both descent and defence.

The single erratic flight pattern is commonly used by fish. The stickleback, *Gasterosteus aculeatus* for example, is a favoured prey of many fresh-water diving birds, surface feeders, and larger fishes, and typically gives a fast single erratic display when chased. Most people who have gone pond-dipping for tiddlers will be aware of the way in which sticklebacks and minnows (*P. phoxinus*) accelerate away from

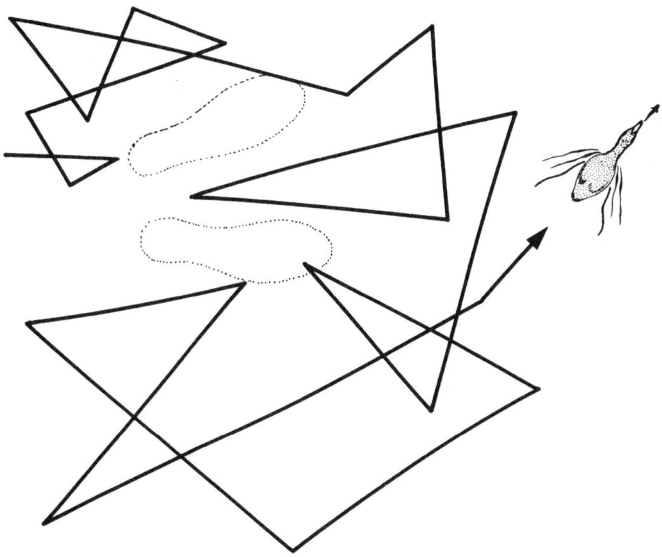

Fig. 4.7. The single erratic display of stickleback chased by red-breasted merganser duckling.

the pursuer in a wild three-dimensional zigzag. And this is true also of the larger free-swimming fresh-water fish such as rudd (*Scardinius erythrophthalmus*), trout (*Salmo trutta*), and perch (*Perca fluviatilis*). Aquarists too will be well familiar with the single erratic display in feeding or territorial disputes between individuals of various species from the smallest guppies, *Lebistes reticulatus*, to the largest cichlids. But the single erratic display of fishes may be more familiar to skin divers, for fishes apparently react to humans under water as they do to most other animals which share their environment, only accelerating away when almost in contact with the intruder. While skin diving we have seen this in many species, particularly wrasse (*Labrus bergylta*), grey mullet (*Mugil chelo*), and spotted goby (*Gobius ruthensparri*). It is apparently the norm for free-swimming fish, whether marine or fresh-water. The extra acceleration made possible by the specialized fish form acting on the watery medium would seem to allow a nearer approach than is tolerable in terrestrial or aerial animals.

A field observation of a typical single erratic display of a stickleback chased by a red-breasted merganser duckling, *Mergus serrator*, is recorded in Fig. 4.7. This is based upon studies of sea ducklings in the Canadian tundra (Driver 1960). The ducklings were imprinted to the experimenter, an ideal condition for the study of their behaviour. They were taken daily to shallow pools where they could find their own food, and frequently they would chase sticklebacks in shallow water in close proximity to the experimenter. Such chases were often successful.

The same display of fleeing fish was seen by the snorkeling Konrad Lorenz off the Florida Keys. Observing a territorial dispute between two individuals of the same species of marine jewel fish he wrote (1966) '. . . the stranger, although he is a little bigger than his adversary, switches around and flees with vigorous strokes in wild zig-zags, trying to avoid the ramming movements of his pursuer . . .'

There are numerous other examples of fish single erratic displays given in the extensive fish literature, some of which are noted in the following chapter where we consider the colour changes which may be associated with single erratic displays. It seems in fact that the single erratic display is more common in fish than in other vertebrates, possibly because of the locomotory advantages inherent in the operation of the typical fish form in the normal medium. It is also of interest that shelter is a limiting factor in most fishes, and particularly in pelagic species.

A significant feature of the fish single erratic display is its similarity to that seen in other animal groups. For example, one can predict that a given fish will accelerate away from an intruder in a three-dimensional zigzag, and that reversals of direction are likely to succeed each other. But beyond this the reactor's predictive ability is blocked by unforeseeable changes of angle, velocity, and duration of movement. In fact the sequence is quite likely to include a zigzig, a zagzag, or a temporary halt, making accurate prediction impossible.

The species of fish, birds, and mammals which are normally fast-moving employ single erratic displays which generally cover a considerable amount of linear space. However, in the slower-moving vertebrates and in many (but not all) invertebrates, the single erratic display is often centred around one spot, involving skipping, hopping, and bouncing behaviour rather than movement away from the trouble spot altogether. The reason for this contrast in form of display is probably not so much the prey's locomotory powers alone, but the relationship between the speeds of predator and prey. Where it can roughly match the speed of the predator the prey incorporates the single erratic display in a straightforward flight path helping not only to avoid attempts at seizure but often preventing the predator from approaching closely enough to launch such an attempt. Where the prey is much inferior in speed the single erratic display cannot keep the predator at a distance but it can disorientate attempts at seizure and can do so more effectively by incorporating more frequent and sharper changes in direction. This inevitably means that less net linear distance is travelled and 'on the spot' displays are created. The significance of

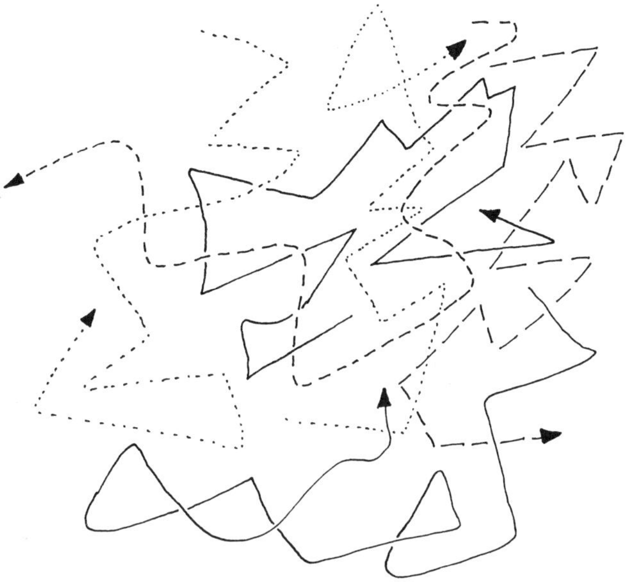

Fig. 4.8. The erratic display of dancing gnats or midges. Each individual path is a single erratic display, but when several insects are dancing together, as here, the effect is as in the multiple erratic display (p.72).

the protean principle is emphasized by this change in display form as the physical limitations on direct fleeing increase.

In the Class Insecta, containing over three-quarters of the one-and-a-quarter million known species of animals, we find that the behavioural uncertainty principle is widely used. As examples we may take mosquitoes, such as *Anopheles*, and gnats, such as *Culex*. Most field workers, from the tundra to the tropics, will be aware of the sometimes overwhelming presence of such insects and will probably be familiar with their erratic dancing flight. The insects dance singly or in groups of varying numbers of individuals but, however many individuals there are, the flight pattern of each is a complex three-dimensional single erratic pathway whch continually turns back on itself. The insect thus stays in the same general volume of space. (Fig. 4.8).

The significance of these aerial dances is by no means completely understood. They are often concerned with mating and might even involve a population-regulation mechanism (Wynne-Edwards 1962). But they certainly are protean in form and must make it very difficult for an aerial predator such as a bird to select and capture an individual insect.

When a mosquito is pursued by a predator with a speed approximat-

Fig. 4.9. The single erratic display: mosquito chased by eider duckling.

ing to its own, a straightforward single erratic display is seen. We have observed this repeatedly when mosquitoes near the water surface have been chased by swimming ducklings, particularly eiders, *Somateria mollissima* (Driver 1960, 1974). An example is given in Fig. 4.9. We must stress that — as in our other examples — the actor's changes of direction were *not* systematically determined by the actions of the predator, as may be the case in dodging.

Like mosquitoes, many other insects move erratically, even when undisturbed. Such behaviour is characteristic, for example, of many syrphids, muscids, psychodids, and ceratopogonids, to name only examples from the same Order — Diptera — to which mosquitoes belong. In fact, if the flight paths of many small insects are compared with the paths of drifting specks of fluff, it will be seen that the insects follow far more erratic and contorted courses. It is difficult to see why such behaviour should be universally exhibited by so many animals other than as a protean insurance against capture by insectivorous birds or other predators. In some cases, for example in aphids, transport of the flying insect by wind is the main dispersal agent and the insects are then particularly vulnerable to predatory attack. We later suggest that the wide gape of aerial plankton feeders is in part a countermeasure to the protean movements of the prey (p.194).

Kennedy and Booth (1963) studying the free flight of aphids in the

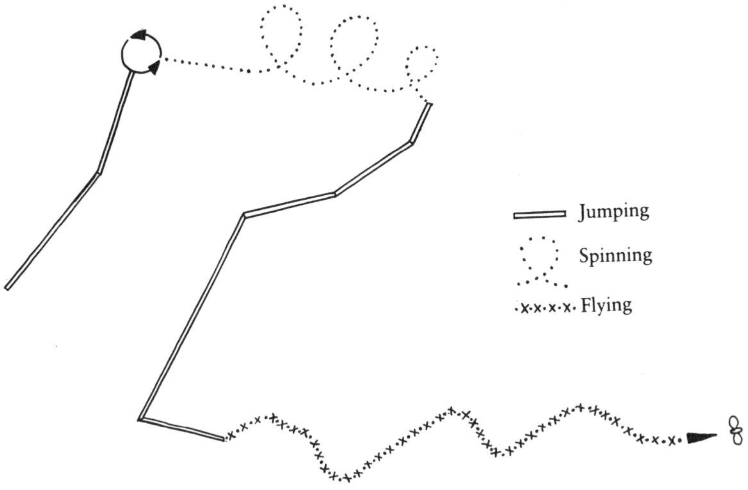

Fig. 4.10. The single erratic display: psychodid fly disturbed by man.

laboratory noted that 'although aerodynamically stable, the fliers behaved in an unstable, sometimes quite "wild" manner at the start'. They note further that 'The type of horizontal movement — looping, criss-crossing, circling, etc., — changed frequently'. They also record that the minimum level of fluctuation in rate of climb was very different from individual to individual. This variation between the degrees of irregularity shown by different individuals is likely to have a further confusing effect. The effects of protean variations between the displays of different individuals are discussed in Chapter 8.

White-flies (Aleyrodidae) show a type of flight even more erratic than that of aphids. Psychodid flies (Psychodidae) also generally fly on an irregular course, although any disturbance results in a dramatic burst of exaggerated protean reactions. *Psychoda phalaenoides*, for example, when disturbed by a sudden current of air or a tactile stimulus, shows a variety of highly protean responses. It may fly in a zigzag, run along with a short series of half-flying hops, or spin round violently like a top, moving in erratic loops over the substratum (Fig. 4.10). There may also be short periods of immobility, and the kind and sequence of responses vary greatly between individuals. It is clear that even though individuals normally show the single erratic display as insurance against possible attack, they are capable of sharply increasing the protean nature of their behaviour if an attack occurs.

This is clearly so in Lepidoptera, many butterflies exhibiting protean elements in normal flight, as described so succinctly in Robert Graves' poem quoted at the beginning of this book. A rarely seen documentary

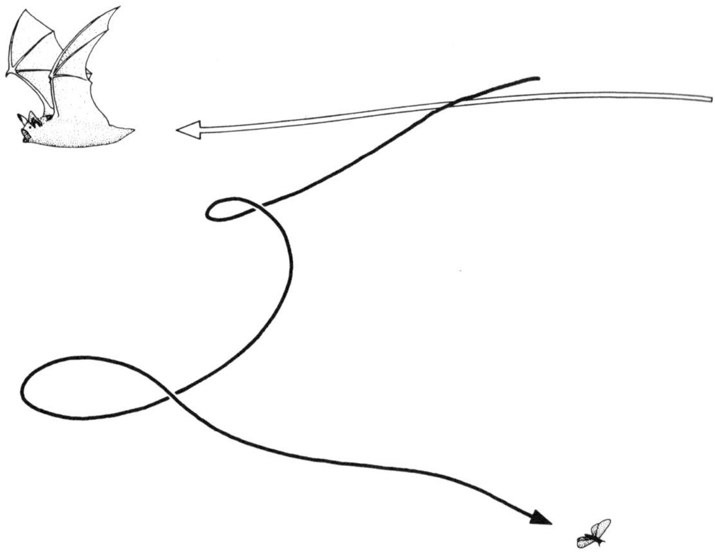

Fig. 4.11. Single erratic reaction of lacewing to bat (after Roeder 1962).

film shows that Graves recognized the erratic flight of the cabbage white butterfly as a 'protean' display, though he did not use the actual term.

This type of flight in which an erratic path is normal is correlated in part at least with the diurnal habit, since many nocturnal insects travel on a much more regular flight path. Many nocturnal moths do, however, show protean reactions when subjected to ultrasonic stimulation of the type used by bats when hunting. Roeder (1962) in a study of the effect of such stimulation on flying insects found that those without hearing organs were unaffected but others, particularly noctuid and geometrid moths reacting within 3 m or so of the ultrasonic source, 'showed a bewildering variety of diving reactions'. He mentions 'a wide variety of manoevres, including passive and power dives, loops, rolls, and one or more tight turns' and comments that 'The variety of unorientated manoevres released by high intensity ultrasonic stimulation continues to defy any attempt at orderly descripton'. In the presence of ultrasound, lacewings (Neuroptera) also dive to cover in an erratic manner (Fig. 4.11).

Roeder and Treat (1961) note that bats soon learn to intercept food moving through the air in a simple ballistic trajectory and they regard the 'random' behaviour of the moths as a natural answer to this predictive ability. We summarize some of Roeder's recordings (1962) in Fig. 4.12.

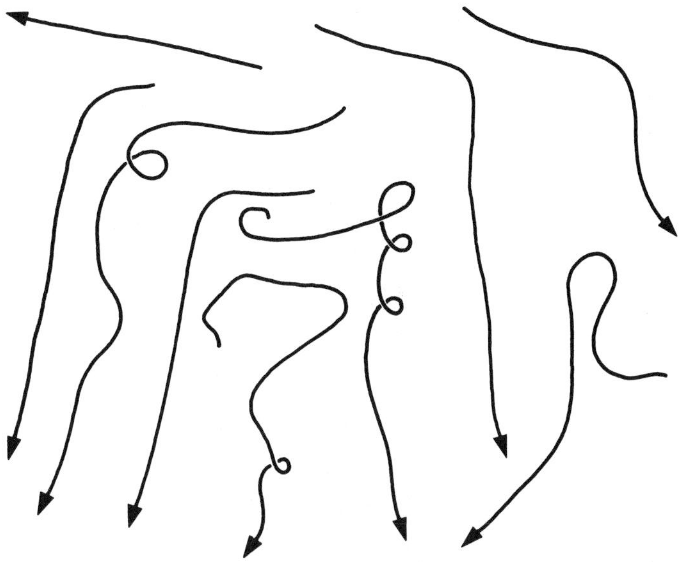

Fig. 4.12. Reactions of moths to ultrasound (after Roeder 1962).

Ground-living arthropods show the single erratic display in various forms. The silver-fish, *Lepisma saccharina* — a primitive wingless insect (Thysanura) — shows the single erratic pattern in its normal movements over the ground, exaggerating the protean characteristics when disturbed. The component 'legs' of the zigzag tend to be more curved than in most other animals, but otherwise the sequence is essentially similar to the others which we illustrate. When proceeding normally, undisturbed, the animal tends to pause between 'legs'; when it is lightly touched, or when a shadow falls on it, the sequence becomes continuous and accelerated (Fig. 4.13).

Springtails (Collembola), bristletails (Thysanura), and littoral Crustacea such as sand-hoppers *Talitrus saltator* and shore-hoppers *Orchestia gammarella* leap erratically to and fro when disturbed by removal of their cover (Fig. 4.14 and 4.15). Anyone who has tried to catch these animals, particularly when several of them are jumping together, will know how effective this display can be. That the erratic movement is not a necessary consequence of specialized locomotory modifications is demonstrated by the characteristics of the diurnal movements of sand-hoppers up and down the beach, when the jumps are fewer, more regular in distance covered, and less hasty, in addition to being predominantly in one direction.

Several species of flea, including *Ceratophyllus gallinae*, *C. garei*, *Ctenophthalmus nobilis*, and *Pulex irritans* also show highly irregular

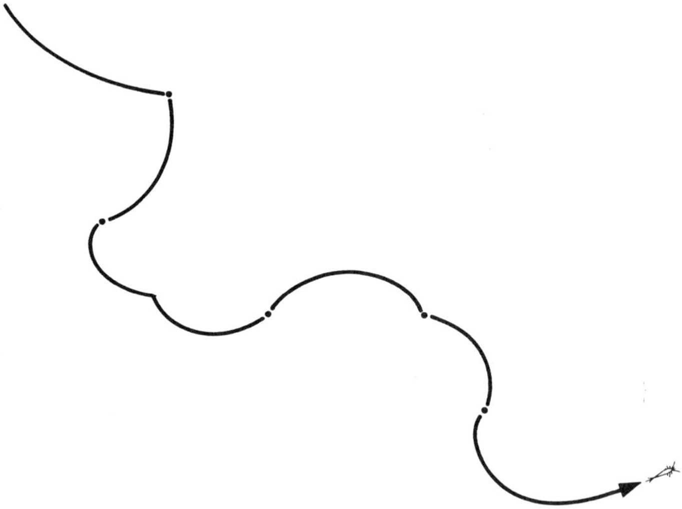

Fig. 4.13. The single erratic display: silver-fish startled by human.

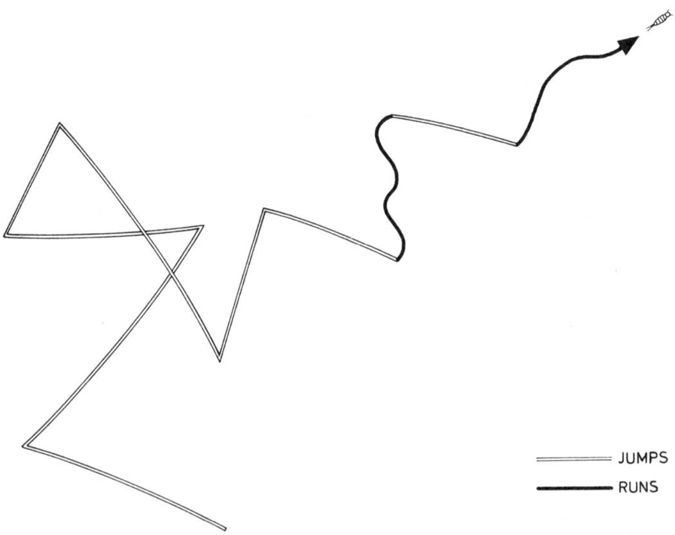

Fig. 4.14. The single erratic display: collembolan disturbed by human.

leaping and scuttling when hunted by their hosts. In *C. garei*, analyses of behaviour after a sharp tactile stimulus has been given show that variations in the direction, distance, and speed of the zigs and zags increase significantly over the variation present in normal behaviour (Humphries 1970).

Many small aquatic Crustacea characteristically move in a jerking or

Fig. 4.15. The single erratic display: sandhopper uncovered on beach.

weaving fashion. In *Daphnia* and *Cyclops* species the protean nature of locomotion becomes very striking if the animal is stimulated by touching it with a fine needle or by gently squirting water from a fine pipette. The animal then moves swiftly and jerkily, zigzagging in three dimensions, and often doubling back on its tracks. Marshall and Orr (1955) record similar behaviour in the small crustacean *Calanus* sp. when attacked by a fish or approached with a pipette: 'The escape reactions may be small darts in any direction away from the fish, but may occasionally be a zigzag flight of a foot or so, or the *Calanus* may whirl around rapidly in a small circle for a few seconds.'

Many of the medium-sized crustaceans also show well-developed single erratic displays when disturbed, for example, prawns (*Leander*) and shrimps (*Crangon*). The former live in shallow pools, resting or scavenging for food anywhere in the pool but particularly amongst seaweed. When disturbed they dart *backwards* into the weed or a rock crevice or simply over the bottom of the pool. In any case they are very difficult to see as they are semitransparent and have pigment spots of changeable size, thereby matching their background very closely. As their escape movements in open water consist of irregular zigzagging darts, they are very difficult to follow visually and have often been lost to view before the movement ceases. In our experience, littoral fishes which feed upon prawns seldom attempt pursuit, depending rather on detecting one as it moves slowly and taking it unawares.

In the case of the bottom-living shrimps the situation is basically the same, with the added factor that the animal is dorsoventrally flattened to maximize the crypsis. If one positions oneself at the surface of a tide flooding in over a shelving sandy beach and watches the bottom a foot or two below one's head, one gradually becomes aware that many species of animals are coming in with that tide, largely cryptic fishes and crustaceans. They are only obvious when they move, which they do in irregular, medium-speed darts. When they stop they disappear unless one concentrates very hard on the moving animal. If one then attempts catch one of these creatures in a net, however skilled one's technique, the prey is likely to escape, largely by a fast single erratic display. The cloud of sand and organic particles stirred up as the animal darts away enhances the effect of the unpredictable movement and the close matching of the background, and one usually has to search for some time before detecting the same animal. In fact, one is more likely to disturb a different animal, which does not become apparent until it moves. This we know by having a colleague observe us in this predator–prey situation.

The whirling behaviour of the smaller crustacea is similar in form to that of the psychodids mentioned earlier (p.57) and also to that of 'waltzing' mice. Certain species of mice and other small mammals, particularly the deermouse, *Peromyscus maniculatus*, show unusual spinning responses when disturbed (Chance and Yaxley 1949). A pigmy shrew, *Sorex minutus*, when attacked by a robin, *Erithacus rubecula*, 'whizzed round and round in circles' each time it was pecked (M.R.A. Chance, pers. comm., 1965). Voles (*Microtus* and *Clethrionomys* spp.) have been observed spinning about in one spot when approached suddenly by an overhead object in a laboratory enclosure (J.C. Fentress pers. comm., 1965). Such gyrations are always confusing to a human observer and doubtless are to non-human reactors also. Probably they act by denying to the reactor any systematic signal to which it can react. They would thus seem to be essentially similar in their effect to the gyrations of foxes, e.g. *Vulpes vulpes*, and stoats and weasels (Mustelidae) which display within full view of their prey — usually rabbits, small rodents, or birds. According to various reports the rabbits seem bemused by the display and allow the predator to approach more and more closely until capture is possible. In the case of the stoat (*Mustela erminea*) Harrison Matthews (1952) writes:

> When several are playing in company they chase each other, box and wrestle together, turn somersaults and leap several feet into the air . . . single stoats make use of such antics when hunting, and it is extraordinary that the intended victims do not take fright, for a stoat cutting these capers is able to approach close to feeding rabbits the nearer of which appear to be fascinated by the strange spectacle, while those at a distance take very little notice. Stoats are said to use a similar performance to get close to birds settled on the ground, making a quick dart at one when within range.

There are several anecdotal accounts of such behaviour in the natural history literature, the dance often ending with capture of the prey.

Professor Tinbergen has kindly allowed us to quote from his own field notes of a stoat 'crazy-dance':

30 May 1966, Ravenglass Reserve. In hide to film wild carrion crows taking Harvey Croze's camouflaged shells; sat 5.30 to 11.30 a.m. . . . 9.15: a stoat suddenly emerged from a small hole in the side of an eroded sand dune on my right and started the famous crazy-dance. It ran up and down the steep slopes like mad, ran in loops over the flat sand in front, making sudden jumps and turns. It had its tail up and made little flicking movement with it, the way a squirrel does (only in 'all directions'). Saw no actual somersaults but it came very near it . . . the movements were too fast to follow.

An example in which a weasel (*Mustela nivalis*) captured a bank vole (*Clethrionomys glareolus*) has been described to us by our colleague J.H. Midwinter. The weasel performed a sequence of bipedal jumps, rolls, runs, and somersaults before pouncing on the vole which had remained motionless throughout (Fig. 4.16).

The crazy-dance of foxes and other Canidae is quite well-known anecdotally but has not hitherto been subjected to serious analysis. Indeed, by its very nature — erratic, unpredictable, and apparently unstructured — it seems to defy logical treatment. Mowat (1963) describes the response of a timber wolf *Canis lupus* to a party of old-squaw ducks on a pond: 'Yipping like a puppy, she began to chase her tail; to roll over and over among the rocks; to lie on her back; to wave all four feet furiously in the air . . .'. C.W. Nicol (pers. comm. 1984) has given us accounts of similar behaviour in the arctic fox *Alopex lagopus*, and the Simien fox (or jackal) *Canis simensis*. In 1968 in the Simien Mountain National Park in Ethiopia, in the Geech Highlands at about 12 000 ft., he watched a Simien fox 'somersaulting, tail-chasing, and side-rolling' on a grassy knoll. In 1961 on a shingly ridge in Devon Island, northern Canada he observed an arctic fox 'jumping on all fours into the air, twisting, and then rolling' in addition to the behaviours mentioned for the Simien fox. In both instances the animals were hunting small rodents.

Anyone who has watched red foxes *Vulpes vulpes* — particularly cubs — at play may have seen similar behaviour. It would seem that when a very high attack drive is blocked from its expression in actual attack, crazy-dance behaviour may be seen as a displacement activity. A tame red fox kept by one of us (P.M.D.) would show all the behaviour described when 'teased' in play with a prey object which continually 'escaped'. We suggest a neurological explanation of the fox crazy-dance in our model on p.185.

Australian aborigines capture kangaroos in the open by performing a display which seems to be similar in many ways to these crazy-dances.

Fig. 4.16. 'Crazy dance' of weasel hunting vole. Here the reactor confusion engendered by single erratic behaviour is used by a predator to confuse its prey.

One individual does a crazy-dance in full view of the kangaroo, the prey apparently being so bemused by the unusual movements that it can be captured by other, non-displaying individuals. One wonders if this aborigine display originated by copying dingos or marsupial predators, or by serendipity.

Some of these convulsoid single erratic displays are similar to deterrent displays which depend on wild incoherent movements (p.182). However, whatever the functional distinctions between them, they all depend for their effect on their highly protean nature. They can all be regarded, basically, as examples of a complex single erratic display taking place within a restricted area.

Single erratic displays are a common feature of play behaviour in young birds and mammals and in humans of all ages. Nice (1962) has

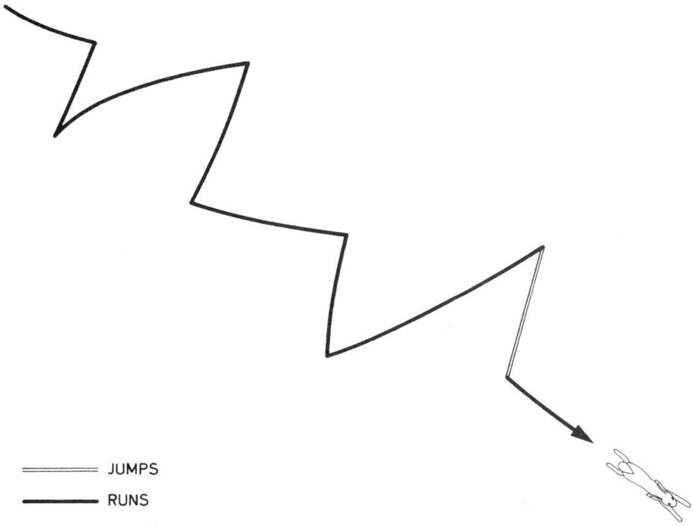

Fig. 4.17. The 'play-fleeing' single erratic display of rabbit.

termed this behaviour 'play-fleeing' in young birds, a term to which we subscribe. She notes that the distinctive character of 'play-fleeing' — which she had previously called 'frolicking' (Nice 1943) — lies in the sharp turns made by the animal. She records this behaviour from a large number of avian species spread over seven Orders. The pattern of this display is like that of other free-flight single erratic displays, including that employed when adult individuals in such species as herring-gull, black-headed gull, mallard (*Anas platyrhynchos*), wigeon (*A. penelope*), rook, and jackdaw chase one another in territorial disputes.

The young of many species of mammals perform play-fleeing displays (Fig. 4.17). This is particularly well known in domestic rabbits, sheep, and goats and, like many others, we have often observed it in wild rabbits. Eibl-Eibesfeldt (1950*a*) records this behaviour in rats (*Rattus norvegicus*) and (1950*b*) in badgers (*M. meles*), roe deer (*C. capreolus*), and horses. Nice *et al.* (1956) note the behaviour for the snowshoe hare (*Lepus americanus*), and Darling (1937) gives details for young red deer (*Cervus elephas*). Play-pronking of springbok is well known and is essentially the same as pronking seen as a predator-reaction. Many other examples could be given (including elephants!, M.J. Coe, pers. comm., 1970). It seems likely that most active terrestrial or aquatic mammalian young demonstrate the single erratic flight path in their play behaviour.

All examples of play-fleeing apparently contain elements homologous with those occurring during the single erratic display seen in predator–prey situations. This clearly supports our suggestion that unpredictable behaviour has been evolutionarily selected for, as play-fleeing in the young may be regarded as the manifestation of innate tendencies which have survival value later in ontogeny. And just as prey-catching behaviours appear not only in the play of kittens, but occur as play in adult cats too, so play-fleeing may occur in adult animals. Lorenz, commenting on Nice's observations on the song sparrow, *Melospiza melodia* (in Nice 1943) writes:

> Most of what you describe as 'frolicking' are emotion-dissociated fleeing movements! The most impressive instance of such frolicking is the pre-bathing play of Anatidae. They do every single stunt that is applied when the eagle is behind them — diving and taking wing at the very moment of emerging and taking a header from the air right under the surface, with a long under-water swim following, and last, but not least, repeated performance of Hakenschlagen — 'hooks' or sudden turns.

The young of our own species also show single erratic play-fleeing behaviour — as may be seen by anyone, any day, in any playground anywhere in the world. This may originally have had an interspecific or internecine fighting practice function, just as it seems to have in other animals (see, for example, Darling 1937). Perhaps even more significant however is the use of the single erratic display in human sports activities. It is, for example, manifested as 'dribbling' in association football and in 'jinking' in rugby football, both of which involve unpredictable changes of speed and direction of movement systematically organized by the actor — who has the ball — but of unsystematic appearance to the reactor — who is trying to get the ball. The protean principle is also of great importance in the plays used in American football and in the feints employed in such combat sports as boxing, wrestling, and judo, but such ploys are not typical examples of the single erratic display and will be considered later. In 'jinking', however, we have an excellent example of the display. Here the actor is deliberately attempting to avoid the clutches of the reactor by means of unpredictable behaviour. The pattern resulting from this conscious attempt to produce a maximal confusing effect is essentially similar to the single erratic display unconsciously used by many free-living prey animals. We see the same unpredictable changes in direction and speed and even the occasional sudden stop. Additionally, the display may occur when the actor is running away from, as well as towards, the opponent. Altogether, when used efficiently the jink is very difficult to deal with indeed and is one of the rugby player's best weapons. It is particular interesting that all rugby players so far questioned (including both authors) agree that although there may be a conscious plan to jink, the detailed parameters *are not under conscious control.* Herein may lie the truth about great rugby players: they are *born* jinkers, not merely the product of intensive training or superior intellect.

The jink in rugby football is of further interest because protean factors other than those in the form of the flight path can be introduced into it. It may be made more complex, for example, by the actor offering the ball to the reactor at a certain point in the display or by vocalizing unexpectedly. This is similar in function to the real or apparent changes in colour, form, or sound characteristics which are introduced into the single erratic displays of many other species. Huxley (in Cott 1940) notes that flash coloration and zigzag fleeing are associated in certain grasshoppers. Nicol (1960) suggests that sudden colour changes, as seen in certain teleost fishes and cephalopods, may 'serve to distract or confuse predators or prey'.

A striking example of the introduction of colour change into a display is provided by the cuttlefish, *Sepia officinalis.* Holmes (1940) describes a 'series of colour changes of remarkable rapidity and completeness', each pattern lasting for a few seconds only. First, a pair of black spots appear, the rest of the animal becoming pale.

> Often this is followed by the animal's contracting its mantle violently and shooting away by the action of its siphon . . . accompanied by a colour change which makes it amazingly difficult to follow, even to the human eye. For while at one moment one's eyes are fixed on a white animal with two black spots, at the next it seems to have disappeared, for its rapid movement is accompanied by total darkening of its body, by full expansion of all chromatophores. This complete colour change, most deceptive to a human observer, must be even more effective in nature in deluding predators.

Further patterns may appear, including longitudinal flickering black lines, black spots, or a zebra pattern, the animal continually darting to and fro. If the predator persists, the cuttlefish is then likely to eject an ink cloud which distracts the predator and behind which the prey makes its escape (Fig. 4.18).

The cephalopods — some of the most advanced invertebrates and the most advanced of the Mollusca — employ this display quite widely. Boycott (1965) describes squid, *Sepioteuthis* spp., switching their colour on and off rapidly when fleeing after disturbance. Deeper water squid such as *Heteroteuthis* can eject either an ink cloud or a luminous secretion which breaks up into sparkling droplets (Nicol 1971). Colour and form changes associated with escape also occur in octopuses, for instance in *Paroctopus bimaculatus* (MacGinitie and MacGinitie 1949)

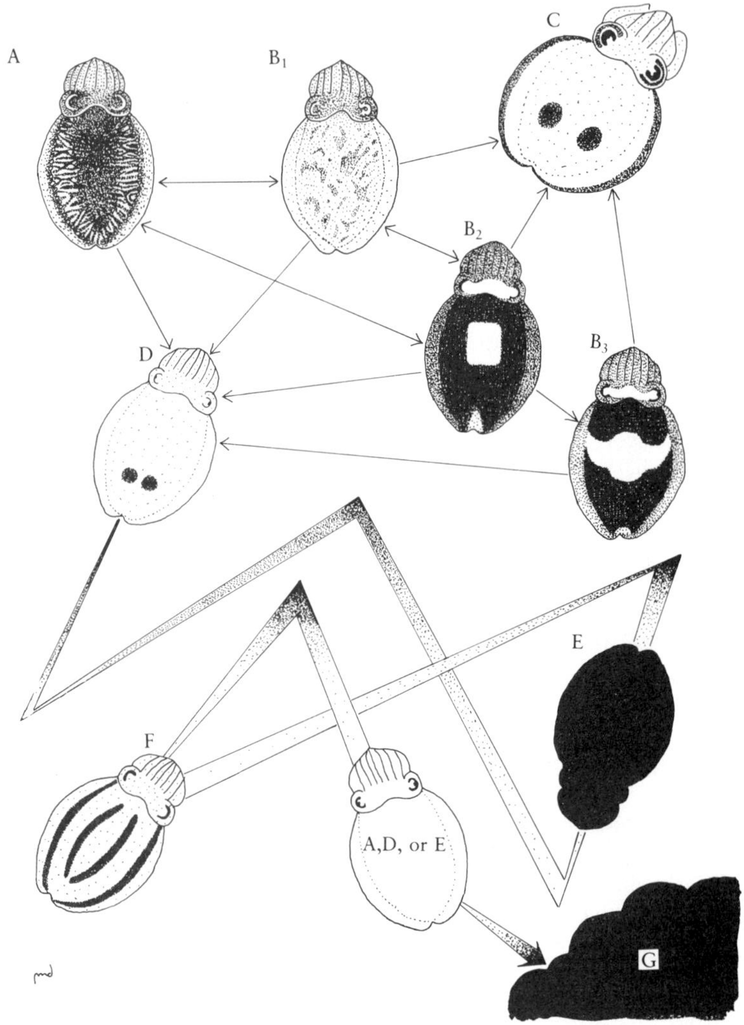

Fig. 4.18. Escape behaviours of cuttlefish *Sepia officinalis*. A, zebra pattern of free-swimming animal (cryptic). B_1, light mottle pattern of animal on sandy bottom (cryptic). B_2, white square pattern and B_3, white stripe pattern of animal on contrasted bottom (cryptic). C, flattened display of cornered animal (deimatic-startling). D, black spot; E, purple black; and F, black striped patterns of attacked animal (protean, possibly deimatic). G, black tone of animal in darkness or under ink cloud (based on Holmes 1940).

and *Octopus vulgaris* (Wells 1962; Packard 1960; Packard and Sanders 1971).

Other marine animals use similar protean escape techniques to the cephalopods. Several prawns for example, including *Heterocarpus*, produce luminous puffs while zigzagging away (Harvey 1952).

Bioluminescence is not uncommon in benthic animals and has long been noted by zoologists of many persuasions so that there is now an extensive literature on the subject. Herring (1979) provides a recent survey.

The protean principle seems to operate in many of these displays. Beebe (1935), for example, noted that the flashing lights he observed were highly varied in pattern; of longer or shorter duration, regular or irregular, and extremely varied in size and juxtaposition. Some were pale yellow, others bluish. Watson (1979) records similar flashing patterns near the ocean surface at night.

Buck (1978) notes that

> sudden changes in direction have been described in [some] animals at the moment of light emission (or, in the fishes *Anomalops* and *Photoblepharon*, at the moment of brief extinction of their mostly continuous emission). Notable examples are the siphonal jetting of the squid *Heteroteuthis*, the darting directional change of the ostracod *Cypridina*... the tail-flip of the decapod shrimp *Systellaspis*..., the positional change of the copepod *Metridia*...

Many teleost fishes change colour swiftly when disturbed, an ability well known in flatfish. Gordon (1977) notes that the summer flounder, *Paralichthys dentatus*, 'can assume a wide range of colour patterns from checkerboard and spotted arrangements to various shades of gray, blue-green, brown, pink, orange and almost black'. In the case of the Nassau grouper, *Epinephalus striatus*, Cott (1940) records that it 'can assume six or even eight different liveries, which may be put on and off within a few moments'. Gordon (1977) writes:

> In one phase the fish is uniformly dark, in another creamy white; in a third it is dark above with white underparts. A fourth phase shows dark bands, the whole fish taking on a light brown coloration. In a fifth phase the fish is pale, with all dark markings tending to disappear. The sixth phase shows a light-colored fish with the whole body sharply banded and mottled with black. The last common phase provides the grouper with a dusky hue above and white below, with a black band in the middle region from head to tail.

He goes on to recount how William Beebe saw a blue grouper with three broad vertical brown bands swim into a clump of coral, to emerge a little later coloured bright yellow with black dots. Even if this were two different fishes it indicates the range of colouration which groupers may display.

Many reef-dwelling fishes share this ability to change colour drastically and rapidly. It basically serves a cryptic function but has acquired protean properties through the almost instantaneous occurrence of the changes when the fishes are provoked. And there may be an acoustic accompaniment. In the Nassau grouper this consists of percussive sounds produced by the action of a specially toughened peritoneum upon the swim bladder. Sound production during the protean display of the squirrel fish, *Holocentrus ascensionis*, is described in the next chapter where changes in visual and auditory appearance are considered in more detail.

The effectiveness of unexpected sound production is seen in the protean screaming of captured birds such as starlings, *Sturnus vulgaris* (Driver and Humphries 1969). The startle, or 'deimatic' effect (Edmunds 1974) may here result in the escape of a prey animal which has already been captured. We may therefore expect it to be of value in a single erratic flight pattern where the predator does not have the advantage of having the prey within its grasp.

The startle effect may also operate in mammals. Rabbits may scream unexpectedly after, or sometimes before, capture. We have not observed the successful escape of such an animal but this behaviour may be effective in the escape of rabbits from inexperienced predators such as fox cubs, or from predators which are not very hungry and therefore have a low 'capture and kill' drive. This could also apply in the case of amphibia which scream before, of after, capture (Campbell 1969).

Novick (1969) recounts the successful startle effect of screaming in the case of a captured vampire bat:

> Crawling to the apex of the cave, I was able to grasp one vampire's wing as it clung to the interior of a crevice. I tugged gently; the vampire tugged back. We had reached an impasse until the vampire, reaching for a more secure hold, momentarily released its grip and I was able to start to pull it from the crevice. The bat responded by swinging around, screeching violently, and trying to take flight. I was unnerved by the activity and let go.

The bat escaped, and it may well be that other mammals escape from predators as a result of the unpredictable onset of such screaming behaviour.

The survival value of forms of behaviour which are rare for any particular type of prey, and which are therefore unexpected, is illustrated by the discovery of the first living specimen of the strombid mollusc, *Trigonia*. This had been dredged from Sydney Harbour by the conchologist Sutchbury and placed on the thwart of his boat. His companion laughed at the suggestion that it was *Trigonia*, saying that all known *Trigonia* were fossils. At this point the specimen put an end to the discussion by jumping back into the sea! Kohn and Waters (1966) have since demonstrated that the leaping of strombids is indeed a means of escaping from predators.

Gastropods such as *Nassarius* and *Struthiolaria*, and bivalves such as *Cardium* and *Spisula* leap away when touched by predatory starfish (Bullock 1953; Ansell 1969*a*,*b*, Feder 1972) The protean element involved is the direction of leap. There is a clear similarity here to the sudden sideways dart of a disturbed shore crab, *Carcinus moenas*. One can be sure that the crab will move, but one seldom knows in which direction.

An elaboration of this behaviour is seen in *Helicarion*, a terrestrial mollusc of the Philippines. When attacked by a bird it vibrates its 'tail' vigorously until it drops off (providing a lure). Simultaneously the animal gives a great leap which may conceal it among the surrounding vegetation (Cooke 1895).

Mammals may also evolve mechanisms for change of appearance during a single erratic display, an outstanding example being the white-sided jack-rabbits *Lepus alleni*, *L. callotis* etc. of Mexico and Arizona. As noted by Nelson (1909):

> By means of muscles the skin of either side can be drawn over the back at will. In this manner the buffy or brown dorsal area is shifted more or less completely to one side and the white on the opposite side is drawn nearly or quite to the median line. This habit has been observed when the rabbits were standing, or moving along at moderate speed, usually after they had been driven from their forms. The enlargement of the white area is always on the side turned towards the chance intruder, and accordingly alternates from side to side as the animals zig-zag away.

This behaviour must introduce another variable into the already complex single erratic display and increase the overall protean effect. Another example is the springbok with its lumbar pouch of white hair which is flashed when the animal is 'pronking'. Such displays must be attention-drawing and occasionally, as in avian injury-feigning, will attract a predator's attention to prey it had not previously observed. This applies to many cursorial mammals (reviewed in Harvey and Greenwood 1978) including rabbits, deer, and gazelles, and one would think it disadvantageous. We discuss this in more detail in Chapter 7, but for the present we would point to one common feature of these back and rump patch displays: they signal to both predators and conspecifics. The predator's attention is drawn to the displaying animal, and conspecifics are alerted with the result that they too may start to display. Thus the predator may become distracted and confused by a multiplicity of displaying animals (Walther 1969; Kruuk 1972).

Finally in this section on single erratic displays and their variability, we refer to the behaviour of the fox, *Canis vulpes*, an animal with a remarkable spectrum of behaviour at its disposal. The fox's convulsive and immobility displays will be referred to under those headings later, but directed fleeing may also have protean features. The flight path itself does not show a typical single erratic zigzag unless the animal is very hard-pressed, but it may be highly unpredictable in other respects. A fox fleeing from hounds may take to water, or run along walls or tree branches. Or it may, deliberately or fortuitously, cross the paths of other foxes. The result is a protean scent trail (D. Driver, pers. comm., 1976) which is just as effective against a pursuer hunting by scent as the standard single erratic display is against a visually-orientated predator (Fig. 4.19). In a similar manner wild hunting dogs may lose wildebeest, *Connochaetes gnou*, when prey individuals rejoin the herd (Kruuk 1972). There is an element of distraction in these examples (see p.133) but, whatever other mechanisms are involved, the unpredictably structured chemical confusion may allow pursued individuals of many species to survive to flee another day.

Derived erratic displays

It is clear that the single erratic display is a common feature of the behaviour of animals engaged in conflict, particularly prey animals. We have indicated the main forms this display can take. We must now consider certain related behaviours which have been largely derived from the single erratic display. Some of these are already known by other names, e.g. convulsive displays and injury-feigning, and for the moment we shall retain those names. In other cases we are describing previously undefined behaviours, and the terminology is original. Some examples involve animals performing the displays in association with other behaviours; others concern animals performing simultaneously. Beneath the diversity and elaboration, however, there is the foundation of the basic single erratic display. We therefore classify these further examples as *derived erratic displays*.

The multiple erratic display

We have seen earlier (p.3) that animals frequently perform their erratic displays in company but not in co-ordination with others, for example in dancing midges (Fig. 4.8), whiffling waterfowl (Fig. 4.6), and bouncing impala and springbok (Fig. 1.2). The result of the increase in the number of actors is a great increase in the confusing effect, for a reactor not only has to attend to different displaying individuals, but also is presented with significant variations between

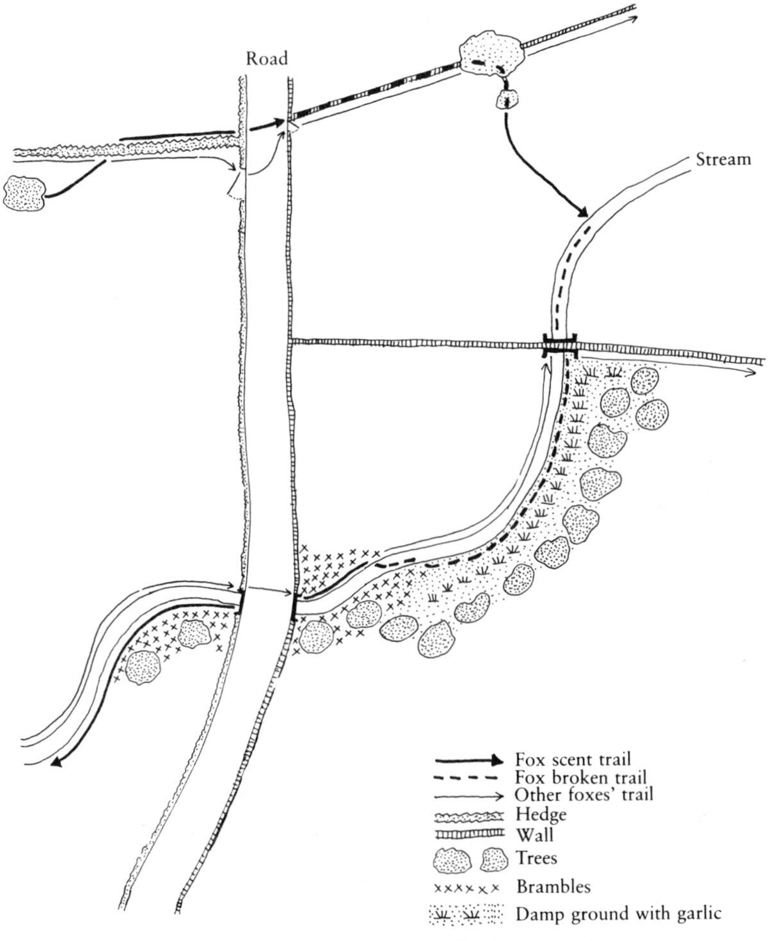

Fig. 4.19. Protean scent trail of fleeing fox.

the displays of those animals. And even if the displays were identical, which is never so, the fact that they are not performed in unison means that at any one moment the different individuals will be doing different things. The effect here is apparently similar to that which we later suggest is achieved in mobbing (p.100).

The multiple erratic display may not always have been selected for in the normal Darwinian sense. It would seem rather that the increased individual protection provided by multiple erratic, compared with single erratic, displays is largely a bonus resulting from the social nature of the actors. But it is *the effect* that is of primary importance, and it is possible that the advantages accruing from the effect of numbers may have operated phylogenetically in reinforcing both the gregariousness and the individual erratic displays which contribute to the multiple erratic display.

The multiple erratic effect is commonly associated with the spatial restriction of a group of animals. In other instances the restriction arises from behavioural requirements of the actors. The multiple erratic effect may be achieved fortuitously when a group of animals is attacked in a physically restricted area, as when a fox enters a fowl house or fish are attacked in a small pool. The panic bouncing of individuals off the walls of the enclosure results in a complex multiple erratic pattern. Normally, the fowls' reaction to attack would be a scatter display (p.85). In a school of fish it would be a united erratic display (p.76). In a restricted space these displays cannot be mounted, and the animals behave in a panic-stricken manner with a multiple erratic effect. Anyone who doubts the effectiveness of this behaviour has only to attempt to seize fowls in a small fowl house. The birds fly in all directions, bouncing off the walls, flapping and scrambling erratically, providing the intruder with an ever-changing stimulus complex. A panic-stricken bird may even fly in one's face, adding a deterrent effect to the confusion. However, as the actors remain within a restricted area, one or more of them will eventually be caught unless further measures are adopted. Thus, when grey mullet are attacked in a small pool, after the multiple erratic display each survivor's flight path will, if necessary, terminate in concealment (Fig. 4.20).

The multiple erratic pattern is seen when several lapwings display over adjoining territories (Fig. 4.21). They swoop and tumble with noisy wings and cries, bounding in the air like midges dancing in a shaft of sunlight. Though not confined by the material walls of an enclosure the birds stay in a loose group, restrained by the invisible bonds of their adjacent territories. Human intrusion results in an increase in the protean nature of the display and the confusing effect of birds displaying erratically all around is considerable.

The dancing midges are also invisibly bound in their communal dance. For reasons not fully understood, the midges stay in a group, each performing its own dance, and the whole forming a maze of ever-changing detail (Fig. 4.8). In both midges and lapwings the display would seem to make the actors less vulnerable to attack. The aerial dances form a protean pattern of confusing detail so that it is more difficult for a predator to fix its sights upon any one animal for an effective period of time. Swatting at a summer roomful of houseflies produces the same pattern, and the same frustration on the part of the predator. Such is also the result of the amazing single erratic/multiple erratic/scatter display of disturbed shore-hoppers (p.92).

The general pattern of distraction and confusion in these displays

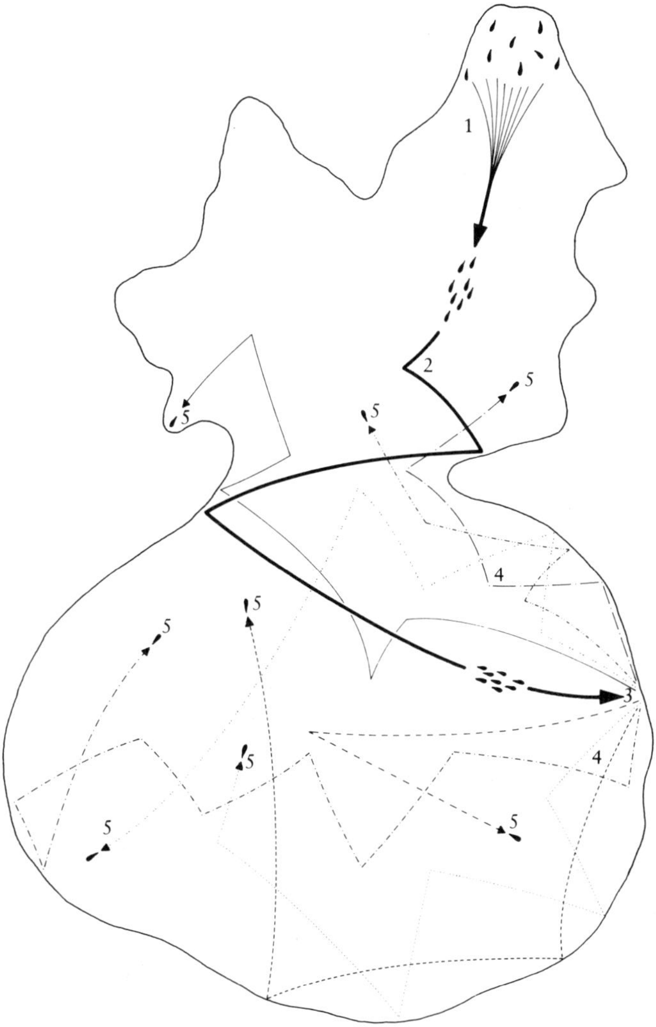

Fig. 4.20. Reactions of grey mullet disturbed in rock pool. 1, bunching; 2, united erratic; 3, scatter; 4, multiple erratic; and 5, single erratic displays.

would seem to be the same as that presented to predators by moving herds of ungulates: it is that quoted for impala in Chapter 1. For the same species Moss (1976) notes how the animals, when disturbed, 'leap explosively' in different directions, while uttering sharp bark-like alarm calls, reaching 10 feet in height and 30-feet span: 'The effect of this defence is to confuse the predator so that it cannot concentrate on an individual'. Groves (1972) makes almost identical comments. J. Hanby and D. Bygott (pers. comm., 1979) confirm this.

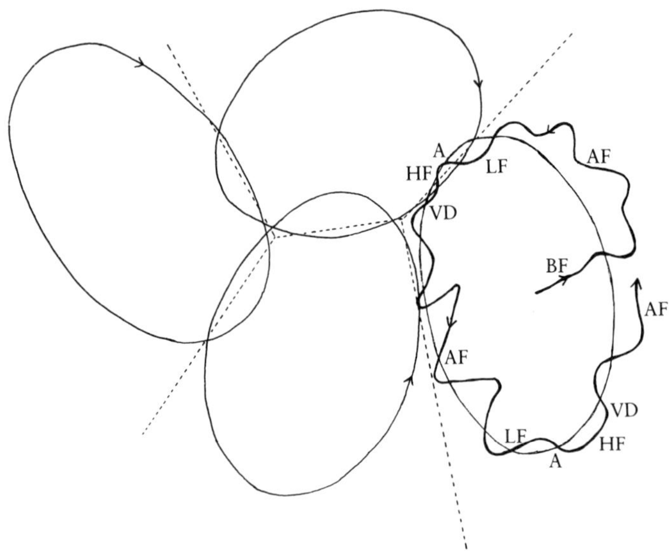

Fig. 4.21. Lapwing: Song flight routes of four territory owners (smoothed). An actual example has been superimposed upon the right-hand route. Broken lines indicate territorial boundaries. Flight type initials as in Fig. 4.5 (based on Dabelsteen 1978.)

The united erratic display

In certain gregarious animals such as schooling fishes and fast-flying bird flocks the pattern of erratic flight behaviour seen so commonly in individuals as the single erratic display is shown by the group moving in unison (Fig 4.22). This is a distinct category of primary protean display involving special mechanisms for the close co-ordination of individuals in united movements. These mechanisms are clearly innate and must have been developed by natural selection. Many species of fishes show this behaviour and in Fig. 4.20 the mullet perform a united erratic display before showing multiple erratic behaviour.

The united erratic display is well-known in birds as it is the typical reaction of gregarious species such as starlings, ducks, and waders, to the appearance overhead of a peregrine falcon. Tinbergen (1951) notes how starlings react in these circumstances by drawing together into a very dense flock which zigzags with close co-ordination (Fig. 4.23). The attacking speed of the peregrine — over 240 km/h in a full 'stoop' or dive — makes it highly vulnerable to damage from collision unless its talons strike first. The close packing in the starling flock, its erratic zigzagging, and the protean shuffling of the individuals' relative positions deter the peregrine so that its attacks are incomplete or not

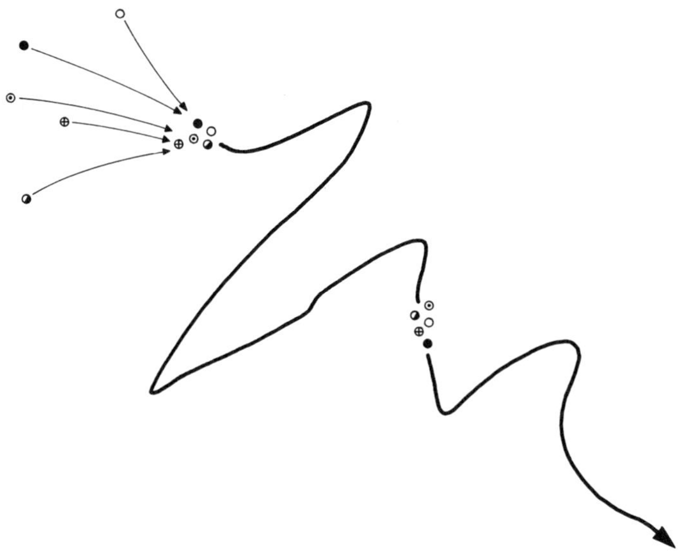

Fig. 4.22. The united erratic display — generalized form.

mounted at all. Only when an individual loses contact with the flock is the peregrine able to select that suitable target and mount a full attack.

Although ducks and waders show the united erratic display in its most strongly developed aerial form — presumably because of their long history as prey of the peregrine — this display is not restricted to these groups. In addition to the starlings already mentioned, other social species such as black-headed gulls display united erratic predator-responses. For the latter species Kruuk (1964) noted that the changes of direction of flight in this display were both synchronous and unpredictable.

Sometimes, a flock performing a united erratic display will separate into two or more smaller flocks, as noted by Witherby *et al.* (1944) in the case of the lesser yellowlegs. Each of the smaller flocks may continue to perform a united erratic display, and will sometimes cross-over one or more of the other flocks in flight. Shuffling of the individuals' relative position also occurs within each flock. All these different behaviours increase the confusing effect. The overall picture is given diagrammatically in Fig. 4.24.

This separation, cross-over, and shuffling is very similar in appearance and effect to behaviour which occurs in the great herds of wildebeest in the Serengeti plains and elsewhere. Kruuk (1972) notes that if a hyena runs into the centre of a herd 'he often causes it to split up into a number of small groups with the wildebeest running very close to one

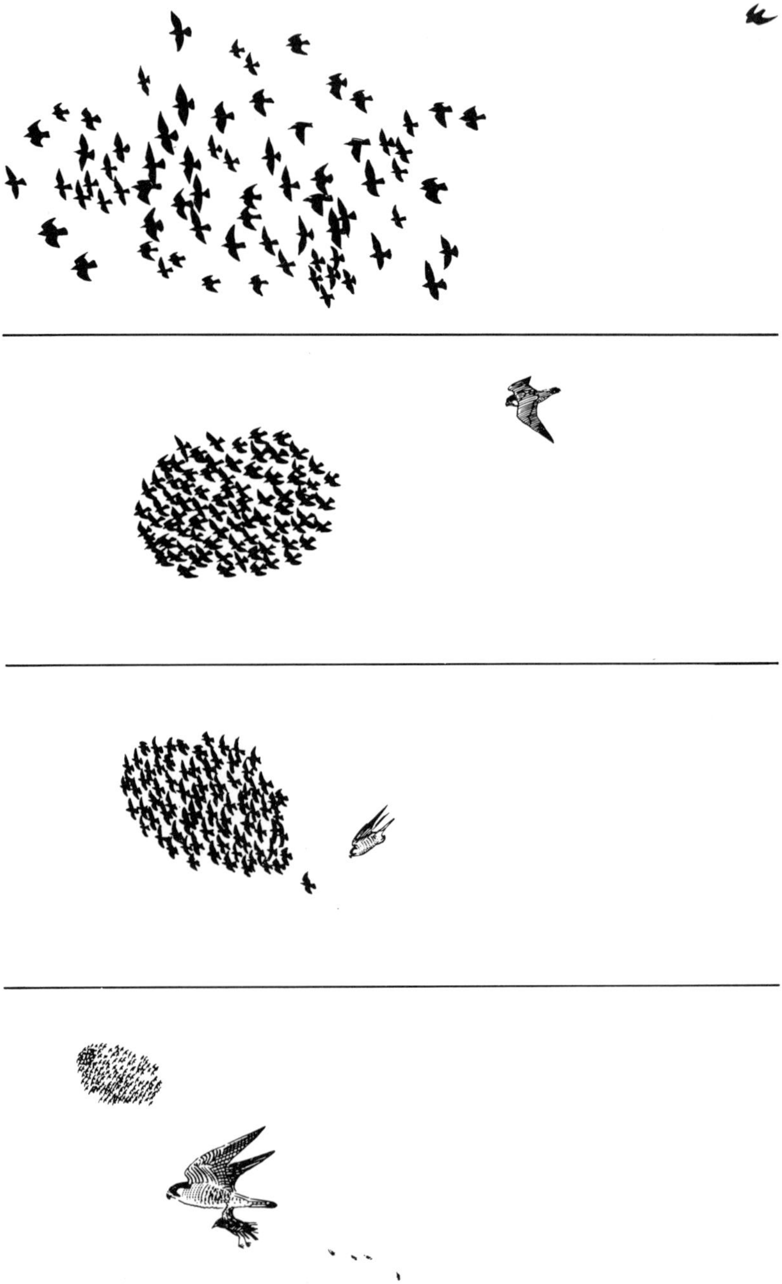

Fig. 4.23. Tactics used in encounter between peregrine and starlings. 1, Starlings in usual loose flock, peregrine appears; 2, starlings bunch to perform united erratic display; 3 and 4, peregrine attacks bird that becomes separated from flock.

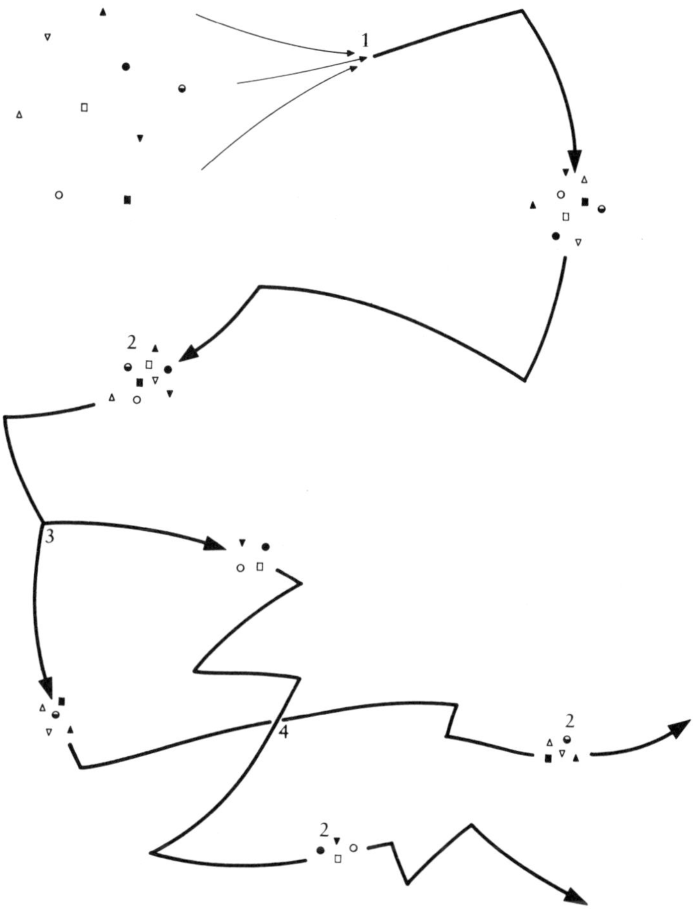

Fig. 4.24. Wader united erratic display. 1, Bunching; 2, shuffling; 3, splitting; 4, crossing-over.

another. These small groups may merge, split up and so on, in one enomous melee'.

Different species often enter into the composition of duck and wader flocks or ungulate herds, and any separation of the displaying group is likely to be into the separate species. This may commonly be seen at resting places favoured by wader flocks on autumn migration. We have noted it frequently in mixed flocks of turnstone (*Arenaria interpres*), dunlin (*Calidris alpina*), and redshank, in any combination. Starlings also, sometimes fly with other species, such as lapwings and corvids.

Many writers have commented on the remarkable co-ordination of the individuals in a starling flock, but this co-ordination is even more impressive in duck and wader flocks, for these two groups have developed fast flight more extensively and the erratic flight pathway is covered at a much greater speed. We suggest that the special wing and tail markings of these two groups — particularly the wing bars and specula in waders and ducks respectively — have been developed at least in part to assist in the co-ordination of flock movements (Fig. 4.25). These markings undoubtedly have species-isolation and/or sexual recognition functions, but the fact that they are situated so that they are displayed optimally in flight suggests an aerial function also. It may be questioned that the sensory, nervous, and muscular elements involved could function quickly enough to serve for co-ordination through observation of neighbours' wing patterns. That birds' visual reflexes may actually be fast enough is suggested by hunters' frequent claims that flying ducks, waders, and others can avoid gunshot by swerving on the flash of a gun. We have suspected this ourselves and have also observed that swimming black guillemots (*Cepphus grylle*) and common loons (*Gavia immer*) can, apparently, escape gunshot by diving. The suggestion is strengthened by a comparison of the average muzzle velocity of 12-gauge shotguns — a little over 300 m/sec — and the mean visual reaction time of starlings under experimental investigation — around 77 msec (Pomeroy and Heppner 1977). Even faster reactions were found by Greenwalt (1955) in the evasive responses of birds in flight to bright stroboscopic light flashes. Titmice (*Parus bicolor* and *P. atricapillus*) reacted within 30 msec. Certainly, visual reaction times of the order quoted should enable waders in fast zigzagging flight to maintain their positions in the flock to well within a metre per course alteration, provided that there is a clear and immediate signal of any change in direction by neighbouring birds.

Evidence of the reaction times that may actually be required in avian united erratic displays is provided by the work of Davis (1980) who used cinefilm to analyse behaviour in dunlin flocks. He found that the average time that elapsed between the first individuals turning and more than half of the individuals turning was 196 msec. He noted however that there was great variability, and that in six of his 10 recorded instances the majority of the flock had turned within 120 msec. Of course, it may be that the changes in orientation recorded by Davis had been set off by some unknown signal preceding the first full turn by any individual; if so, reaction time would certainly not be a limiting factor, but it is a simpler hypothesis to assume that the first turns themselves provide the signal to the rest of the flock. Although this requires swift reactions, the times cited by Pomeroy and Heppner and by Greenwalt are short enough to be consistent with the time-course of events in the dunlin flocks.

Davis (1980) notes that during straight flight by the dunlin flock

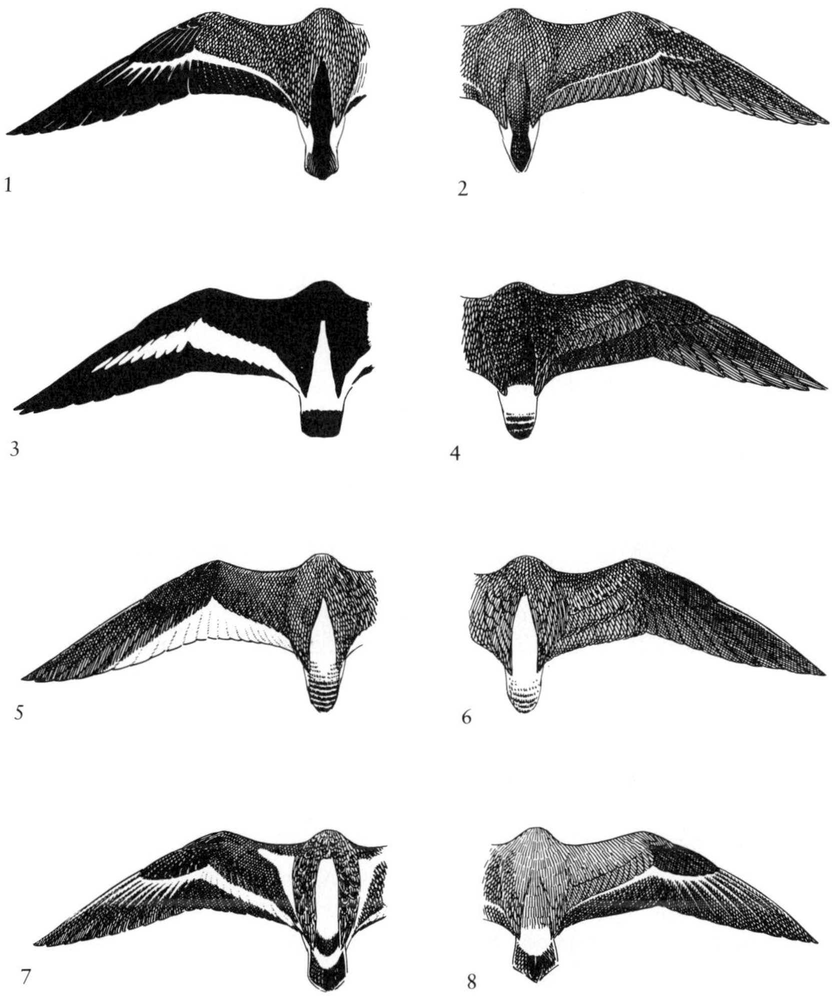

Fig. 4.25. Wader wing and tail markings: principal patterns, with some examples (species illustrated are italicized). (1). *Dunlin*, ruff, sanderling, Temminck's stint, red-necked phalarope, grey phalarope, ringed plover, Kentish plover, least sandpiper, semi-palmated sandpiper; (2) *little ringed plover*, little stint, pectoral sandpiper, purple sandpiper; (3) *oystercatcher*, black-tailed godwit, willet; (4) *green sandpiper*, stilt sandpiper, lesser yellowlegs, Wilson's phalarope; (5) *redshank*; (6) *greenshank*, dowitcher, bar-tailed godwit); (7) *turnstone*; (8) *curlew sandpiper*, white-rumped sandpiper, knot.

there are momentary changes in orientation by individual birds, and he suggests that they might be indicating tentative reversible decisions to change direction; only if sufficient birds make the same 'vote' at the same time would the rest of the flock follow.

We note that such a group-decision mechanism would demand the rapid reaction times cited above, and might therefore also require that the voting signal be clear in both onset and withdrawal. Contrasts of pattern dependent on orientation would fit this requirement, and it is significant that those species with the most clearly-developed wing and tail markings are prominent among those with a highly-developed united erratic display.

The whole matter of wing, rump, and tail markings in birds, and its relationship to flash displays in insects (p.125) and mammals (p.182) is worthy of more extensive consideration. However, for the moment suffice it to say that the protean principle seems to have influenced the development of nearly all such markings.

Schooling behaviour in fishes — particularly those aspects which we consider to be protean — has been the subject of considerable attention by fisheries biologists and others. Eibl-Eibesfeldt (1962) and Kuhlmann and Karst (1967), *inter alia*, have suggested that certain co-ordinated group movements have evolved as antipredator devices. Wilson (1975), however, has quite properly noted that 'It is difficult... to judge to what degree these group patterns stem from co-ordination and to what degree they are the mere outcome of selfish evasive maneuvering by individual fish'. This comment is especially pertinent in view of recent developments in the theory of the 'selfish herd' (Williams 1964, 1966; Hamilton 1971; Dawkins 1976*a*) — which, more properly though less elegantly should be the 'herd of selfish individuals'. This idea depends upon the fact that an attack by a single predator entails a much greater risk to an isolated individual than to an individual closely accompanying others of its kind. In a group the risk from the single attack is shared and is therefore much lower for each individual. This effect is likely to be an important factor in the evolution of grouping behaviour. However, the simple numerical division of risk among alternative targets is not the whole story. The overall risk is not only divided in this way, it is also *reduced* through the operation of largely protean mechanisms. Thus the closely similar appearance and the sharply changing relative positions of individuals tend to delay or prevent attack because the predator finds it difficult to maintain a particular individual in its sights for long enough to orientate and launch an effective attack.

Hobson (1968) noted that 'A number of studies have favoured a "confusion effect" as being an important factor in the school's role in reducing predation' (referencing Allen 1920*a*,*b*; Eibl-Eibesfeldt 1962; Mantiefel and Radakov 1961; Hobson 1965; Starck and Davis 1966). He continues, 'The theory contends that a predator required to direct its attack at an individual fish experiences difficulty when confronted with the numerous targets presented by a school, finding members of the school difficult to distinguish. This theory is generally consistent with my own observations'. It is also consistent with our thesis. In fishes, as in birds, the bunching effect prevents the predator from selecting an individual prey to attack, the confusing image being greatly, perhaps necessarily enhanced by shuffling within the group and the united erratic movement. Shuffling is of course an inherent aspect of the united erratic display, for it would be almost, if not quite impossible for the individual animals in a fast-moving zigzagging group to maintain an exact position in the group relative to all their neighbours.

Hobson (1968) also records that the several prey species of shore fishes which he observed in the Gulf of California frequently tightened up their schools when threatened. He notes that similar reports are widespread and references Allen, (1920*a*,*b*), Blaxter and Parrish (1958), Breder and Halpern (1964), and Eibl-Eibesfeldt (1962). He comments: 'If, as the confusion theory suggests, a predator is confused by multiple targets, then the more of such targets that can be placed in the predator's visual field, the greater should be the effect'. This suggestion is entirely consistent with our thesis. It is further supported by the work of Neill and Cullen (1974). Working with squid (*Loligo vulgaris*), cuttlefish (*Sepia officinalis*), pike (*Esox lucius*), and perch (*Perca fluviatilis*), they found that the predators did best when hunting single prey fishes, less well with small shoals, and worst with large shoals.

Heller and Milinski (1979) found that sticklebacks (*G. aculeatus*), feeding on swarms of water fleas (*Daphnia magna*), only attacked dense swarms when they were very hungry. The authors provide experimental evidence in support of the suggestion that it is the confusing effect of numbers which is the controlling factor. Milinski (1979) notes that 'it does not seem likely that training can totally overcome the confusion effect in sticklebacks'. Earlier observations by Allen (1920*a*) and Hobson (1968) also suggest that experience is of little use to predators attacking fish schools.

Baerends and Baerends-van Roon (1950) consider that a rapid change in the density of the group in itself has a confusing effect. This may be so, but we suggest that the concentration into a more tightly-packed group has as a primary function the establishment of a formation which makes the united erratic display optimally effective. It would appear that schooling behaviour in fishes increases hydrodynamic efficiency (Weihs 1973*a*) and so the displaying animals will be able to move faster and more erratically with minimal increase in energy expenditure. It may even be that the drag-reducing mucus produced by fishes to improve their swimming efficiency gets washed off in small amounts and thus benefits fishes swimming close behind (Breder 1967). The fact that schooling of prey fishes seems, in some cases, to make them more vulnerable to predatory attack (Bullis 1961; Williams 1964) does not necessarily detract from the protean thesis. As Hobson (1968) points out, some predators may have 'developed methods of attack that capitalise on what is generally an advantageous maneuver of the prey'. We are reminded of the natural variability and complexity of biological systems.

Once the necessary minimum group density for a fast-moving united erratic display has been established, the problem then is to carry out the display effectively. This involves bringing into operation certain of the elements of unpredictability outlined in the next chapter.

The protean nature, and the confusing effect, of the united erratic display apparently depend on two main factors: the erratic structure of the united flight path and the unpredictable shuffling within the group.

There is abundant evidence of the occurence of unpredictable zigzagging in united erratic displays. As for the shuffling, this may be seen by anyone who is able to watch flocks of ducks or waders when they are disturbed. For fishes Hobson (1968) again provides information, noting that the species he observed, after packing closer when threatened 'also increase swimming speed as individuals begin to weave in and out about one another. Such behaviour is certainly effective in making the individual more difficult to distinguish'.

Many species of schooling fishes have a silvery colouration, as noted by Breder (1959), and in the next chapter we see how apparent changes in colour are introduced into the erratic display by the movements of such fishes, and other animals too. Hobson (1968) notes

> The effect of masking the individual is greatly enhanced when the schooling fish is silvery, like *Harengula thrissina*. When such fishes are alerted, they, like the other schooling species, increase their swimming speed as they begin to dart in and out. Because of the irregular, constantly changing course they follow, sunlight reflecting from their silvery bodies at rapidly changing angles is manifested in a large school as a bewildering shower of brilliant flashes.

B.N. Walker (pers. comm. 1980) suggests that a predator may fix its attack orientation on one of these flashes, thus missing the herring, which has moved on. The protean movements of the fishes make this flashing display quite unpredictable in time and space. The flashes have no detectable connection with the speed, direction, or changes in the fishes' movements, and the overall confusion effect must be very considerable.

The confusing effect of this unpredictably varying pattern of flashes is similar to that seen in a number of other animals, including the 'jewel flies' (Dolichopodidae), the flashing of which is described on p.127. No predators seem to be able to overcome the confusing effects of unpredictable movements combined with unpredictable colour flashing. The highly reflective surfaces of many species of both fishes and insects, together with the frequent development of iridescent colouration, seem to have evolved largely through the consequent enhancement of the confusing effect of erratic movement. Here we have another example of the influence of the protean principle on convergent evolution.

Many schooling species of fish are boldly striped or barred, and Starck (1966) has pointed out that 'the shifting maze of lines makes individuals difficult to distinguish'. Such an effect is readily seen in shoals of the tropical aquarium zebra fish, *Brachydanio rerio*. This effect has also been noted by Spinage (1962), and others, in the case of running zebra. Kruuk (1972) writes that a zebra's stripes 'may well have a confusing effect on a predator at close range, particularly when the predator is confronted with many zebra at the same time'. Or, as Professor Hardy has suggested (pers. comm., 1964), 'the varying widths of stripes may create an optical illusion about the direction in which the animal is moving...'. D. Bygott tells us however that the black and white stripes at long range make the zebra look grey and provide excellent camouflage against the plains background. In a mixed herd, wildebeest are noticed first.

Finally we may note that — as in the avian united erratic display — there is remarkable co-ordination in the schooling behaviour of fishes, particularly in the united erratic movements. As stated by Shaw (1960) 'The most conspicuous features of schooling fish are their constancy in orientation and their synchronization of speed and direction of movement'. Such synchronization is essential for the success of a united erratic display and we may assume that it has been developed, at least in part, to that end. As in the case of birds, the necessary equipment seems to be present. Fishes' lateral-line systems, for example, enable them to assess the vortices set up around nearby individuals — an obvious prerequisite for co-ordination within the school.

Scatter

When a closely-packed school of fishes, flock of birds, herd of deer, or almost any other group of active gregarious animals is subjected to an attack by a predator which suddenly appears at close quarters, the almost invariable reaction of the prey is to scatter. When an attack is pressed on the middle of a group, scatter involves the centrifugal fleeing of the individuals. Scatter is like an explosion — the same

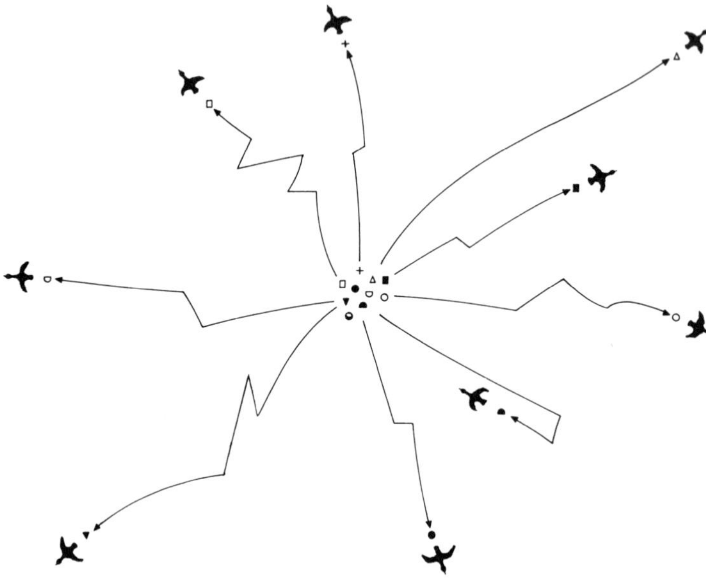

Fig. 4.26. The scatter display — typical appearance, seen here in the crash dive and scatter of an eider duckling brood.

number of basic units suddenly coming to occupy a much greater volume of space. The effect is to greatly reduce the chances of more than one individual being killed during a single attack, this effect being much enhanced by the protean elements involved. These are, principally, the multiple choice suddenly presented to the predator, and the several separate single erratic displays which vary in form in an unpredictable manner within and between themselves. The individual flight paths often terminate in concealment but, generally, fleeing does not proceed so far that the group is unable to re-form once the danger is past. But the flight paths followed result in a protean dispersion; an important phenomenon which we consider in more detail in Chapter 8. The basic principle is that knowledge of the position of one prey animal does not allow correct prediction of the position of others.

The scatter display and its effects constitute one of several examples of protean phenomena which have been widely commented on by naturalists and others familiar with the ways of animals. Joshua Slocum (1900), for example, with his considerable experience of the behaviour of pelagic animals at close quarters, records scattering as the mode of escape used by dolphin-fish fleeing from a shark.

Scatter in bobwhite quail (*Colinus virginianus*) is depicted in one of Audubon's famous bird paintings (1872–38), and this reaction to

Fig. 4.27. Covey of partridge scattering from sparrowhawk.

predatory attack is commonly seen in other members of the pheasant family Phasianidae, in grouse Tetraonidae, in ducks Anatidae, and others (Fig. 4.27).

In ground birds, scatter takes the form of flying or running away from the point of encounter. Kittiwakes (*Rissa tridactyla*) rise and scatter in response to the sudden appearance of skuas (Stercorariidae), whether on land or water. Diving birds on water, such as broods or adult parties of the various species of sea ducks Mergini and Somateriini, respond to attack by instantly crash-diving. This involves immediate concealment before the group spreads out over a wide area under water. In all cases observed by us, or reported to us, the individual flight paths have been unpredictable in their detailed form, including their direction. All that one can be sure of is that the great majority of the actors will radiate from the centre of attack. As a result, some individuals of an eider duck brood, for example, may surface after the crash-dive closer to a stationary human than they were before he appeared (Driver 1960). This has also been observed in tufted ducks, *Aythya fuligula* (D. Goodwin pers. comm., 1964). But this does not mean that such individuals are putting themselves at greater risk by this behaviour, for most predators will not remain stationary, as may a human observer, and will not stay at the point of first encounter.

In some cases a persistent predator may wait for the surfacing of an individual bird which has crash-dived and force it to dive repeatedly, tiring it out (Meinertzhagen 1959). And when groups of gulls, *L. argentatus*, as opposed to individuals, prey upon family parties of shelduck, *T. tadorna* (Van Dobben 1934) or crêches of common eider (Munro and Bédard 1977) the duckling mortality is very high. But such cases are relatively rare. And they do show that protean defence displays may require considerable escalation of an opponent's armaments, if they are to be overcome. Normally, scatter is of distinct advantage to the individuals in group. The spatial organization of an unthreatened nidifugous brood is a fairly loose one. This provides a good target for a gull or raptor, as an individual target can be easily selected. Young nidifugous birds are unable to perform a united erratic display and, if the parent cannot protect them, the only alternative for such a brood is to hide or disperse, and scatter involves both. Similarly, the nidicolous young of perching birds (Passeriformes), towards the end of the nestling period, may explode from the nest on the close approach of an intruder. This is mentioned in a cautionary note in the British Trust for Ornithology *Instructions for ringers* (N.D.). In mammals, active young or adult rodents will explode from the nest when suddenly disturbed. All species of mice and voles seem to do this under natural conditions, as do grey squirrels (Chance 1957). We have observed it in captive field mice (*Apodemus sylvaticus*) and domesticated mice (*Mus musculus*). To a human, at least, the explosion is most confusing, and we suspect that natural selection has moulded the behaviour to achieve just such an effect.

Scatter reactions are clearly innate. Eider ducklings, for example, will crash-dive and scatter if attacked on their very first outing in the water. Very young hamsters and field mice will scatter from the nest in slow motion. J.C. Pernetta tells us that this behaviour is characteristic of the British shrews *Sorex araneus*, *S. minutus*, and *Crocidura surveolens cassiteridum*, and also farm kittens. It is, after all, natural for animals to move from a discovered nest to a place of better concealment, and moving along separate routes to separate refuges is easier, and more efficient in such cases, than moving all together.

Bowen (1931) and Spooner (1931) have described the scatter reactions of fish schools when disturbed. Hobson (1968) writes, 'as a predator rushes into a school, the tightly packed members then radiate out in all directions before the advancing predator, probably contributing to any confusing effect that might be operating'.

Konrad Lorenz whilst snorkeling (1966*b*) observed small prey fishes, when attacked, darting 'at lightning speed in all directions, upward, downward, and even leaping above the surface, producing in a few seconds a large clear space of water, which only gradually fills up again when the predator has passed'. The general pattern of this scatter display in fishes seems to be the same as that in other groups.

The efficacy of fish scatter and single erratic displays is reflected in the occurrence of predator countermeasures. A number of marine and fresh-water predators, particularly birds, may encircle a school of prey fishes, thus restricting the scatter. The thresher shark, *Alopias vulpinus*, achieves the same effect by circling a school at speed, thrashing the water surface with its tail as it goes. Alternatively, various piscine predators upon fish schools charge into the schools striking in all directions, then returning at a more leisurely pace to feed upon the dead and injured. These, and other antiprotean devices are reviewed at the end of Chapter 7.

Invertebrates also show the scatter reaction in appropriate circumstances, that is, when a group of individuals of a gregarious species is suddenly attacked. Boycott (1965) for example, describes a combination of single erratic, scatter, and distraction displays by the squid, *Sepioteuthis*. When a shoal of squid is disturbed, each individual becomes brown, with green spots: 'Like most squids they may switch this uniform colouration on and off rapidly and since the shoal breaks in all directions any one individual is hard to follow. The difficulty of following the animals may be enhanced by the distracting ejection of a dense blob of drifting ink.' The central feature of this behaviour is the scatter display. This is made more effective by the complex variations in the individual flight paths and the introduction of the ink cloud.

Within the arthropods, groups of young spiders, such as the garden spider, *Araneus diadematus*, show a scatter reaction to sudden interference. For this species Bristowe (1958) notes that the young 'cluster together in a neat ball which quickly disintegrates into its component parts in response to any disturbance'. A simple way of demonstrating this is by blowing gently on the ball of spiders. A similar reaction to attack occurs in the tree-living South American bug, *Psocus venosus*. As recorded by Howard (1901) huddling groups of this species, in the adult form, scatter 'when seized with sudden panic', the individual flight paths terminating in concealment.

A well-nigh incredible example of scatter combined with crypsis in the defensive armament of an insect is seen in the tropical flattid bugs which, in the cryptic phase, aggregate on a plant stem in such a way as to resemble flowers — even to the extent of different-coloured individuals taking up different positions (Cott 1940; Wickler 1968). When such an aggregation is disturbed the insects take to the wing, scattering in all directions, but eventually settling and quickly re-grouping to form an 'inflorescence' (Fig. 4.28).

Most mobile predators seem to find multiplicity of prey confusing. Schaller (1972) notes 'When attacked, gazelle and impala typically scatter, an effective method of confusing a predator. On several occasions I saw a lioness stand undecided in high grass looking quickly back and forth at gazelle leaping around her without being able to select an individual'. This has also been noted by D. Bygott (pers. comm., 1979). We have frequently observed this effect in the case of a dog hunting rabbits.

The examples we have cited show not only that the basic scatter display is widespread among animals, but that its detailed form varies between species. For instance, the speed with which scattering occurs depends not only on the velocity attainable by each individual, but also on the extent to which the individual's flight path shows single erratic characteristics. Where motion is fast but jerky, with extreme changes of direction, scattering from the original area may take some while, and initially the display closely resembles the multiple erratic form. Perhaps the best illustration of this, and one which anyone can see for themselves, is the reaction to disturbance of the sand-hoppers, *Orchestia gammarella* and *Talitrus saltator*. These amphipod crustaceans live around the high-water mark in countless millions, clustered in hiding under seaweed and other debris and burrowing into the sand. When their cover is removed they hop, apparently in any direction, with great vigour and variability, each animal performing a single erratic display in which direction of movement alters sharply and drastically, frequently being reversed. The effectiveness of this display in preventing the tracking of a single, particular animal is much enhanced by the fact that many other single erratic displays are taking place at the same time and are overlapping (Fig. 4.29). The human would-be captor may know perfectly well that there is a systematic aspect to the display — namely that the hoppers will tend to scatter from the point where their cover has been disturbed, and that they will stop at some point where cover is again available — but this knowledge is of no assistance at all in dealing with the display. Scores, or even hundreds of hoppers may be jumping all at the same time and apparently at random. The protean features of this activity create confusion in the reactor that is very considerable indeed.

The effectiveness of this sand-hopper display was clearly demonstrated when — while this paragraph was first being composed — we went to collect some of these crustaceans for further study. On turning over the seaweed washed up on Bardsey Island beach we were so confused by the combination of protean features that a considerable

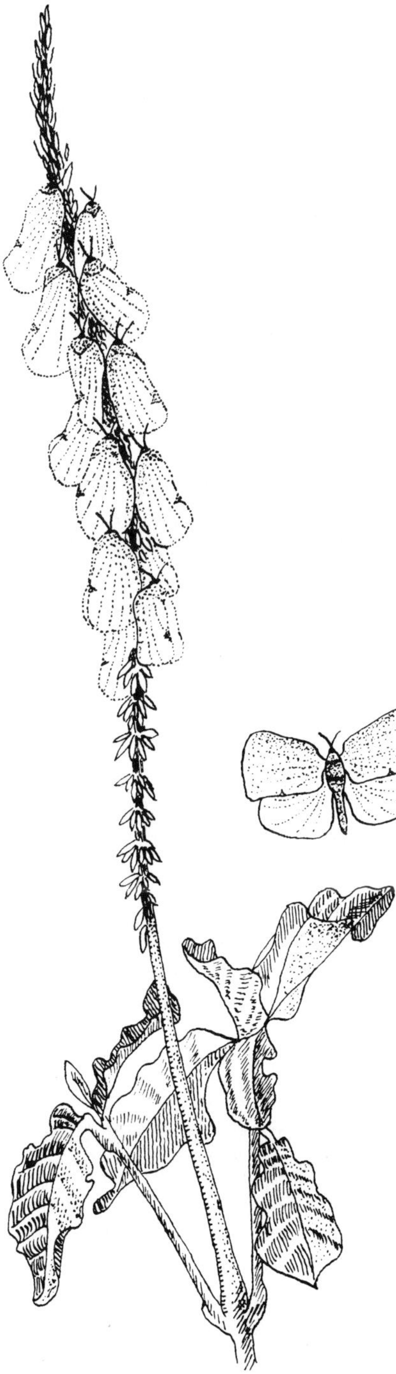

Fig. 4.28. 'Inflorescence' of flattid bugs (redrawn from Wickler 1968).

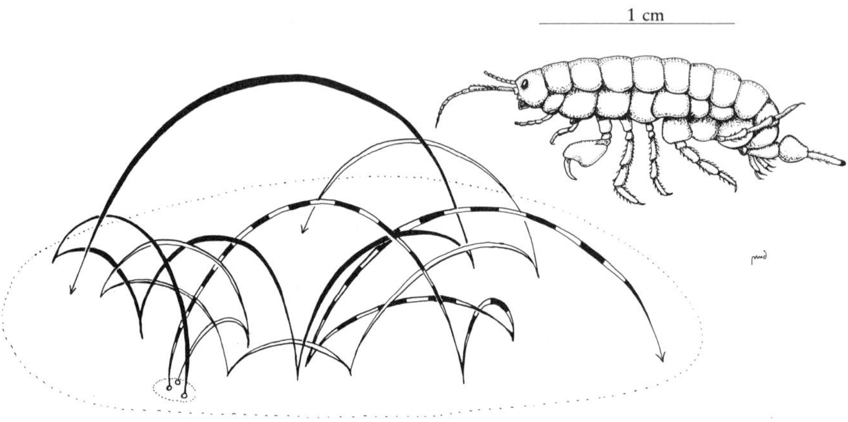

Fig. 4.29. Simplified diagram of shorehopper display when cover removed with *Orchestia* sp. to show burrowing and leaping appendages. Three individuals are indicated here. In a more typical field example, when a stone of approx. 700 sq. cm. (the area indicated here) was lifted, 300 + hoppers displayed thus and disappeared into the substrate (over a wider area) within 20 secs.

effort of concentration was needed for the focusing of attention on one individual. Only then did a capture attempt become something more than a random grab. The abundant hoppers on this, and any similar stretch of shore, constitute a favoured prey of turnstone, chough, and other birds which may often be seen moving the beach debris and digging in the sand for them. Presumably, the combination of protean behaviours is an effective defence. For, although millions of prey animals may be present, observation indicates that birds have to work extremely hard to capture enough to meet their needs.

Distraction

The phenomenon of distraction display has attracted the attention of many biologists, particularly ornithologists (Armstrong 1947, 1949*a*, *b*, 1956; Simmons 1952, 1955; Brown 1962; Gramza 1967; Barash 1976). We included distraction displays in our first broad survey of protean defence in prey animals (Humphries and Driver 1970) for no one appeared to have realized that the effectiveness of such displays often depends to a considerable extent on protean principles. We describe here more fully the forms of distraction display in which protean elements play a major role, and in Chapter 7 we discuss aspects of the mechanisms and evolution of these displays.

Wilson (1975) defines distraction display as 'any distinctive behaviour used to attract the attention of an enemy and to draw it away from an object that the animal is trying to protect'. Whether or not a conscious attempt at protection is involved, the basic characteristic of distraction is the direction of a predator's attention away from one individual or group to another, less vulnerable or biologically less important individual or object. The predator's attention to a *prize* is confused by the introduction of one or more *lures*. For example, in the classic example of injury-feigning in birds, the parent acts as the lure while the eggs or young form the prize. The same principle applies even in more complex examples involving several actors.

Protean behaviour is seen both in the individual, for example in zigzagging, and as a group property when the numbers, positions, and behavioural roles of the group members change unpredictably, or when an animal, by vocal mimicry, gives the impression that there are more animals present.

The release of lures In its simplest form, distraction display may be seen in an interspecific encounter involving a single prey animal only. There are numerous accounts in the literature of an animal releasing a part of itself as a lure which distracts the attention of the predator from the actor proper — the prize. Lizards commonly shed the tail when grasped thereby. Cott (1940) summarizes the situation admirably:

> Different species, belonging to families including Gekkonidae, Teiidae, Gerrhosauridae, Lacertidae and Scincidae, have . . . the power of self-mutilation or 'autotomy'. Owing to their speed and agility, or to their skulking habits, and to the nature of their surroundings which afford numerous hiding-places, they are difficult to capture and enemies often succeed only in seizing them by the tail. When this happens, the tail is rejected and left executing the most violent wrigglings and contortions in the grasp of its pursuer, whose attention is thus diverted while the lizard is running off into a safe retreat. Subsequently the lizard grows a new tail which, however, differs from the original organ in its aberrant scaling and absence of vertebrae. This habit of autotomy — which involves special structural modifications in the tail and special reflex activites on the part of its owner — is clearly adaptive.

In support of this Edmunds (1974) notes that in the flying gecko, *Ptychozoon kuhli*, 'the tail is more conspicuously coloured than the rest of the body, presumably so as to divert an attack towards the tail rather than at the body'.

Cott (1940) writes: 'Every one who has collected lizards in the tropics knows how abundant such individuals are; and the evidence is confirmed by the finding of amputated tails in the stomachs of birds and snakes'. Among such records are two lizards' tails from the common coucal, *Centropus sinensis* (Mason and Maxwell-Lefroy 1912); a skink's tail from the snake, *Lycophidion capense*, and four skinks' tails from the bush snake, *Philothamnus semivariegatus* (Loveridge 1933). Numerous birds of several families include lizards in their diet and this has doubtless formed a major selective pressure on these reptiles. The stomach of a buzzard, *Buteo rufofuscus*, was found to contain five skinks and four chameleons (Peters and Loveridge 1936), and a secretary bird, *Sagittarius serpentarius*, had eaten eight chameleons and 12 lizards (of two species), as well as a tortoise, three frogs, an adder, two locusts, two quails, and other animals (Stark and Sclater 1900–06).

Autotomy is more frequent in the invertebrates, where regeneration of lost parts is less of a problem. MacGinitie and MacGinitie (1949) describe how sea cucumbers (Holothuria) and annelid worms (*Echiurus echiurus*), when disturbed, discard parts of the viscera and proboscis respectively. The Mollusca provides many examples of autotomy, summarized by Stasek (1967) and Edmunds (1974). Various eolid, dorid, and sacoglossan sea slugs for example, when attacked, autotomize the attention-drawing dorsal papillae which contain defensive glands and/or stinging cells as well as branches of the gut (Edmunds 1966*a*,*b*). These papillae wriggle for some minutes after detachment, keeping predators occupied while the prey animal escapes. The well-named sea slug, *Discodoris fragilis*, autotomizes sections of its mantle when attacked (Alder and Hancock 1864; Edmunds 1971), and species of the sacoglossan *Oxynoë* autotomize the tail in similar circumstances (Stamm 1968; Warmke and Almodovar 1972). The associated regenerative abilities are common in marine invertebrates, including sand-dwelling molluscs and worms where projecting parts are vulnerable to predation by free-swimming fishes.

In a number of species several lures may be released successively. This occurs in the brittle stars, *Ophiocantha eurythra*, which progressively shed portions of their arms or disc; in jack-knife clams or razor shells (*Solen* spp.) which release successive parts of the siphon; and in scale worms (*Harmathoë tuta*) which break off successive portions of the body. Certain polynoïd worms, when provoked, cast off their luminous segmental scales which form glowing lures. In such cases the multiplication of prey objects will cause confusion through numbers, as in multiple erratic and scatter behaviour. There are obvious similarities with the release of inky or luminous clouds by cephalopods (Holmes 1940; Hall 1956; Boycott 1958).

Many decapod Crustacea autotomize a portion of a cheliped or claw limb (Huxley 1884; Yonge 1949; Edmunds 1974). As in lizards this involves the contraction of special muscles across a plane of weakness.

In these cases part of the function of autotomy may be to release the animal when it becomes mechanically trapped by its stony environment (Street 1952). The same function of mechanical release will apply in the more likely circumstance that the trapping is achieved by a seizing predator, and autotomy in both lizards and decapods is probably related primarily to this effect. However, the organ released will inevitably exert some distracting effect upon the predator, and in several species of crabs autotomy is combined with defensive attack. Crabs and other crustaceans with chelae will flee or freeze when approached in the open. If further provoked, however, many species attack, and a chela may become autotomized while pinching the predator. Robinson *et al.* (1970) note that more than 50 per cent of experimental attacks on *Gecarcinus* and *Pomatocarcinus* resulted in the autotomy of an actively pinching chela. The attackers in these experiments were teddy bears. When the attacker was the otter, *Lutra annectens*, the attack autotomy resulted in the predator having its attention totally diverted from the prize to the lure while it removed the painful object. A prey animal could hardly bring about a stronger attachment between lure and predator!

In the instances noted above, the lure performs a relatively limited amount of movement and the prey must rely also upon the protean unexpectedness of the display. Except in cases of attack-autotomy, such displays are unlikely to effectively distract advanced avian or mammalian predators. In more advanced lure displays therefore the lure is more complex. Hall (1966) for example describes how the leading male of a disturbed group of patas monkeys, *Erythrocebus patas*, runs violently about, bouncing erratically off bushes and moving gradually away from the rest of the group, which crouches. The conspicuous white hindquarters and blue scrotum of the male undoubtedly add to the distracting effect. A number of the more social species of primates are known to perform similar displays.

Injury-feigning We define **injury-feigning** as

'that behaviour of birds — usually adults in the breeding season — which gives the actor the appearance of injury and which results in reactors being led away from eggs or young in pursuit of apparently easy prey.'

Other animals, particularly reptiles, may appear injured when they are not when faced with or captured by a predator, but this is seldom in a parental context. And some birds may show other kinds of distraction behaviour in an imminent capture situation. But injury-feigning is the principal distraction display to predators. It is particularly well-demonstrated in ground-nesting birds such as waders, ducks, and gamebirds. We do not believe that the bird is consciously feigning injury but normally the behaviour is so similar to the disorganized movements of an injured animal that it cannot be distinguished from the real thing.

This behaviour has intrigued people from early times. It is commented upon in Book IX of Aristotle's *Historia Animalium.* Poets too have noted it, sometimes with interesting detail as in Robert Bridges' *The Testament of Beauty* (1929):

It is pretty to mark
a partridge, when she hath first led forth her brood to
run among the grass-tussocks or hay-stubbles of June,
if man or beast approach them, how to usurp regard
she conterfeiteth the terror of a wounded bird
draggling a broken wing, and noisily enticeth
or provoketh the foe to follow her in a vain chase;
nor wil she desist from the ruse of her courage
to effect her own escape in loud masterful flight,
untill she hav far decoy'd hunter or blundering hoof
from where she has bid her little ones to scatter and hide.

The amount of information in this passage is remarkable, for it tells us about the form of the behaviour and its context in some detail. It is essentially similar to field observations of many naturalists and biologists. For example Saunders (1937) records a wood duck, *Aix sponsa*, successfully distracting a red-shouldered hawk, *Buteo lineatus*, from her nine young. The mother duck lay on her side in the water with ncek outstretched and one wing flapping in the air, paddling in circles. She avoided the hawk's attacks by taking cover under bushes, and at one instance her display even distracted the hawk from two of her young which had reappeared from cover. Chapman (1897) flushed a black-throated diver, *Gavia arctica*, from hatching eggs. After flying some 10 m, the bird fell to the water, apparently with a broken wing, and paddled in narrow circles on one side as if paralysed. Dewar (1913) experienced the display of 20 or more pratincoles (Glareolidae) simultaneously. Some were simply lying as if wounded. Others were floundering as if in pain. Still more were fluttering along with one wing extended as if broken; some apparently with both wings damaged. We illustrate such behaviour in the black-winged stilt, *Himantopus himantopus* (Fig. 4.30). Derek Goodwin (in Simmons 1955) noted the reaction to a pike of a female great crested grebe, *Podiceps cristatus* — swimming with large young on a lake:

Suddenly with loud cries . . . it jumped vertically out of the water, about 3 feet into the air, or so it appeared. It fell, or half fell half flew back and as it hit the water it violently sped away, thrashing over the surface like a young duck when chased.

Fig. 4.30. Some injury-feigning postures in black-winged stilt.

The female nighthawk, *Chordeiles minor*, attracts an intruder's attention by flying near in a conspicuous manner and, after landing near the intruder but away from her nest, maintains her conspicuousness by spreading the wings. This gives the appearance of flight disablement and optimally exposes the white flashes on the primary feathers (Gramza 1967).

Injury-feigning is greatest in intensity and ease of elicitation when the bird has eggs near to hatching or newly-hatched young — 'when she hath first led forth her brood to run'. In terms of natural selection, this is what one would expect for it is then that the brood is most vulnerable to predation. This synchronization of hatching with injury-feigning has been noted in a variety of species including the little pratincole, *Glareola lactea* (Hume 1890), the killdeer plover, *Charadrius vociferus* (Friedmann 1934), the ringed plover, *Charadrius hiaticula* (Laven 1940), the dunlin, *Calidris alpina* (Saxby 1874), and the common eider (Driver, field notes).

Typically in injury-feigning the actor remains close to the reactor — the predator or human observer — this being necessary to keep the reactor's attention from the brood. The predator is usually led further and further away from the prize until eventually the lure — i.e. the displaying bird — breaks contact or the predator loses interest. This display has been noted as being successful against a variety of pre-

Fig. 4.31. Wing-thrashing injury-feigning of old-squaw duck.

dators including the red-throated diver, *Gavia stellata* (Nicholson 1939) as well as the red-shouldered hawk already mentioned, and several mammals including the arctic fox, *Alopex lagopus* (Huxley 1925), the common fox (Brown 1939), the coyote, *Canis latrans* (Hochbaum 1944), the eskimo dog (Hohn 1958), various other domestic dogs (Woodward 1944; Jourdain 1936–7; Hochbaum 1944), the domestic cat (Beven 1946), the otter, *Lutra lutra* (Pitt 1931), the stoat, *Mustela erminea* (Lynes 1910), and man (Serventy and White 1944; Armstrong 1947; Pickwell and Smith 1938; Taverner 1936).

We summarized the essential features of injury-feigning in Chapter 1 but we must emphasize here the central importance of the protean principle in maintaining the reactor's attention while also avoiding capture. It is natural for an injured bird to behave in an erratic manner and the irregular avoidance manœuvres are therefore not seen as such by the predator. The display thus, while seeming to put the actor at great risk, has inbuilt protection.

When an intruder refuses to be led away by injury-feigning, as sometimes happens if the reactor is a field naturalist, the actor may return to the point of encounter and 'try again'. We have observed this repeatedly in the ringed plover and also in the old-squaw duck (Driver 1960). We use the latter observations for our illustration (Fig. 4.31) and later analysis (p.166). In one typical instance an old-squaw female was discovered on a small pond with a brood of recently-hatched young. On the observer's approach the parent bird performed a complex and erratic sequence of diving, flying — both efficiently and, apparently, inefficiently — swimming half-submerged as if sinking, and thrashing the surface of the water with the wings as if unable to fly at

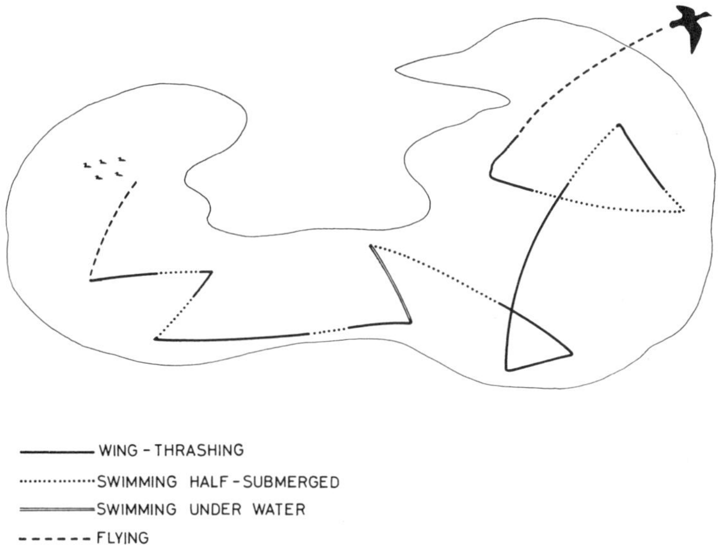

Fig. 4.32. Old-squaw injury-feigning — sequence of events in actual example.

all (Fig. 4.32). After displaying for some 10 min the duck flew away but soon returned. The whole procedure was repeated four times (with varying detail), after which the duck alighted on another pond and called the brood to her. At no time during the proceedings was the observer able to predict what the bird would do next — as might have been possible had the display been an epigamic one.

Detailed consideration of the mechanism and origins of injury-feigning we leave until later (p.163). It is generally considered to involve the operation of conflicting motivations: those concerned with defence of the young on the one hand, and with fleeing on the other. Whatever the mechanism, evolutionary pressures seem to have operated to ensure the consolidation of effective randomness, enabling the actor to display without significant danger to itself. This has considerable significance with respect to discussions of altruism (p.253).

We have so far considered examples in which only one actor is involved. When several actors participate that protean effect is greatly enhanced through the multiple choice phenomenon. This occurs with injury-feigning when both parents display, as is not uncommonly the case, and is even more significant when there is a high density of the adult population, as noted for several species by Armstrong (1954). In these instances the presence of more than one lure is fortuitous, but there are forms of distraction display in which the multiple-lure situation is a central feature. Such displays include 'mobbing'.

Mobbing Harvey and Greenwood (1978) define *mobbing* as 'the hustling of predators by potential prey'. This is very well developed in a number of social mammals including primates (Kortlandt and Kooij 1963; Crook and Gartlan 1966; Stoltz and Saayman 1970; Eisenberg *et al.* 1972). It may also be seen in ungulates (Schaller 1967) and hyenas (Kruuk 1972). And it has been studied widely in birds, where it is known to occur in many different families (Altmann 1956).

The majority of mobbing studies to date have been concerned with motivation, causation, ontogeny, and specific occurrence (Hinde 1954*a*,*b*; Andrew 1961*a*–*d*; Curio 1959; Altmann 1956). We consider the evolution of mobbing later (p.170). No one, however, seems to have applied themselves to the problem of how animals which normally are vulnerable prey become, during mobbing behaviour, aggressors upon their own predators. It is not simply weight of numbers, for many mobbing species usually flee from their predators, whether alone or in numbers. It is the operation of the protean principle which makes mobbing relatively safe for the mobbers.

Mobbing may involve actual attacks upon the predator, so the more mobbing animals there are, the better. In many avian species mobbing behaviour is therefore associated with the production of locatory calls which can attract birds of a number of species. These calls are convergent in form (Marler 1959). They are usually loud clicks lasting 0.1 sec or less and covering 2–3 kHz of frequency or more in the range 0–8 kHz.

However, even though the locatory calls are usually successful in attracting a 'mob', mobbing can be a dangerous occupation. The smaller and faster raptors such as sparrowhawks do occasionally kill birds by which they are being mobbed. But generally, mobbing is effective. Kruuk (1964) and Hoogland and Sherman (1976) showed that predators are effectively deterred by the mobbing behaviour of black-headed gulls, *Larus ridibundus*, and bank swallows, *R. riparia*, respectively. A similar conclusion was reached by Andersson and Wiklund (1978) who demonstrated experimentally that the mobbing behaviour of fieldfares, *Turdus pilaris*, significantly reduced predation on eggs in artificial nests located near real fieldfare nests.

'Typical' mobbing occurs when perched raptors such as the tawny owl, *Strix aluco*, and the buzzard, *B. buteo*, are pestered by groups of small birds which may occasionally form their prey. Hinde (1954*a*,*b*) and Marler (1956) made detailed analyses of this behaviour in the chaffinch, *Fringilla coelebs*. The small birds do not attack *en masse*; each individual alternates with the others in making brief mock attacks on the predator, whose attention is thereby repeatedly distracted so that a counter-attack can seldom be launched. Owls and buzzards are perfectly capable of killing small birds — though they cannot easily catch them — but only seldom is one of the mobbing individuals caught in the talons of a rapidly striking foot when it ventures too near. On the whole, the unpredictable alternation of attack from one mobbing individual to another is enough to confuse a predator's capture mechanism.

A more complex example of this kind of distraction is given by Meinertzhagen (1959). He observed a flying tawny eagle, *Aquila rapax*, attacking a group of houbara bustards, *Chlamydotis undulata*, on the ground. As soon as the eagle attempted to press home its attack on one bustard, some other bustard would run across in front of it, distracting its attention (Fig. 4.33). Eventually the unsuccessful and exhausted eagle abandoned its attacks. Other observers have noted this mobbing technique for this species, and Meinertzhagen records that it deals with the imperial eagle, *Aquila heliacus*, in a similar manner. Apparently, this bustard, which is approaching the size limit for efficient flight, has developed this protean running display as a standard defence mechanism.

In houbara bustards it is the flight or defensive aspect which is the more prominent. In other cases the approach or offensive aspect may dominate. This seems to be the case when ground hornbills *Bucorvus cafer* deal with snakes (Gurney 1861; Meinertzhagen 1959). A small group of hornbills gathers round a snake with wings spread down and forwards for protection (Fig. 4.34). Each bird in turn makes incomplete attacks on the snake, which is distracted from one member of the group to another until eventually it is overcome and killed. In this way a small group of birds is able to overcome a large snake, which would be impossible if they attacked singly or in a regular and systematic fashion.

The protean element in these mobbing displays is the erratic succession of attacks from different angles. This is even shown by some small mammals. The dwarf mongoose, *Helogale undulata*, for example uses this technique against predators (Rasa 1977). The well-known snake-capture technique of the Indian mongoose, *Herpestes mungo*, is similar, although only one actor is involved, since the rapidity of the actor's movements and its unpredictable changes of position duplicate the multiple-lure system described above. The mongoose, bristling up its fur, leaps over and around the snake so that the reptile is threatened from angles which change rapidly and unpredictably (Fig. 4.35). Kipling's classic account (1894) of the battle between Rikki-Tikki-Tavi and the cobra would seem to be an accurate description of this behaviour. An eyewitness report of encounters between captive individuals of the species in question tallies in all essential details (D. Ilsley,

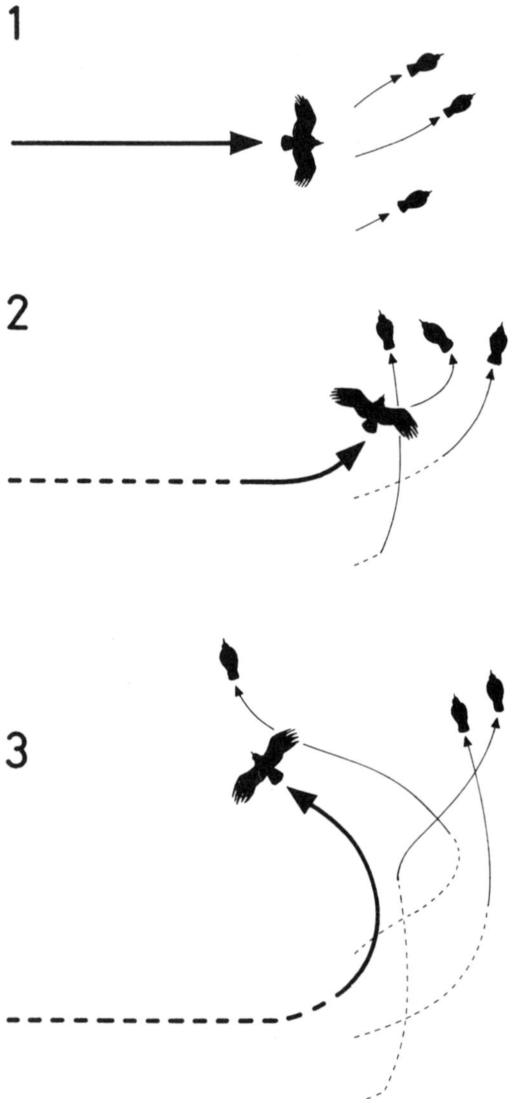

Fig. 4.33. Distraction display of running bustards attacked by flying eagle. (1) The eagle attacks one individual; (2) another individual crosses the eagle's path, causing a diversion; (3) the pattern is repeated by the 'third' bird.

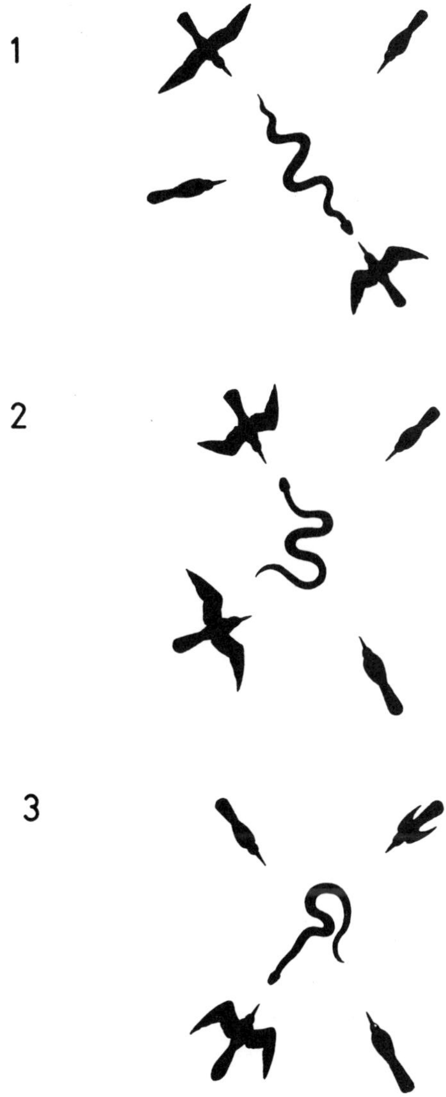

Fig. 4.34. Mobbing distraction of snake by hornbills. (1) Snake attacks one bird and is attacked from behind by another; (2) as the snake turns its attention to the second bird, it is attacked by a third; (3) the procedure is repeated.

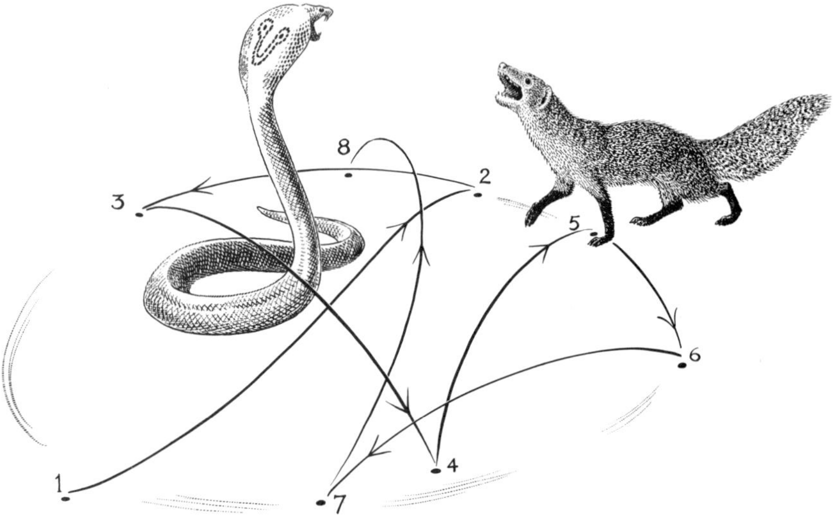

Fig. 4.35. Mobbing-type behaviour of single Indian mongoose.

pers. comm., 1984). The crucial and culminating event is when the mongoose's protean movements put it literally one jump ahead of the snake. Hudson (1892) owned a cat which dealt with snakes in a similar manner.

Individual birds also can overcome dangerous predators if they use this technique. Meintertzhagen (1959) described two instances where a hawk in the air was beaten — by a lapwing in one case and a black-headed gull in the other. The 'prey' dived at the hawk repeatedly, at the last second veering up, down or sideways unpredictably. In both cases the hawk eventually left the field.

Even humans may be intimidated by single birds in this way if the encounter occurs in a situation which is already potentially dangerous for the human. We can testify to this from our experiences with birds on sea cliffs and in tree tops. There can be little doubt that human lives have been lost in such conditions as a result of unexpected attacks by gulls, crows, or raptors.

W.J. Hansen (pers. comm., 1980) makes the point that, in aerial mobbing of a single predator, the turbulence set up by the succession of close encounters must have a very disturbing effect upon the predator. The vortices resulting from successive stoops will repeatedly, but unpredictably, destabilize the reactor's wings so that the bird may be wallowing in turbulence in otherwise clear air. Often the only means of escape is to dive and gain speed. A single erratic display may then enable the erstwhile predator to escape.

Fig. 4.36. Mobbing: the unpredictable attacks of a pair of gulls can divert foxes from the nest area.

Anyone who doubts the effectiveness of this buffeting need only approach the nest of a great skua, *Stercorarius skua*, when it has eggs close to hatching or very young chicks. A typical attack consists of a power dive or low level approach in which the flight speed is enhanced by the ground effect, terminating in a steep climb perhaps 1 m from the intruder.

Two actors are, of course, more effective than one, as we have often observed when choughs, mistle-thrushes, swallows, and other birds have driven away carrion crows, buzzards, and ravens from their nest areas. The effectiveness of the double attack has been recorded on film by Professor N. Tinbergen in the case of carrion crows taking a herring gull's eggs, and also by Professor O.S. Pettingill in the case of dolphin gulls (*Larus scoresbii*) so pestering a penguin (*Pygoscelis papua*) bringing crustaceans to its young that it is forced to give up its catch.

A well-known form of mobbing is the 'dive-bombing' performed by birds such as gulls and tern (Laridae) and skuas (Stercorariidae) where the birds swoop upon a ground predator singly or in numbers (Fig. 4.36). Typically, the attack is pressed very close to the intruder and humans at least are sometimes struck by wing or foot. The changes in attack direction add to the distracting effect, as does the occasional defecation of the actors. Anyone who has worked in gull colonies will testify to the effectiveness of this display. The best remedy is to keep one's head down and concentrate on the work in hand. Even then the occasional blow on the head is distracting, for there is no way of telling from which direction it will come, or when. That this display is functional in distracting the birds' normal predators is shown by the work of Bergman (1941) and Kruuk (1964). The same basic technique may be used offensively by herring-gulls and greater black-backed gulls preying upon crêches of eider duckling (Munro and Bédard 1977). We have experienced such attacks in colonies of many species of gulls, terns, and skuas.

Even birds as formidable as eagles may be overcome by aerial mobbing. At Gibraltar the spectacular views of migrating birds — particularly raptors — attract many ornithologists and it is well-known from observations there and elsewhere that mobbing by gulls, particularly in large numbers, can be a serious problem for birds of prey. Indeed, species such as the short-toed eagle, *Circaetus gallicus*, may even be driven into the water and drowned. This effect of numbers operating on the protean principle is also seen when starlings drive away sparrowhawks — on rare occasions even killing them.

Vocal mimicry We have included the release of lures, injury-feigning, and mobbing under 'distraction display' because they all confuse a would-be predator's primary object by causing it to switch attention from prize to lure or from one attacker to another. A similar principle seems to operate in avian *vocal mimicry* and so we consider this behaviour here (see also p.241 where we consider avian vocal mimicry from the point of view of display diversification).

Bower birds (Ptilinorhynchidae) are outstanding vocal mimics. Indeed, the spotted bower bird, *Chlamydera maculata*, is probably the most accomplished of all avian mimics. This species incorporates all kinds of sounds into its repertoire and reproduces them so faithfully that — to human ears at least — they are indistinguishable from the real thing. Not only does this species, like others in the family, mimic other birds in its vicinity, and mammals too, it also reproduces many kinds of non-vocal sounds. Marshall (1954) lists: the noise of cattle breaking through scrub and of emus crashing through wire fences; a mallet striking a wedge; sheep walking through dead branches; the chopping of wood; the crack of a stock whip; and various other sounds of camp and bush. A recording of a captive individual of this species (Driver and Midwinter, N.D.) includes the sound of horses' hooves, knocking on a door, shovelling coal or gravel, water sloshing in a bucket, and shutting a drawer, as well as a variety of human vocalizations and bird calls of a number of species — a truly astonishing performance from one individual.

The elaborate mimicry of the spotted bower bird is particularly noticeable during the breeding season and Marshall quotes a number of sources of evidence for its use 'as a distraction display when the nest with eggs or young are menaced'. Mimicry of predatory birds may be included and it may occur with injury-feigning. Campbell (1901) noted that the spotted bower bird could mimic a whistling eagle, *Haliaster sphenurus*, so faithfully that a hen and chickens would run for cover at the sound.

There can be no doubt that the spotted bower bird's mimicry is accurate and diverse and that from time to time it will fortuitously produce the call of some much more formidable animal. Whatever other functions this mimicry may have it is clear that it may act as an antipredator device, for a would-be predator cannot but be taken aback, if only momentarily, by the sudden unpredictable production of the call of a much more formidable opponent than the one it has in view. This is analogous to the flashing of eye spots in Lepidoptera (p.125).

It seems reasonable to suggest that calls resembling those of alternative prey, or extraneous sounds may likewise disrupt a predator's attention and decrease the effectiveness of its hunting.

This elaborate combination of distraction and deterrence has presumably evolved because of the great vulnerability of ground-nesting bower birds with their long breeding season. The males have to maintain their females' attention to themselves and their territories over the long period of waiting for rain with its unpredictable termination soon resulting in the production of insect food for the young. The vulnerable male therefore produces protean vocalizations as a form of insurance against predatory attack.

Doubtless this principle has operated in the development of other examples of vocal mimicry in birds, for example starlings (p.242).

Immobility

So far we have considered protean armaments which involve certain specific kinds of locomotor activity. Now we are to consider behaviour which consists largely of inactivity. Immobility obviously can have protean properties only in relation to associated events or to other instances of immobility. We are here considering those examples of immobility which deny to the predator the stimuli which would release capture or killing reactions, in other words, cryptic concealment, and 'shamming dead' or thanatosis. We are not concerned with immobility which involves aggressive resemblance or other forms of deterrence.

Fleeing is frequently terminated by cryptic immobility, particularly where the hunted animal is less swift than its pursuer. Several examples have been given in earlier parts of this chapter. In such cases, the appearance of the animal during immobility may have different kinds of protean relationship to its appearance during fleeing, preventing accurate prediction of the animal's appearance, position, or behaviour. These relationships basically involve motion, colour, and form. Thus the onset of immobility varies in relation to the duration of fleeing and the particular type of movement immediately preceding immobility. For instance, young ringed plover become immobile after varying lengths of running flight path, while the dipterous fly, *Psychoda phalaenoides*, which has a single erratic display including irregular sequences of zigzagging, hopping, and spinning, may become immobile after any one of these forms of motion. Thus the motion context in which immobility suddenly occurs varies between different instances of the display.

Changes of colour from brightly-hued while in motion to cryptic with the onset of immobility are common in butterflies and indeed in many other insects. Changes in form may also accompany immobility; Cott (1940) gives several examples. Our description of the single erratic display of the small tortoise-shell butterfly (p.37) illustrates all three kinds of protean relationship between fleeing and immobility.

We conclude that the effectiveness of many instances of defensive immobility is due — in part at least — to the nature of its relationship with preceding events, a relationship involving protean variation and transformation.

A further protean element in the general phenomenon of immobility displays is that individuals in a population are often polyethic for immobility: not all will become motionless in any given circumstances. Ornithologists who have handled birds such as starlings will be familiar with the screaming and catalepsy which sometimes, but not always, occur. It is impossible to predict which individuals will scream or become immobile when captured. And the unpredictability involves the cessation as well as the onset of immobility. We have pointed out elsewhere (Driver and Humphries 1969) that if one is handling a silent and immobile bird and it suddenly screams, there is a startle response which sometimes results in the bird's escape. Moreover, the unpredictable onset of vocal activity is sometimes accompanied by a similar transition from catalepsy to violent struggling. It is well-known that immobility in a captured prey animal deprives the predator of the major external stimulus for attack: prey movement. If such a captured prey animal suddenly and unexpectedly becomes active, vocally or otherwise, then it has a greater chance of escape than if it had been continuously active or indulging in regular, and therefore expected, bursts of activity.

Tinbergen (1962) has recorded the successful use of this behaviour by a bullfinch, *Pyrrhula pyrrhula*, when captured. He remarks that this behaviour would have definite survival value in the wild even if predators only occasionally relaxed their attention to the prey, and that it is found in many small birds. It is also found in larger ones, for Lees (1964) records the escape of a mallard, *Anas platyrhynchos*, from mongrel dogs in this way. And we have been given an eyewitness account (M. Evans pers. comm., 1980) of a pheasant, *Phasianus colchicus*, escaping thus from a domestic cat. The bird was carried with its head hanging, and apparently dead, and placed on a lawn. It then sprang up and escaped. We have seen, or know of, similar encounters between house mouse and cat, rabbit and cat, and blackbird, *Turdus merula*, and man. The escape of a well-known explorer from a lion (Livingstone 1865) seems to have involved the same processes in both predator and prey. Fortunately, the prey was able in this case to subsequently record his feeling during the encounter. We comment further on this in our analysis on pp.188–92.

This tactic of immobility followed by sudden action must have a protean basis if it is to succeed. If all individuals of a population showed immobility, and if immobility were of consistent duration and its cessation easily predictable, a predator could develop countermeasures. Thus, captured mice do not always 'play dead'; many of them struggle until killed. But this means that those individuals which do not struggle have a chance of escaping if they suddenly 'come to life'. Struggling from the outset may sometimes lead to escape, but such continued struggling is a tactic which, like 'playing dead', inevitably has relatively low success in the prey's desperate position. However, its level of success would not seem to depend to any great extent on its incidence in the population. The success of limp immobility on the other hand is likely to depend very much on the expectations the predator has developed from earlier encounters. Immobility will presumably be more successful when rare in the population, when its duration is variable, and when its ending is a sudden transformation to violent motion. The genetic aspects of this behaviour would make a most interesting study. It may well provide an example of frequency-dependent selection (Clarke 1962*a*). One certainly need not postulate the operation of 'altruism' or any other self-sacrificing behavioural mechanism on the part of strugglers or those occasional mice which during immobility receive such a disabling mauling that escape becomes impossible.

In a number of species apparent catalepsy occurs as the ending of a behavioural sequence in which, to human eyes at least, the actor appears to be dying. Thus, reptiles such as the hog-nosed snake, *Heterodon platyrhinos* (Ditmars 1954) and the grass snake, *Natrix natrix* (D. Goodwin pers. comm., 1964) perform convulsive 'death throes' before becoming cataleptic and 'dead'. In the grass snake, at least, this behaviour is polyethic, some individuals not showing it in apparently suitable circumstances.

It would seem from Hudson's careful accounts of life in the Argentine (1892) that the fox (*Canis azarae*), the opossum (*Didelphis azarae*), and the spotted tinamou (*Nothura maculosa*), behave in a very similar manner when captured, not only 'feigning death' but also preceding this with highly naturalistic 'death throes'. Hudson states that the tinamou was able to escape from man in this way, and the fox from dogs. However, this behaviour is far from being perfected in all cases and in some species may be exposed as deception by the automatic, stereotyped aspects of the reaction. For example, Watson (1973) notes that

> both the Cameroon toad *Bufo superciliaris* and the hog-nosed snake *Heterodon platyrhinos* sham dead, turning over and lying on their backs with their tongues hanging out when threatened. But the mechanism is not perfectly developed in them yet, because they make the hilarious mistake when put the right way up, of immediately turning upside down again.

It is revealing stereotypy of this kind that protean phenomena are elsewhere so often employed to mask.

An example of apparent lifelessness in a fish is seen in the reaction of a sucking loach, *Gyrinocheilus aymonieri*, to being chased with a net in an aquarium (A.J. Driver pers. comm., 1978). Fish were being removed from a tropical aquarium so that it could be cleaned; something which had occurred to the same group of fish twice before. On this occasion the loach performed a very vigorous single erratic display — even leaping from the water — before being caught. When captured it collapsed, sinking to the bottom of the net and then to the bottom of the bucket, lying on its side and showing irregular respiratory movements. As a result of this collapse it was replaced in the aquarium where it sank to the bottom again and stopped breathing. A swirling current caused by the removal of water from the aquarium carried the fish up to the surface, and then it sank again, apparently lifeless. It was assumed to be dead and was again removed and placed in the bucket, whereupon it recommenced feeble respiratory movements, soon righted itself, and recovered. We, and many aquarist colleagues, have observed such reactions in a number of species of fish.

Catalepsy is not necessarily advantageous every time it is used in conflict between animals. The feeding drive of a predator may be so high that it devours a cataleptic prey animal without the usual release, by the prey's struggles, of killing and consuming behaviour. More significantly perhaps, Armstrong (1947), in a survey of many instances of 'shamming dead', mentions individuals of tinamou, song sparrow, and field sparrow, *Spizella pusilla*, actually dying through 'fear' in an interspecific encounter. We have several records of a similar nature for the blackbird, *Turdus merula*, in which species this reaction is quite well-known — at least among bird-ringers. Armstrong also quotes an account by Hudson of a jaguar, *Felis onca*, which apparently was so paralysed by fear that it allowed itself to be burnt alive. One may compare this with the human reaction of becoming 'petrified with fear'. Doubtless this has led to many deaths when a prompt and more appropriate reaction would have resulted in survival. And, apparently, if lizards such as *Anolis carolinensis* and *Phrynosoma cornutum* use the immobility response too often, they actually die (Hoagland 1928). But these examples do not negate the basic protean thesis, which suggests that if — in the desperate situation of capture by a predator — unexpected behaviours are occasionally effective, they will add an important weapon to the actor's armoury.

Convulsoid behaviour

Catalepsy and convulsoid behaviours are often connected contextually and may be connected neurally. Both involve apparent malfunctions which may in fact be adaptive and of survival value. 'Convulsions' in individual prey animals when attacked typically consist of short bursts of protean activity — largely wild, undirected, and apparently uncoordinated movements — frequently followed by complete immobility. Numerous examples are known. The skipping of the small tortoiseshell butterfly has already been described (p.37) and similar skipping occurs in many moths. It is interesting to reflect — and this is the only time we will make this point in the present work — that, as with many other categories of protean behaviour, millions of examples are doubtless occurring in nature while this sentence is being read.

A good example of convulsoid skipping in moths is seen in the yellow underwing, *Triphaena pronuba*, a common species in many parts of its wide range over Europe, Asia, and North Africa, where it occurs in grassland habitats. There are several similar species with the same geographical range. When a yellow underwing moth is disturbed, it usually flops around on the ground — frequently but variably on its back — occasionally flashing its hindwings, in an apparently uncoordinated manner. Blest (1957*b*) describes how this skipping behaviour, which he terms 'flopping', replaces more systematic displays when moths of another family, the Saturnidae, are stimulated particularly strongly. This skipping behaviour has a considerable startle effect upon man — even if he is expecting it — and makes the moth difficult to handle. Similar behaviour is shown by a large variety of insects and other terrestrial arthropods. Spiders, for example, may bounce about on all 'fours' with great vigour and rapidity, as in tarantulas, or they may bounce and somersault in all directions and orientations, as in many of the smaller species.

It can readily be observed that mice, in captivity, have difficulty in dealing with such a convulsing animal. Lizards and snakes also frequently perform convulsive displays when cornered and are then almost impossible to catch with the hands alone. This response may even be seen in clawed toads which are commonly kept in laboratories (Chance and Russell 1959).

It is possible that rapid colour changes may have startle and confusion effects similar to that induced by convulsions. For instance, Bullock (1958) quotes Buck (1954) on a luminescent response of sea pansies, *Renilla spp.* On being disturbed this colonial coelenterate produces a series of waves of luminescence spreading from many loci to involve the whole colony of polyps which form the flower-like head. Bullock notes that these waves 'arise so frequently and with such a shifting of loci and development of new loci that one gains the impression of frenzied activity'. The display lasts for a few minutes and has been termed 'berserking', because of its similarity to the supposed 'mad frenzy' of ancient Norse warriors. Nicol (1955) describes this display in detail and notes great variation in excitability between different specimens. According to Panceri (1872) the display also occurs in the sea pen, *Pennatula spp.*, a close relative of the sea pansy. Members of this Order, Pennatulacea, are noted for their ability to luminesce — upon stimulation — by exuding a secretion containing luminous granules. The colours produced are blue or violet in most species, but may be yellow or green (Hyman 1940).

Buck (1978), referring to our suggestion that the berserking of sea pansies may be a protean antipredator device (Humphries and Driver 1967), comments that the main natural predators of *Renilla* are the nudibranch mollusc, *Armina*, and the starfish, *Astropecten*, against which luminescence is ineffective. Sometimes, indeed, the luminescence may not occur at all, even when the colony is being eaten alive (Bertsch 1968). However, Kastendiek (1976) considered that light emission in *Renilla* protected the colonies against single fishes because of the large number of such colonies with only one bite mark. And Morin (1976) concluded that in other alcyonarians the sudden luminous waves passing over the colony may deter fish and crustacean predators.

We suggest that the *Renilla* display is likely to be effective against fishes and crustaceans but not against predators with different response modes. Any prey defence against one predator type is likely to be less effective, ineffective, or not employed at all against other kinds of predator attacking that same prey.

Escape reactions may be released by the colour displays of cuttlefish, as mentioned by Chance and Russell (1959). Holmes (1940) was the first to suggest that this kind of effect is induced by sudden changes in visual patterns. Special orange, orange-red, and brown-black chromatophores, revealed to varying degrees and in varying combinations, give the cuttle a wide range of possible variations in appearance. Similar colour displays occur in other cephalopods such as *Octopus* (Packard and Saunders 1971) and *Argonauta* (Young 1959).

It would seem that convulsoid behaviours have come to be used by otherwise defenceless animals, when they are threatened with imminent capture, through the effectiveness of the protean elements of the display. The startle effect is a result of the unpredictable onset of the display and of changes therein, and the difficulty of handling a convulsing animal stems from the confusing effect of the single erratic and multiple erratic displays.

To suggest, as we do, that convulsions have a positive role to play in an animal's behaviour repertoire is not new. Chance (1957) has already made the point, and readily-acceptable examples are coming to light as a result of field observations. For instance, it would seem that convulsions act as begging displays in some species of pelicans (Schaller 1964; Burke and Brown 1970).

A number of observers, including the present authors, have noted that some epileptics may be more prone to fits when they are placed in situations which they find unpleasant — the condition known as hysterical epilepsy. The result of the fit is that the sufferer is removed from that particular situation. Watson (1974) writes

> There is no evidence that the *grand mal* or most severe fits are in any way hereditary, so the possibility exists that they might . . . be psychosomatic reactions produced in response to certain threatening conditions. This suggestion takes epilepsy out of the area of brain damage and neural disorder and makes it a behaviour pattern that can be produced in any normal brain and might in certain situations have survival value.

Over the years we have become convinced that convulsions in man sometimes function in this way. There is in fact no fundamental biological reason why convulsive characteristics should not have been affected by evolutionary pressures as have other characteristics, and it may be that convulsions, at least at the neurological level, are more common in everyday behaviour repertories than we have suspected. Their erratic overt structure predisposes them for a protean display function.

Seen in this light the convulsive displays of those predators which employ them are more easily understood. As we note elsewhere (pp.184–7) the convulsive gyrations of foxes and stoats seem, at the same time, to draw the reactor's attention and confuse it by avoiding any systematic signal content. The prey is therefore denied the initiation of escape behaviour. This mechanism could also operate in the capture of kangaroos by Aborigines (p.64), and may even indicate a sound biological basis for the anecdotes of pythons bemusing their prey by means of erratic gyrations. A classic and, we now suggest, pertinent example is 'Kaa's Hunting' in Kipling's (1894) *Jungle book*:

> Then he began making loops and figures of eight with his body, and soft, oozy triangles that melted into squares and five-sided figures, and coiled mounds, never resting, never hurrying, and never stopping his low humming song.

For some time we were hesitant about including this quotation because of its highly anecdotal flavour and its content in a work of fiction. However, we have now noted that a number of herpetologists have described convulsive behaviour as a protective device in a number of species of reptiles.

Parker (1963) records a complex, rapidly-gyrating and coiling display in the saw-tailed vipers and asps of northern Africa and western Asia, and also in the African egg-eating snakes, *Dasypeltis spp.*, which produce sound by the rubbing together of rows of specially-serrated scales. He comments that the performance is uncanny to watch, since adjacent parts of the body continually move in opposite directions with a thoroughly confusing effect. After, or during, this display the snake will strike. Snakes of the genus *Dasypeltis* are in fact harmless, but the most widespread species *D. scabra* mimics, in different parts of Africa, various poisonous species. Both colouration and swirling display match that of the carpet viper, *Echis carinata*, in the northern part of the *Dasypeltis* range, that of the night adder, *Causus spp.*, over much of its range, and that of the Cape horned viper, *Bitis caudalis*, in south-west Africa (Gans 1961; Edmunds 1974). These swirling displays may have inspired Kipling's accounts, though it seems that they are used primarily when the snake is itself potential prey.

At first sight it may seem curious to associate colour flashing in sea pansies, skipping in Lepidoptera, rapid colour changes in Cephalopoda, 'convulsions' in amphibians and reptiles, and swirling displays in snakes all under the one heading. Yet they all seem to have the same effect upon a reactor: the inhibition of capture behaviour through the confusion of signals. We therefore categorize convulsoid behaviours as any rapid and apparently uncoordinated changes in the appearance or movements of an animal, occurring within a restricted area so that the actor displays without moving away. Such categorization could include crazy-dances, but the latter seem to be composed of action patterns 'borrowed' from systematic behaviours such as play, and we discuss them later, after single erratic displays (pp.184–187). Clearly, much further study will be necessary to formalize the analysis of these more complex protean phenomena.

General comments

We believe that this outline of protean displays and their effects provides abundant evidence that the irregularity or protean principle has been an important factor in the evolution of the behavioural weaponry of very many species throughout most groups of animals, including man. We would stress that we have endeavoured to be selective in our choice of examples, always attempting to assess the validity of the data. We are familiar with many of the examples ourselves, and we believe that our thesis is strengthened by the way in which, time and time again, experienced field naturalists, when first acquainted with the idea of protean behaviour, confirm our examples or provide new ones.

The classification we have adopted arranges the various protean displays into more or less discrete categories according to their appearance to an observer and their effect upon a reactor. But this does not mean that these categories of display have all evolved independently. It can readily be seen that the single erratic display, with suitable modifications in its form, or with a multiplication of displaying individuals, can be developed into any of the other categories of protean display. Thus the single erratic display performed by a number of animals in unison becomes the united erratic display. When performed by a number of animals at the same time and place, but not in unison, it becomes the multiple erratic display. A number of animals performing their single erratics when fleeing from the centre of a cluster produce the scatter display. The single erratic display performed with injury simulation becomes injury-feigning, or when carried out on the spot is seen as a convulsion.

In all cases it is the *elements* of unpredictability which have been moulded by evolutionary pressures within the context of other behavioural armaments to produce a specific category of protean display. We consider these elements in some detail in the next chapter.

Summary

We distinguish between *primary protean displays*, which are the result of natural selection, and *secondary protean displays*, which are chance occurrences.

The escape behaviours into which the protean or irregularity principle has become incorporated as a major feature fall into a number of more or less discrete categories. These we regard as 'protean displays'.

The principal form of protean display, found throughout the animal kingdom, is the *single erratic display*. This involves the erratic fleeing of a single prey animal, for example by flying, swimming, running, or jumping, separately or in combination. Typically, the actor varies its flight path by zigzagging, changing the path by unpredictable variations in speed, angle of turn, direction of turn, frequency of turning, and other means. Running hares, flying snipe, and swimming fish all provide common and well-known examples of this display when they are disturbed. The single erratic display is so widespread that it must be regarded as a major feature of the behavioural armament of all principal groups of animals. Most of the following protean displays seem to be derived from the single erratic display and may therefore be regarded as *derived erratic displays*.

When the single erratic display is performed by a number of animals simultaneously, but not in unison, a discrete category of display is seen which we name the *multiple erratic display*. This is more confusing to a reactor because of the multiplicity of actors and the increased difficulty of selecting a target. A dancing swarm of mosquitoes is a typical example. Another is the bounding display of groups of antelopes.

In many species of gregarious animals the *united erratic display* is seen. In this case the actors display in unison, typically bunching together, making it increasingly difficult for a predator to select an individual prey animal. The closely-packed group follows a flight path similar to that of the single erratic display, with the added confusing factor of shuffling within the group. Splitting, crossing-over, and regrouping may also occur. Flying flocks of waders and ducks and swimming fish schools provide excellent examples.

Scatter is a group display in which the individuals in a quite closely-knit group explode in individual single erratic movements and thus spread suddenly over a much larger area. The explosion also results in a more or less random dispersion of the individuals. Typical examples are seen in broods of game birds, each individual flight path usually ending in concealment; also in the underwater scatter of groups of diving ducklings, the individual flight paths being hidden beneath the water surface; and the explosion of young from the nest in many species of perching birds and small mammals. This is another display seen in many different groups throughout the animal kingdom.

Distraction displays involve the distraction of a predator's attention from the 'prize' by means of one or more 'lures'. In its simplest form this is seen when certain reptiles and many kinds of marine animals shed parts of themselves and escape while the predator's attention is re-directed to the shed organ which usually shows some degree of movement. In some animals several lures are released successively. Protean elements are introduced in the unexpectedness of the occurrence and the movement of the lures.

Avian injury-feigning is another complex example of distraction behaviour in which the protean movements of the lure — in this case the parent, usually mother bird — are a central requirement for the success of the display. The predator's interest in the lure must be maintained, yet the lure must avoid being captured, so behavioural mimicry and protean movements are operating together.

In *mobbing* behaviour, the actors alternate roles between lure and prize, the protean effect being based upon the unpredictable switching of role. This behaviour is seen widely in the higher vertebrates, as in the mobbing of owls and other birds of prey by smaller perching birds; also in the dive-bombing of humans and other intruders by gulls and terns in territory; and in the overwhelming of snakes by hornbills.

Vocal mimicry may also bring about a predator's switching of attention. Bower birds, for example, may mimic birds of prey, and potential predators may be distracted by this unexpected and unpredictable occurrence.

Immobility displays may be strongly protean by virtue of their unpredictable onset and duration. As in injury-feigning, mimicry of another mode of behaviour is often involved, many examples of immobility being known as 'shamming dead'. This denies to the predator the releaser for attack, and the actor thereby has an increased chance of escape. The irregular occurrence of this behaviour prevents the predator from developing learned countermeasures. Several bird species are known to perform immobility displays, and the phenomenon is also common in several other major groups, both vertebrate and invertebrate.

Protean elements are very apparent in *convulsoid behaviours*. These are used by certain predators as well as by many prey animals. Convulsions confuse by the unpredictable changes in the behaviour of the actor. It may flop around in an uncoordinated manner, or show unpredictable colour changes, all of which deny to a reactor any systematic information. This is seen, for example, in many species of moths and other arthropods, which may flop and skip around when disturbed, and in many reptiles, rodents, and other vertebrates when cornered. Foxes, stoats, and some other predators show convulsoid behaviours which block the escape reactions of their prey. The rapid and unpredictable changes of colour in sea pansies, cuttlefish, and other marine animals seem to have the same effect and must be regarded as convulsoid displays.

The categories of display outlined here do not necessarily represent the principal avenues of selection pressure. Rather they show the main areas of biological armament in which protean phenomena have played an important evolutionary role. The effectiveness of protean displays depends ultimately on the elements of unpredictability outlined in the next chapter.

5
Elements of unpredictability

> He who knows the properties, the changes, and the processes of matter, must of necessity understand the effects and sum of what it does, has done, or can do, though his knowledge extends not to all the parts and particulars thereof.
>
> *Bacon:* Essay on *Proteus, or Matter*

In Chapters 3 and 4 we have described the general forms of escape behaviour and shown how protean elements may be added to them in various ways in different species. Even intrinsically protean displays such as scatter and distraction are open to further complexification and variation, and seemingly static and regular escape mechanisms like catalepsy and other immobility devices can be made more effective by being linked with protean phenomena like single erratic colour changes or convulsive displays or by being protean in the frequency of their occurrence.

Although protean displays are, in detail and in total pattern, as diverse as the species which use them, it is possible to dissect them into relatively few types of component. A zigzag escape path, for example, may be protean in a number of ways: in variation in the length of each zig and zag, in variation in the angles of turn, in changes of speed, and in variations in the sequence in which any of these three components occur.

These elementary components of protean displays we may call the 'elements of unpredictability'. In addition to the three basic elements just mentioned, the elements of unpredictability include individual changes in shape, colour, sound production, and form of movement; differences in timing of the onset of a behaviour; changes in number, behavioural role, and relative position within a group; and differences in the displays of various individuals of the same or different species. We shall examine each of these elements and describe some of the ways in which they become woven into a great variety of escape displays through variations in their forms and combinations. It must be kept in mind that operation of any of these elements will increase reaction times and thereby benefit the actor.

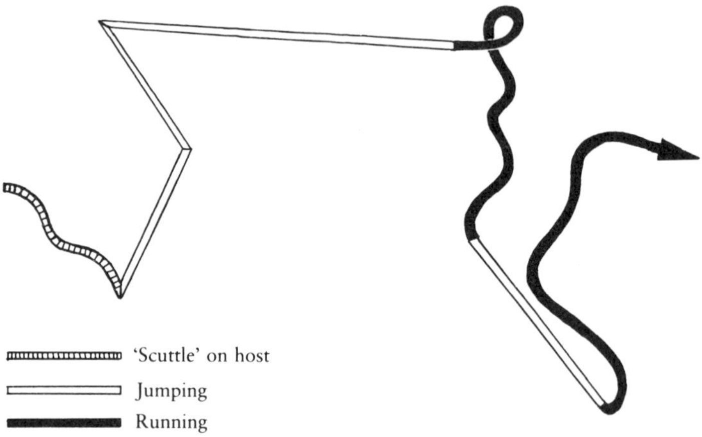

Fig. 5.1. The single erratic display: duck flea disturbed by man.

Change in speed

Changes in the speed at which the actor is moving are commonly seen in single erratic displays, not only at the onset when the actor accelerates away from the point at which it detected the predator, but also during the display itself when variations in speed form an integral part of the sequence. One leg of a zigzag for example may be covered at a different speed from adjacent ones. This can readily be observed in the escape behaviour of many species of fish and birds, in adult hares and rabbits and in play-fleeing of the young of various mammals. It also occurs in convulsive deterrent and united erratic displays, in the individual components of scatter, and in injury-feigning. The change in speed is not necessarily the direct response to an action by the predator, as is the case in dodging, but often appears to be endogenous. For example, when injury-feigning is performed to a stationary observer, it still commonly involves changes in speed.

The most extreme kind of variation in speed is that occurring as a result of momentary pauses during convulsive deterrent and single erratic displays. In insects particularly, brief pauses of variable duration are incorporated into the single erratic display, usually at those points where changes in direction or form of movement occur. For example, the single erratic display of *Psychoda phalaenoides* (Fig. 4.10) in response to a light tactile stimulus is usually liberally punctuated by momentary cessations of movement, some lasting only a fraction of a second, others being of two or three seconds' duration.

A somewhat similar pattern of interrupted movement occurs in many species of fleas (Fig. 5.1) during escape from the antiparasite activities of their avian or mammalian host. Pauses occur in the 'still' or 'stance' postures, the latter being the attitude which the flea must adopt in order to store energy in its pair of metathoracic resilin pads as it winds up for a jump. Brief moments of immobility are thus essential to the continuation of a sequence of jumps, although neither the incidence nor the duration of the pauses are determined entirely by the mechanical demands of the mobile phases (Humphries 1970).

Similar considerations apply to many other hopping insects, and to small aquatic crustaceans whose progress is necessarily jerky since they propel themselves by the synchronous beating of a single pair of antennae and/or other appendages.

Again, changes in the form of movement during the single erratic display from running to flying, or vice versa, or from hopping or spinning to running, inevitably require that the insect must briefly cease moving at the point of changeover in locomotory pattern. The duration of such pauses is however, frequently much longer than is necessary from this point of view, and is variable. Also, pauses occur when there is no change in the pattern of movement and when there is no mechanical necessity to stop.

The interruption of movement during the convulsive and single erratic displays is usually easily distinguishable from a full immobility display, either by the relatively longer duration of the immobility display, or by a difference in the posture assumed, or by a tendency of some species to become cataleptic and unresponsive during an immobility display. For instance, the flashing skipping of the yellow underwing moth and the small tortoise-shell butterfly necessarily includes brief pauses as the insect falls back to the ground at the end of each skip. At this point the wings of the small tortoise-shell close tightly, hiding their bright colour exactly as in the posture of cryptic immobility which so often terminates the single erratic skipping and fleeing displays in this species. The difference is that wing closure is momentary, functioning to allow a vigorous downbeat which propels the insect upwards again. In many other insects the posture of cryptic immobility is different from those assumed during pauses in other displays. In fleas for example, cataleptic immobility invariably occurs in the 'leg' posture which closely resembles the attitudes usually assumed by fleas in death. 'Leg' is never used as only a brief break in single erratic display; it is always a relatively long-lasting cryptic posture during which the flea is unresponsive to most kinds of sensory stimulation (Humphries 1970). Darwin (quoted by Carrington and Meader 1911) noted such differences in a variety of insects.

A protean change of speed of an extreme kind is seen in the young of birds such as the ringed plover, *Charadrius hiaticula*. These chicks have a highly cryptic downy plumage and normally freeze when disturbed. However, in certain circumstances they run when disturbed and can then be seen against their background. But this visibility is destroyed when they stop, for they stop unexpectedly, coming to an apparently instantaneous halt. The eye of the observer, not expecting this, has followed-through along the expected flight path, and has great difficulty in finding the bird, which is now highly cryptic again. This is functionally similar to the flight and unexpected landing of a number of birds and insects which show flash colouration in flight.

Finally, it is noteworthy that football players employ unexpectedly changes of speed to considerable effect.

Change in direction of movement

Alteration of direction is an obvious and basic component of most forms of protean display. It is the main feature of the single erratic display, is fundamental to the deterrent shuffling of a coherent group, and is found in many convulsive, scatter, and distraction displays. A sudden change in direction frequently precedes cryptic immobility in fleeing lepidopterans and makes location of the camouflaged insect very difficult.

In small animals where the inertial resistance to a swerve is low, the angle of change can be sharper and the rate of variation in direction is often very high. The larger power-to-mass ratio intrinsic to the small size of arthropods may well be the reason why it is possible, in terms of biological and mechanical efficiency, for so many insects and small crustaceans to move habitually in a protean fashion. Where the predator is larger than its prey, its relatively greater inertial resistance to change in direction may well play a significant part in making the single erratic escape pattern more effective. Presumably this factor operates at all size levels through rabbit and dog, up to antelope and lion. These comments also apply to the changes in speed which have already been discussed and which we have seen to be particularly rapid and striking in the escape displays of small arthropods.

In the classical zigzag, performed by a running terrestrial animal such as the European hare, change in direction is intermittent and fairly simple to describe. But in some forms of single erratic display, change in direction is almost continuous and the animal moves in a series of curves rather than straight lines. The silverfish, *Lepisma saccharina*, provides the simplest example, where the animal is essentially limited to movement in a single plane and the curves tend to be of approximately the same radius, though varying in length (Fig. 4.13). More complex displays result when movement is three-dimensional, as

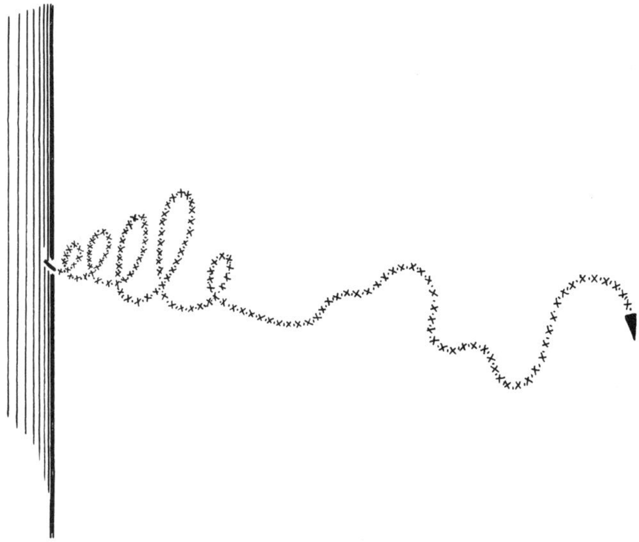

Fig. 5.2. The single erratic display: flight path of chironomid fly disturbed from its resting place on tree.

in the hopping and bouncing displays of collembolans, shore-hoppers, fleas, and antelopes. Even more complex are aerial and aquatic erratic displays, for here there is the relatively greater freedom of three-dimensional movement. The changes of radius of curvature in the whirling of chironomids disturbed from their resting place on tree bark (Fig. 5.2) and the spiralling of noctuid and geometrid moths when exposed to the ultrasonic probes of hunting bats (Fig. 4.12) are specialized examples of this commonly-occurring feature of change in direction of movement.

An example of change in direction of movement which has connections with this section and the next, and is not readily catalogued in Chapter 4, is the well-documented instance of flying fish escaping their predators by leaving the water. This must be as near to a disappearing act as it is possible to get. Many species of the marine family Exocoetidae have enlarged pectoral fins on which they glide after leaving the water, and an enlarged lower lobe of the caudal fin which enables them to taxi along the surface. Some species can even maintain height by vibrating the 'wings'. Flying fishes have little manœuvrability in the air, but when there are suitable breezes, or up-draughts from waves, may travel for up to 400 m at heights up to 9 m. They may take to the air for a number of reasons but the principal one is escape from predators (Gordon 1977). These predators are numerous and include tuna, bonito, albacore, swordfish, marlin, porpoise, and, particularly, the dolphin-fish or dorado which, it is often claimed, is the fastest fish in the sea. Although flying fish show typical single erratic movements when pursued in the water, this does not protect them against the fastest predators and they then take to the air. Although, in calm water, a predator may follow beneath the surface, we are assured (W. Hansen pers. comm., 1979) that a flying fish above the surface is very difficult to see from below it, and when the water surface is broken is invisible. Most skin divers will appreciate this situation. It is also seen in various species of squid (Mollusca; Suborder: Decapoda) which, when pursued, may jet-propel themselves through the water surface and glide on their fins.

The only other 'disappearing act' we are aware of at present is the behaviour of snowshoe hares, *Lepus americanus*, when pursued by lynx, *Lynx canadensis*, in deep snow. If the single erratic display is of no avail and the lynx pounces, or is about to, the hare may dive into the snow and disappear from view. The lynx dives into the same spot, but the hare has burrowed away beneath the surface and, after a brief interval, reappears unpredictably and continues its flight. The whole thing may be repeated several times, and the hare is often successful.

Changes in form, colour, sound, timing, and type of movement

Transformations of shape, though the most striking ability of the Proteus of Greek mythology, are uncommon as protean devices in biological reality. Thus the wing closure and folding associated with the adoption of a cryptic attitude in moths and butterflies undoubtedly involves a considerable change in apparent shape, but is of little significance as a protean element since it is almost entirely subordinate to the cryptic function.

Indeed, many transformations of shape which occur in escape displays are not protean, for they are not associated either with rapidly repeated changes in the pattern presented to the predator or with irregularity of behavioural sequence. Such systematic transformations include the taking in of air or water to increase in size, as in globe-fishes, porcupine-fishes, chameleons, and many species of frogs and toads. In many mammals and birds, increase in size by the erection of hairs or feathers is sometimes incorporated into protean display, for the transformation in size can be achieved much more rapidly in this way than by inflation. The mongoose, using a bouncing single erratic display to avoid capture when attacking a snake, fluffs itself up, almost doubling its apparent bulk — particularly in the tail which expands into a great brush. The position of the vulnerable parts is thus made even more uncertain, decreasing the chances of an accurate strike by the snake. In protean displays by prey animals, increase in size by fluffing occurs during some convulsive deterrent displays such as the sudden and unexpected struggling, screaming, defecating display of a previously passive captured passerine.

Warning devices which involve the sudden display of hidden coloured structures often include movements or posturing which change the apparent shape of the animal. Again, some of these displays are largely systematic, like the protrusion of whip-like processes from the hind end of the puss moth caterpillar, or the startling eye-spot displays described by Cott (1940) in the mantis, *Pseudocreobotra wahlbergi*, and in the eyed hawk-moth, *Smerinthus ocellatus*. Such displays probably rely largely on a startle effect and the deterrent effect of predatory resemblance. Simple flash displays, like those of the many lepidopterans which suddenly open the wings to show areas of bright colours, or that described by Carpenter (1921) in the green African grasshopper, *Phymateus viridipes*, which when attacked by a predator lifts its wings vertically to display their red and black colour, may have a startle effect. Or, when the prey has been in view of the predator prior to the flash, they may have an anomalous effect (Sargent 1978) which will also delay the predator's response. The unexpected, and largely unpredictable occurrence of flash displays would seem to be an essential part of their selective significance. These remarks will also apply to the sound displays heard in a number of grasshoppers which buzz or click during flight.

The unpredictability of onset of a behaviour must have widespread significance, for it is seen in combination with a wide variety of displays and in many different kinds of animals.

In some cases the suddenness of activity itself seems to be the principal antipredator component of a display. The sudden unexpected screaming of a captured bird or mammal is a case in point (p.70). Inglis *et al.* (1982) who apparently ignore or overlook our belief (Driver and Humphries 1969) that it is the unpredictable occurrence of distress-calling which is the main survival factor in captured passerines, demonstrate considerable interspecific variation in the occurrence of distress-calling in British passerines (27 species). There is also wide variation in such calling among North American birds (Norris and Stamm 1965; Rohwer *et al.* 1976; Perrone 1980). There is considerable intraspecific variation in the timing and occurrence of calling in all cases. All of these studies support our thesis, for the greater the variation, the greater the surprise effect on predators.

Another excellent example of sudden activity is provided by the 'skipjack' beetles of the Family Elateridae. *Athous haemorrhoidalis* is a British species, around 1.4-cm long, which may be found in summer at

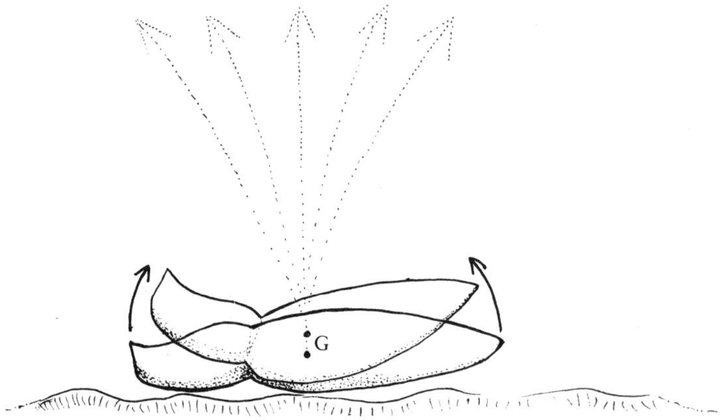

Fig. 5.3. The jump of the skipjack or click beetle *Athous haemorrhoidalis*. The two profiles, taken from cine film, are 64 msec apart. G shows the change in centre of gravity. The dotted arrows indicate possible trajectories, according to substrate irregularities (after Evans 1972).

the tops of grass stems and other herbage. When disturbed, it normally drops to the ground — like many other beetles — and takes cover. However, if it falls upside down, particularly when on a smooth surface, it has difficulty in righting itself. It then performs a prodigious leap which may carry it 30 cm into the air, usually somersaulting several times before it falls back to the ground (Evans 1975). The leap is executed by means of a special mechanism which throws the prothorax violently upwards with an acceleration of up to 700 *g* in about one two-thousandth of a second. This force is transmitted via the wing cases to the ground, and the animal is propelled into the air (Fig. 5.3). It may leap in any direction and may, of course, land the right way up. If it does not, it leaps again.

In such species the time lapse between landing on the back and executing a jump is highly variable, as anyone may see by placing a specimen on the hand. One can never tell precisely when the animal will jump. Such variation in onset of the display will clearly have survival value in nature through its effect on an actual or potential captor. It is thus analogous to the unpredictable timing of the recovery from apparent death in death-feigning animals (p.109) where the display can only be effective against experienced predators by virtue of its unpredictability.

The unpredictably erratic occurrence of colour flashing is particularly well demonstrated by certain flies of the Order Dolichopodidae. These insects, sometimes called 'jewel flies', have iridescent colours

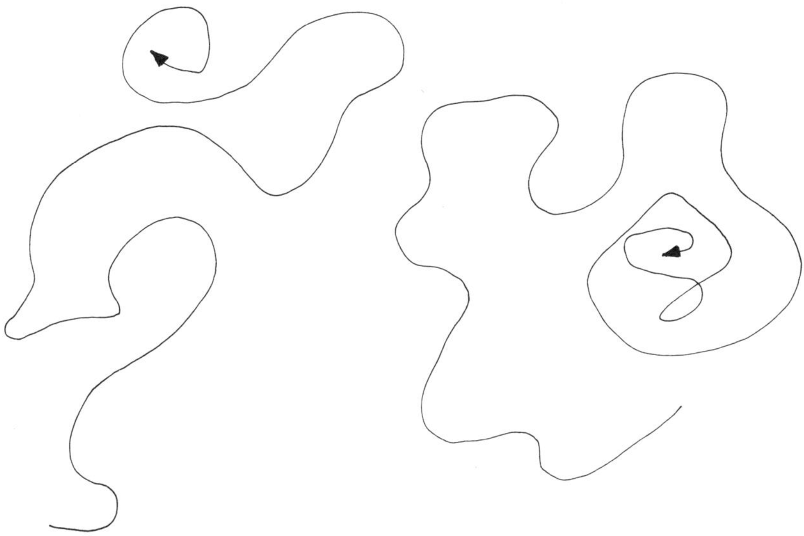

Fig. 5.4. Flight paths of jewel flies (Dolichopodidae). As the insects turn in the sun their silvery surfaces flash intermittently.

which change with the angle of view and with the amount of incident light. As a result they gleam and flash as they fly in the fast erratic manner so typical of the true flies. In *Argyra diaphana*, for example, with its deep green-blue thorax covered with a silvery sheen 'the effect is very striking, little flashes and gleams of silver alternately appearing and disappearing with great rapidity' (Colyer and Hammond 1951). When several of these insects are flying together back and forth in different directions against the complex background of foliage and ground debris, the effect is confusing in the extreme (Fig. 5.4).

Fishermen in many parts of the world will be familiar with the intermittent appearance and disappearance of an iridescent animal if they have observed kingfishers, such as the Old World *Alcedo atthis*, flying rapidly in and out of shade, twisting and turning as they follow watercourses and avoid obstacles.

In many instances regular warning displays may become so complex that they acquire protean characteristics. Shelford (1903) described the behaviour of the cryptic brown and grey mantis, *Hestiasula sarawaka*, when disturbed. It rears up and spreads the forelegs sideways to expose crimson coxae, black-edged yellow femora, and jet black on the lower surface of the thorax. It also opens the wings and elytra showing their striking yellow-on-black mottling. The display is emphasized by side-to-side swaying, an up-curving abdomen, quivering and rustling of the wings, and a rhythmic snapping sound produced by the mutual impact of the fore tibia and femur. In a single display we therefore have all the components considered in this section — a sudden and complex colour change, repetitive sounds, an apparent change in shape due to the special posture, and a distinctive quivering movement.

Such behaviour has much in common with convulsive deterrent displays which derive their startling effect from the sudden onset of a highly complex and novel stimulus situation. Convulsive deterrence usually relies primarily on the visual effect of rapid, irregular, and unusual forms of movement as in the deermouse, *Peromyscus*. But sound production and colourful effects may also be added: for example convulsive deterrence in the conspicuously black-and-yellow Siamese tree-snake, *Dipsadomorphus dendrophilus*, includes hissing, a drumming sound produced by beating the tail on the ground, wild writhing, gaping, and striking at nearby objects (Annandale 1900). The skipping display of the small tortoise-shell butterfly (p.37) is punctuated not only by bright flashes of colour but also by a clicking sound produced by the vigorous wing beats. In extreme cases, the deterrent effect can be produced by a static animal which creates, in humans at least, the illusion of movement and continuous change of form by means of complex rapidly changing colour patterns — the berserking of sea pansies (Bullock 1958; Panceri 1872) and the similar effect produced by swift colour changes in cuttlefish (Holmes 1940) are well known. These displays are described in detail in Chapter 4.

In the same way that changes in colour are effective against a visually-hunting predator, the production of pulses of sound during escape is presumably of greatest importance in instances where hearing is the main sense used by the hunter. In this connection, it is interesting that some species of arctiid moths are able to foil the attacks of bats (*Myotis lucifugus*) by producing bursts of clicks at the rate of over 1000 per second (Blest 1964; Dunning and Roeder 1965). The clicks are produced by microtymbal organs on the metathoracic episternites (Blest *et al.* 1963) and are largely ultrasonic, having in *Halysidota tessellaris* a dominant frequency of about 60 kilocycles/sec (Dunning and Roeder 1965). The frequency sweeps over the range 30 to 90 kilocycles/sec some 40 times every second. Blest (1964) recorded 47 arctiid and ctenuchid species which will produce these sounds in response to the tactile stimulation of experimental handling. Such tactile stimulation gives some idea of the moths' probable reactions when seized by a predator. A predator–prey situation in which the clicks are highly functional was revealed by Dunning and Roeder (1965). They found that in *H. tessellaris* and *Pyrrharctia isabella* the clicks are elicited in response to an impinging train of ultrasonic pulses such as those emitted by hunting bats, and that tape recordings of the moths' sounds when played back to *M. lucifugus* engaged in the task of capturing mealworms (*Tenebrio*) tossed into the air considerably reduced the bats' success rate (see p.151). Flying arctiids responding to the presence of bat ultrasonics by clicking also show the looping and diving pattern described in noctuids (p.59); their acoustic display is thus combined with a downwardly-directed single erratic escape path.

Percussive display is again combined with single erratic behaviour in the squirrel fish, *Holocentrus ascensionis*, which produces volleys of thump-like sounds. In the Nassau grouper, *Epinephalus striatus*, the percussive elements form a background to a kaleidoscopic single erratic display. Such a display seems admirably suited both to deter or confuse the visually-hunting predator and possibly to 'blind' temporarily the predator's lateral-line system much as brilliant flashes of light would, for a short period, impair vision.

In another reef dweller, *Haemulon flavo-lineatus*, the visual display accompanying the single erratic is even more striking. The colour of the fish changes from pale yellow to deep bronze, with numerous blue stripes which vary in intensity, further longitudinal and transverse stripes flickering transiently and repeatedly on this background (Pycraft 1925). This colour display, like those of squids and cuttlefish, depends on the swift muscular control of chromatophores. But changes in colour associated with zigzag fleeing usually occur as an automatic result of the form and orientation of movement. Thus fishes may present a different appearance according to their angle in relation to an observer. Denton and Nicol (1965*a*,*b*, 1966) show how silvery-sided fishes in their normal orientation are well camouflaged, the amount of light reflected from any part of the body matching closely that coming from the background against which it is viewed in nature. This effect is produced by oriented reflecting crystals which lie in the skin at different angles relative to the surface of the fish. If, however, the fish intermittently moves out of its normal orientation, as during a vigorous twisting zigzag, a series of silvery flashes is produced, the fish being extremely difficult to locate when it resumes its usual camouflaged orientation.

Hemmings (1966) noted the loss of crypticity when silvery-sided fishes move out of their usual orientation, and has since described to us an example in which a whole school of sardines gave a brilliant silver flash and swam down for 1–2 m, after which normal orientation was renewed, and the school disappeared. This illustrates how, when an escaping fish swims downwards, it becomes suddenly conspicuous until it resumes the usual cryptic orientation. The flash may have functions in addition to the production of confusion in a pursuer. It may, as suggested by Hemmings, act as a social signal permitting close co-ordination in the school as a whole. But this too may subserve a protean function (see p.133).

Many reef fishes are brilliantly and variously coloured, displaying different colour patterns according to their angle in relation to an observer; again the appearance of the fleeing prey changes from moment to moment, a protean device which is particularly appropriate against the coloured chiaroscuro of a coral reef.

Intermittent colour changes are automatic too in many insects, including grasshoppers, mantids, cicadas, and many lepidopterans, and in birds such as the turacos (Musophagidae), all of which reveal colourful patches on the wings when taking flight. In these cases there is an additional confusing effect in the sudden disappearance of the patch of colour when the animal closes its wings and alights, much as a silvery fish 'disappears' as it resumes a cryptic orientation. The same principle will apply in the case of the various species of frogs which have brightly-coloured patches on the inside of the thighs. These patches too are suddenly revealed, and concealed, when the animal flees.

The significance of colour changes during fleeing may well include their effect in concentrating the predator's attention on the most conspicuous part of the display, rendering it more difficult to predict the flight path as a whole, and more difficult to find the actor once its colourful labels are no longer to be seen. Edmunds (1974) has also made this point, suggesting that 'predators may be induced to follow this bright colour instead of including other features of the prey in their searching image, and so when the bright colour vanishes the predator gives up the search'.

This point is particularly well illustrated by the turacos, which are mostly pigeon-sized birds of tropical forests. A brilliant patch of red on the wing is seen in most of the 22 species. The few, duller-coloured species are found in more open situations. When a turaco takes flight, the red flashes — almost iridescent in the gloom among the trees — attract and hold the attention. But when the bird alights it snaps the wings abruptly shut a little way before it lands, and the disappearance of colour at a distance from the landing point results in the whole bird disappearing from sight. The position of the bird is unpredictably related to the point of disappearance, for the eye of the observer has travelled on along the flight line. Closing the wings replaces flash colouration with crypsis and the bird is able to run and climb rapidly from view with its highly-developed feet and legs.

It would seem that adaptation to niche in the colourful turacos owes much to the protean principle, for not only do the specialized hind limbs allow for unusual agility in the trees, but also the feather pigments are specially developed to subserve the disappearing trick. The red colour is due to turacin, a copper porphyrin unique in the animal world. The green pigment, which gives such excellent crypsis, is turacoverdin, also found only in the forest turacos. Behaviour and structure are thus woven together into an integrated complex which allows the bird to disappear from view in a confusingly unpredictable manner.

The addition of colour to a display sometimes makes capture of a lure more likely; this is true of some of those distraction displays in which an individual sacrifices parts of its body to save the rest. Certain polynöid worms shed scales which flash luminescently after they become detached (p.94). If cut in half, these worms luminesce brilliantly — but only in the posterior section, the anterior half remaining cryptic and having the ability, if overlooked by the predator, to escape and regenerate the missing parts. In many other autotomous distraction displays, the lures are made conspicuously different from the prize in form or by their active and unusual movement. The violent lashing of a lizard's newly autotomized tail is well known (p.93), and the ejected viscera of holothurians and fragmented siphon of the jack-knife clam in no way resemble the animal which has sacrificed them. If, as suggested by Tinbergen (1960) for insectivorous birds, predator's hunting is guided by specific 'searching images' based on their recent successful experience, autotomous distraction displays should indeed have the lures as different from the prize and as conspicuous as possible. The significance of this difference is in no way reduced if the 'searching image' theory proves to be a mistaken one, for there are other ways in which it may affect the reactor. In this connection Vitz (1966) and Berlyne (1960) have discussed the attention-drawing properties of stimulus novelty and stimulus complexity.

In distraction displays in which the lure is itself a whole organism which must avoid capture, this avoidance is sometimes facilitated by the illusion of transformation into another type of creature. This can be achieved by a special posture or type of movement. In the injury-feigning types of distraction display, special kinds of movement may enable the lure to change its apparent form considerably. Certain injury-feigning waders and some passerines may appear so like rodents that the display has been termed the 'rodent run' (Williamson 1950; Simmons 1955; Rowley 1962). A predator must to some extent have different capture techniques for birds and mammals, respectively, and its efficiency is likely to be reduced by an apparent transformation of one type of animal into the other.

Apart from such unusual forms of movement, protean displays frequently involve an increase in the variety of everyday modes of movement, like running, scuttling, jumping, flying, spinning or walking. This is well illustrated by the display of the old-squaw (Fig. 4.32) and has also been demonstrated quantitatively in the protean behaviour of the duck flea, *Ceratophyllus garei* (Humphries 1963, 1970).

Variation in the distance between points of change

Changes in colouration, form of movement, speed, or direction are likely to be more effective in deceiving or confusing a predator if their exact onset is unexpected or unpredictable. Most protean displays include this element of unpredictability, and we have illustrated (Figs. 4.3 to 4.18) variation in the distance travelled between the exact points where changes occur. In some cases this variation may be governed by movements of the predator. Whenever the predator approaches within a certain limit, the prey will dodge. But variation frequently occurs without direct control by predator approach as in the zigzag take-off of the snipe and probably in the great majority of single erratic patterns of invertebrates. In the single erratic display of a rabbit, *Oryctolagus cuniculus*, pursued by a dog, we have observed on a number of occasions that close approach results in a burst of zigzagging which is much more complex than would be needed for dodging. Even in species which can be seen to dodge in response to an action of the predator, the varied changes of the single erratic display give the impression of it being partly endogenous in origin, since they continue to occur even if the predator remains stationary. Anyone who has attempted to catch fish in a pond or rock pool will be aware that a rapid movement of hand or net in pursuit of the fish results in a wild zigzagging dash which is a directed dodge in the sense that the fish avoids the instrument of capture but is endogenously protean in the sense that most of the zigzagging occurs after the capture movement has ceased. We have also observed this closely in the case of sticklebacks pursued by red-breasted merganser ducklings in open water. The problem of the internal or external determination of protean displays is dealt with in detail in Chapter 6 where we consider their causation.

Change in number, behavioural role, or relative position within a group

Real or apparent changes in the numbers of prey objects are a central feature of scatter and of distraction by autotomy. Bowen (1931) and several later workers supposed that the protective effect of schooling in some fish species might indeed derive from the bewildering effect of the scatter displays which the initial organization into schools makes possible. Baerends and Baerends-van Roon (1950) suggest that in fish schools it is sudden changes in density, either by scatter or by close packing, which are effective in bewildering attackers.

Close packing into a conspicuous mass sometimes occurs as an aposematic device, as in the caterpillars of the cinnabar moth on ragwort, and in the small black longitudinally white-striped reef-fish, *Plotosus anguillaris*, contact with the skin of which is intensely painful (Mortensen 1917). But in other instances the contraction of the group into a dense mass is not associated with distastefulness or other physical deterrents and the protective effect must derive from the transformation in group organization. Small porcupine-fishes for example will cluster together closely on the appearance of a predator such as the garfish, the cluster according to Beebe (Norman 1931) giving the appearance of a single large round prickly fish. Individuals becoming separated from the cluster are immediately seized by the garfish. Spooner (1931) has remarked on the similar sudden close packing seen in sprat and pilchard schools when attacked by seabirds. As remarked by Baerends and Baerends-von Roon (1950), there is some resemblance between these fish displays and the deterrent bunching of starlings and other birds; in both instances, it seems possible that the close packing of individuals renders it impossible for any one of them to act as an adequate releasing stimulus for predatory attack. Also, even if an attack is launched, the close packing is likely to make it much more difficult for a predator to capture an individual prey animal. These suggestions have been reinforced by a number of recent experimental studies including Neill and Cullen (1974), Milinski and Heller (1978), and Heller and Milinski (1979).

In the multiple lure situation (Fig. 4.33) the changes are rather in the behavioural roles of the actors, which alternate as prize and lure, or as background mob and attacker. This type of mobbing display, and deterrent shuffling, also involve changes in relative position. Bouncing and weaving in groups of fleeing ungulates likewise create change in the relative positions of the group members, producing a multiple-lure effect against pursuing predators such as lion. This is well-documented in impala, springbok, and wildebeest.

Number and relative positions of luminescent points are again the main variable features employed in the berserking of sea pansies. These two characteristics are also important in the chorusing of bush crickets, where the changes in numbers and relative positions of the sources of sound have a 'ventriloquial and confusing effect on potential predators' (Broughton 1965).

The example of bush crickets is particularly interesting since it has been found that the simultaneously produced courtship songs of two individual males of the same or different species are often modified in such a way that the combined effect closely resembles the song of one individual although produced from two spatially-separated sources. This can be achieved by one individual suppressing parts of its song, or by modifying and adding to its own song until this closely resembles that of the other species (Broughton 1965). Regen (1908) working on the bush cricket, *Pholidoptera aptera*, was first to study this behaviour in detail, finding that males would alternate their chirps, if within hearing range of each other, to give the effect of a single song with indefinite location. Baier (1930) showed in *P. griseoaptera* that, if the males in a group alternate, they do so in no particular order. This is the acoustic equivalent of the unpredictable change in behavioural role seen in visual multiple lure displays such as mobbing. A similar effect was noticed too by the physicist G.W. Pierce when the bush cricket, *Pterophylla camellifolia*, replied to a series of squawks, emitted by Pierce's assistant, with an equal number of bursts of sound (Pierce 1948). Dumortier (1964) has reviewed the alternation and synchronism of singing in male insects.

In the bush cricket, *P. griseoaptera*, Jones (1966) showed that actively chirping males would alternate on a one-to-one basis with each other's chirping or with artificial signals such as those made by a Galton whistle, hissing between the teeth, or pure tones from a loudspeaker. Alternation occurred up to repetition rates of over 60 per minute. Each chirp lasts about 100 msec. Occasionally two males synchronize chirps. Walker (1964) has shown that stridulation, though necessary to attract females, may have the disadvantage of allowing location of the insect by a predator, and he cites an instance of a cat stalking bush crickets by sound. Localization by the predator is likely to be hindered by the brief high-frequency nature of the chirp, its irregular occurrence, the alternation with individuals in other positions, and the fact that disturbance by close proximity of a predator causes the individual to cease singing for several minutes while males elsewhere continue to chirp, providing an effective multiple-lure distraction effect, as many entomologists will testify.

We regard it as highly significant that much of what we have said about bush crickets can also be applied to certain groups of birds, particularly some of the barbets, *Capitonidae*, and bush-shrikes, *Laniidae*. In these species duetting or antiphonal singing is common, two mated birds singing antiphonally to produce a single song. In many species the illusion that only one bird is involved is complete, and this includes such varied forms as the marbled Guiana quail (*Odontophorus gujanensis*, the Central American wrens (*Pheugopedius hyperythrus* and *Thyrophilus modestus*, African barbets (*Trachyphonus* and *Lybius* spp.), and tropical bush-shrikes (*Laniurus* spp.). Most, if not all, of the antiphonal singers are birds which sing in dense cover, but the barbets, at least, duet when the two singing birds are visible to each other, so maintenance of contact cannot be the only function of this type of vocalization. As Payne and Skinner (1970) have suggested for the barbets, duetting may function in territorial advertisement and maintenance of the pair bond. But to human ears the effect of these antiphonal songs is highly protean, and it is likely that the predators of these birds also find it so.

Differences between the displays presented to the predator on different occasions

The escape displays which a predator must attempt to overcome often vary in pattern between individual prey and between different elicitations in the same individual. Except in the most simple and regular cryptic displays, the relatively inexperienced predator is likely to meet new problems each time it attempts to make a capture. We suggest that the variation found in escape patterns within the same prey species may come about in at least three ways: first, by differences in the predator's behaviour, or other environmental influences, resulting in different patterns or details of response; second, by polyethic differences, that is, genetically-governed variations such as those shown by deermice (Dice *et al.* 1963; Chance 1959) with respect to convulsions; and third, if the nervous system itself can generate a degree of randomness of behaviour within a given response pattern (see p.177) then successive elicitations of that response pattern must differ in the details of the behaviour shown.

A particular prey species is generally not limited to a single type of escape response; it has a repertoire of basic escape patterns, though not all of these patterns are necessarily available to a given individual, even under suitable circumstances. Rothschild and Haskell (1966) studying the defensive reactions of the garden tiger moth (*Arctia caja*) found that when the moth is at rest, a sudden disturbance will elicit one of three quite distinct forms of aposematic display. It may fall down 'shamming death' with wings folded and legs drawn up exposing the maximum amount of red. Alternatively, it may release a strong-smelling defensive vapour from its cervical glands, raise the patagia to reveal the red hairs beneath, and spread the wings to expose the scarlet hind areas. The moth then tilts forwards to direct its red surfaces towards the disturbance. In the third kind of reaction the moth again raises the patagia and secretes defensive vapour, crawling forwards slowly, at intervals flapping its wings, the movements of which are accompanied by a rattling sound. Which of the three kinds of display is elicited depends partly on the nature of the stimulus — the first tends to occur mainly to ventral tactile stimulation, the second to dorsal tactile stimulation on the thorax, and the third to disturbance at a short distance. However, Rothschild and Haskell found that the second and third types were never both elicited in the same individual, and that the stridulatory wing-flapping display once given is repeated to further molestation whatever form this takes. They suggest that a genetic element is involved and that it is possible that in *A. caja* the two displays are a form of behavioural polymorphism.

Chance (1959) coined the term 'polyethism' to cover such behavioural polymorphisms and cites as an example the sporadic occurrence, in populations of deermice (*Peromyscus maniculatus*) and some other rodents, of individuals which react with bursts of wild convulsive seizure-like behaviour when normal routes of escape are blocked, often biting anything within reach (Chance 1957; Chance and Russell 1959). It must be emphasized, however, that polyethism is not necessarily always a protean phenomenon — as would be the case if it served to hinder learning processes in the predator — but might be maintained by other kinds of selective advantage.

A similar point must be made concerning the diverse repertoire of escape patterns in any prey individual; where a prey population is subjected to attack by several different species of predators it can be observed that in many instances the prey will employ a specific mode of defence against each kind of predator, as illustrated by Kruuk (1964) in the black-headed gull, *Larus ridibundus*, where different defences are employed by breeding birds against aerial and terrestrial intruders.

Protean effects must however arise where different displays are on different occasions directed against the same individual predator by prey individuals which appear identical to the predator. Certainly, success in the use of fleeing after catalepsy (Tinbergen 1962) or in the use of sudden struggling and screaming after immobility by birds and small mammals (Driver and Humphries 1969) as devices for escaping after capture by the predator, depends on the predator having little or no expectation of the response, a point which correlates well with the fact that these devices are only used sporadically.

Protean effects arising from variation within a population are also seen as colour polymorphisms in displays which are, individually, quite systematic. For example, numerous insects and molluscs exist in a number of polymorphic colour forms.

Owen and Wiegert (1962) suggest that in the meadow spittlebug, *Philaenus spumarius*, such diversity may confuse predators that rely on past experience to tell them whether or not an insect is edible. It is interesting and suggestive of a protective function for the polymorphism that in *Philaenus*, and several other species, colour variations are found particularly in the females, which from a reproductive point of view require a somewhat longer life span than the males.

There are other examples of colour polymorphisms which do not fit into these established categories, yet which probably act as protean devices against predators. These are discussed in detail in Chapter 8.

General comments

In attempting to define the elements of unpredictability it is necessary to guard against the tendency to view these displays exclusively through human eyes. It is abundantly clear from the work of Burtt and Catton (1966) on insect vision, Lettvin *et al.* (1959) on what the frog's eye tells the frog's brain, and Young (1961) on octopus vision, that the perceptual world — the *Umwelt* — of animal species other than ourselves is sometimes greatly different from our own. Displays which are not effective as protean devices against man or other mammalian species may thus be effective in confusing species which analyse the visual qualities of the environment in a different way, or which have less complex and flexible central nervous control of the mechanisms for seizing the prey.

We may now consider to what extent randomness in escape behaviour is relevant to the rapid transformation aspect of protean displays. If relative or apparent randomness of sequence is to be regarded as an essential criterion of protean behaviour, then it might appear that displays which involve regularly repeated changes should not be termed protean. This argument would, for example, exclude simple flash coloration in which a drab cryptic insect changes into a bright colourful object and back again. The attempt to separate the random aspect from the transformational aspect, however, is probably unrealistic, since transformation of stimulus presented to the predator is an essential prerequisite for any kind of irregularity in sequence of the stimulus units. The transformations can of course take place at a variety of levels, occurring merely in direction of movement or, at the other extreme, involving changes in an individual's entire appearance. Transformation of stimulus pattern at one or more levels is therefore an essential part of a protean display. Some degree of randomness of sequence is characteristic of most protean displays for, as explained in Chapter 6, it lengthens the time during which the transformations continue to be effective in confusing the predator. Essentially, the random aspect of the display renders it impossible for the predator to learn display details and make accurate predictions. A sufficiently complex but regular transformational display might work equally well against predators of limited learning ability.

In attempting to tease out and separate the elements of upredictability from the displays in which they occur so that each could be isolated and considered in detail we may have given the impression that these elements are naturally separate in nature, exerting isolated, exactly specified effects on predators. But protean elements not only interact with each other, they are normally woven into other kinds of display, so that the action of these displays against predators is likely to arise from a combination of protective effects.

So far we have said little about the exact nature of the action of protean phenomena against predators, although it has been necessary in places to refer to the evidence for certain ways in which particular displays may be protective. In Chapter 6 we consider this problem and bring together certain of the published facts and theories of ethology and experimental psychology which, when applied to the protean phenomena already outlined, form the basis for a general explanation of the effects of many protean displays.

Summary

The basic elements of unpredictability in protean displays are changes in speed and direction of movement; changes in shape, colour, sound, timing, or form of movement; variations in the distance between points of change; changes in the number of individuals, their behavioural role, and their position within groups; and behavioural polymorphism (polyethism). All operate by increasing a reactor's reaction time.

Changes in speed are common in single erratic displays and convulsive displays, often interspersed with periods of immobility. Changes in direction of movement are also common and form a basic component of most forms of protean display. These two elements, united in varying combinations, result in a high degree of unpredictability of movement in many prey responses to predators.

Changes in shape, colour, sound, timing, and form of movement also occur in an unpredictable fashion in many protean displays. Occasionally they will all occur together, as in certain convulsive displays. Alternatively, an illusion of movement and change in form can be brought about by rapidly changing colour patterns, or by changes in the display of differently coloured parts of the body. Changes in sound production may jam a predator's auditory detection system in a protean manner. Changes in the form of movement are particularly common in distraction displays, especially in avian injury-feigning. Protean displays also frequently involve an increase in the variety of movements performed by the animal in any one behavioural sequence.

The elements considered above are most likely to confuse a reactor if their onset is unpredictable, and most protean displays include this factor as variation in distance or time between points of change. This may be seen from the simplest single erratic display to the most complex convulsion. In some cases the unpredictable onset of a display seems to be the principal protean weapon.

In group displays, confusing elements are added by apparent changes in the number of individuals involved, or by changes in their behavioural role or position within the group. One or more of these elements enter into scattering behaviour and its opposite — closer packing as in fish schools. Changes in position within a group occur in schooling fishes and flocks of ducks or waders in the presence of a predator. Changes in behavioural role are seen in mobbing situations where the individuals alternate roles between lure and prize. Changes in apparent number of actors may be seen to operate also in an auditory context in the case of the singing of bush crickets and the duetting of bush-shrikes.

Differences between the displays presented to a predator on different occasions may be the result of environmental influences, or of differences in the internal controlling factors of a single individual on successive occasions, or of genetically governed — or 'polyethic' — differences between different individuals in a species. Any, or all, of these factors may increase the protean effect in an animal's reactions to its predators.

6
Conflict and confusion

> 'Confusion is a word we have invented for an order which is not understood.'
>
> *Henry Miller* (*Tropic of Capricorn*, 1938)

The validation of protean theory

In this chapter we bring together much of the information which is central to the validity of the theory of protean behaviour. The common feature of the diverse protean patterns described so far is to be found in their biological function.

The validity of the theory that there exists a definite biologically coherent class of protean behavioural phenomena rests firmly on three major points.

First, it is based on the evidence for the widespread occurrence of rapid, complex, and apparently unsystematic transformations arising — very largely — in one kind of ecological context, viz. the direct confrontation between a predator and its prey. We have presented and systematized a selection of the evidence for the occurrence of protean behaviour in such situations, and have shown that the basic designs and elements of many protean patterns appear again and again in similar form in widely separated taxonomic groups. Such convergent evolution, cutting across the diversities of taxonomy and habitat, is strong evidence for the existence of an ecological principle of general validity.

The immense variety of forms of protean displays and the many different behavioural and physiological mechanisms and anatomical structures which are pressed into service to create the displays are no obstacle to this line of reasoning. As Pantin (1950) has pointed out, there is no necessary correspondence between a particular behaviour pattern and one particular class of physiological mechanism. He says: 'The common feature of such patterns is that they meet some particular environmental requirement. In different organisms this may be met by machinery utilizing different properties of the material structures which compose them'. The common feature of the protean pattern, the environmental requirement which it meets, is an organism's continuing need to foil attack by an alternative method to systematic forms of defence or escape. Thus if the behavioural mechanisms of a sufficiently large number of predatory species are vulnerable to escape displays incorporating irregular transformations, then we may expect, in different species of prey animals, the development of diverse behavioural patterns and structural devices whose common theme is the satisfaction of this particular environmental requirement.

Second, our thesis depends on protean displays being of survival value and therefore open to the forces of natural selection. So we may enquire whether there already exists in the literature evidence that, under natural field conditions or laboratory simulations thereof, protean displays are effective in increasing the chances that a prey individual or group of individuals will escape from a predator.

Third, the action of protean displays against predators must ultimately be capable of explanation in terms of their effects on the operation of the perception and response systems of the predator. We must therefore enquire how the known elements of the behaviour control systems of predatory animals might be expected, on theoretical grounds, to respond to displays of a protean nature. In particular, it becomes important to know whether such action against the predator comes about in several quite different and unrelated ways, or whether there exists one general mode of action in reducing predatory efficiency. In this connection we shall also consider whether protean displays are likely to be effective against all predators, or only against certain classes of predators, for as Holling (1965, 1966) has shown, there are fundamental differences in the behavioural ecology of vertebrate and invertebrate predators.

It is the second and third points, both concerning the biological function of the displays, which must now be examined in depth. Certain established properties of vertebrate perceptual and response systems are highly relevant to the problem of the action of protean displays against predators. Since these properties have been described and measured almost entirely under controlled laboratory conditions, we shall discuss their translation into natural field situations, and cite, where possible, field evidence for the occurrence of the theoretically-predicted events.

Protean displays as inducers of conflict

Protean displays differ fundamentally from more regular escape behaviour in their action on the predator. Whereas regular displays like crypticity, or the exposure of deterrent eye spots, systematically deny or falsify information to the predator, protean displays in contrast appear to overload the predator with ever-changing information. In short, the predator becomes confused. It was the recognition of this distinctive feature of protean displays which led Chance and Russell (1959) to assign them to a new and separate category of escape behaviour. In their paper they point out that, although the three protean displays (of *Sepia*, *Peromyscus*, and *Xenopus*) with which they were familar were all subjectively confusing to human observers and made the displayers difficult to handle, nothing was known about the nature or mechanisms of their postulated effect on predators. They comment that 'what happens to the instinctive mechanisms of a predator faced by such a display is a new and interesting question which may have important implications for ethology'. Driver and Humphries (1970) suggested an answer to this question, and in the present chapter we shall develop and discuss our answer in greater depth.

Our suggestion is that protean displays act as inducers of conflict and confusion, in part or parts of the reaction system of reactors. In other words, at some point or points, probably in the reactor's brain and brainstem, the more usual patterning of regular, predictable impulses resulting from regular, predictable stimuli, becomes disturbed. The development of information theory (Shannon and Weaver 1949) has supplied the theoretical framework within which the idea of 'confusion' can be understood in precise terms. Berlyne (1957*a*, 1960) and Attneave (1959) have applied the concepts of information theory to psychology, treating stimulus and response as mathematically equivalent to the input and output of a communication channel. Information, in the technical sense used in communication theory, refers to an input which facilitates the prediction of an event, and so enables one to select a response to the signal with greater likelihood of this response giving a desired result. Cherry (1957) technically defines the quantity of information input in terms of probability; a signal of low probability has high information content. The probability is assessed from past incidence of the signal. Reception of information is limited by channel capacity, in this case the capacity of the sensory–perceptual–response system of the predator. It is important to note that the limit is informational, and not simply dependent on the number of simultaneous stimuli (Broadbent 1958; Miller 1956). Thus the ability of a predator with a given channel capacity to respond appropriately to the escape behaviour of its prey will depend both on the number of impinging stimuli and on the information content, or probability of each. It is hence of significance that protean displays usually involve not only an increase in the number and variety of stimuli presented to the predator, but also the appearance of unusual, or unexpected, forms of behaviour. In general terms we may suggest that the predator's communication channel becomes flooded with irrelevant information, amongst which the systematic message 'prey object and its location' becomes lost.

The informational approach is likely to be a useful tool for understanding the action of protean displays, since it provides a common, precise, and quantitative language for dealing with a wide range of events, but little is known of the information encoding and processing capacities of predators under natural conditions. Indeed, it is only relatively recently that the channel capacity for man himself has been determined: Briggs and Swansea (1971) have shown that the human brain can encode information at a rate of about 6.7 bits per second and process information at about 13.5 bits per second. It is reasonable to assume that other vertebrate species will not exceed these rates, and will be no better than humans at handling protean situations.

It is helpful to talk in terms of the predator's limited channel capacity, and the overloading of that capacity by protean displays, but it is necessary to look at this process in more detail and to see in what specific ways the animal's information-processing abilities break down. In particular, we may ask what long- and short-term effects the breakdown will have on the behaviour of the predator, and whether these effects will assist the prey to escape.

Since the recognition of protean display as a device against predators has been relatively recent, very little research has yet been done to analyse directly the nature of the effect on the predator. Fortunately, however, much relevant evidence exists already, scattered throughout the literature of experimental psychology and ethology. In any circumstance in which an animal is faced with a complex stimulus situation and the necessity of rapidly making a correct response chosen from two or more alternatives, features similar to those naturally operating in protean displays are likely to be present.

When an animal is simultaneoulsy subjected to two or more stimuli, each usually associated with a specific response, some degree of conflict between the alternative different stimulus–response processes is likely to occur. In the wild state such conflict is by no means limited to the predator–prey interaction but occurs in many situations, most conspicuously in forms of social behaviour such as territorial and other agonistic displays, and in courtship, but also less obviously in most, if not all other kinds of activites, like feeding and exploration. In many cases the conflict is expressed behaviourally and striking response patterns characteristic of conflict situations can be recognized. These patterns include ambivalent postures and movements, repeated alternation between the tendencies in conflict and compromise activities (Andrew 1956), redirection of a response to a different target (Bastock *et al.* 1953), and the intervention of apparently irrelevant aspects of behaviour such as displacement activities (Armstrong 1947; Tinbergen 1952). The application of conflict theory is demonstrated in Moynihan's study of agonistic behaviour in gulls (1958).

The occurrence of these conflict patterns in predator–prey interactions has been demonstrated experimentally by Blest (1957*a*), who was primarily interested in investigating the possible deterrent functions of the large eye-spot designs which are commonly found on the wings of butterflies and moths. In order to discover if the eye-spot pattern is indeed a deterrent and if it is more effective than other novel stimuli (for example, a pattern shaped like a cross), he constructed an outdoor semi-natural situation in which insectivorous birds were presented with a mealworm flanked by an artificial eye spot or novel stimulus display. In this simple conflict situation of systematic type, Blest recorded behaviour falling into all of the categories of conflict responses mentioned above, and it is not difficult to see that in the wild such reactions by a predator delaying and possibly giving warning of its attack would be likely to increase the chances of a prey animal escaping. Further, Blest also noted that birds confronted with the display would frequently simply hesitate for a second or two without any special activity being performed, then collect the bait. In connection with this we may note that Berlyne (1957*a*) has provided abundant evidence that a consistent effect of conflict is to increase reaction time. In mild conflict the delay in the reaction is slight with little overt behavioural effect, but in extreme conflict the delay is much longer and the animal's behaviour may take the form of an orientation or startle reaction (Sokolov 1963; Berlyne 1960). It is hardly necessary to point out that slower reactions by a predator are likely to assist the survival of the prey. A fraction of a second may make the difference between life and death.

The above experiments on insectivorous birds investigated a deterrent display which is simple, regular, and predictable. Protean displays are much more complex and provide a far greater number of stimuli likely to arouse diverse incompatible responses in the predator.

It is almost certain that protean displays induce conflict in several ways, depending on the timing, number, and novelty of the elements which we outlined in Chapter 5. First, they often present the predator with a number of stimuli simultaneously, each requiring a different response. This occurs, for example, in the scatter display of a group, where a single target explodes into several alternative targets differing in the speeds and directions of motion. In this instance, conflicting directing responses are aroused, and the predator's attack will be disoriented unless it can rapidly cut off input of information from all but one target. Even if it is able to do so, the processing of information necessary for this target selection will take valuable time. In prey survival, milliseconds can be decisive.

Conflict may be induced in a somewhat similar way by the presentation of a very rapid sequence of stimuli, each arousing a response which cannot be completed before the next stimulus has appeared and set up competing processes. A very rapid zigzag or whirling display is likely to disorientate a predator in this way, and multiple-lure displays may work in a similar manner.

In other instances, conflict may be induced not only in the orientation mechanism of the predator, but also in its basic motivation. Such effects on mood occur for example in multiple-lure displays of the aggressive-mobbing type, and in individual displays which include aggressive, threatening, or deterrent components, as, for example, the bouncing aggression of *Peromyscus* and the production of bright flashes of colour or bursts of intense sound during erratic display by certain insects, cephalopods, fishes, and birds. The work of Hogan (1965) on the Burmese junglefowl, *Gallus gallus spadiceus*, and of Leyhausen (1956) on cats has shown that, even where the prey does not employ such mood-shifting components, a predator's approach to its prey is often ambivalent and includes a proportion of what may loosely be termed 'fear'.

A second basic way in which protean displays induce conflict arises from the erratic nature of the display, which defeats anticipation or prediction by the predator. Predatory behaviour may be considered as a skill, developed by the dual processes of maturation and learning. An important feature of a skill is that the lag between input and output is minimized. This is achieved by anticipation of the latter part of a series of stimuli, where the beginning of the series contains sufficient information for action to be taken on the rest of the series. A bird pecking at a moving insect will peck in advance of the position at which the prey was perceived; the prey's course is being predicted. Similarly, hunting dogs and other carnivores learn to cut off a prey animal fleeing from them at an angle (Kruuk 1972; Schaller 1972; Hanby 1983). Further, it is a well-known principle in experimental psychology that the more frequently two stimulus events are associated in the input to an organism, the greater will be the tendency for the central correlates of one to activate the central correlates of the other (Osgood 1957). Thus one stimulus will often evoke preparatory responses which allow an optimal reaction when the second stimulus occurs. This process is certainly useful when the organism is dealing with systematically-ordered stimulus systems. But the protean display is highly unsystematic and consequently many preparatory responses will be wasted. Further, they will also often conflict with the behavioural processes required by the new situation. Such conflict is likely to occur, for example, where the predator is attempting to follow a prey animal giving a single erratic display, and also where there is rapid change in behavioural role or relative position within a group of displaying prey as in the multiple-lure and united-erratic patterns.

In its extreme form this kind of conflict appears as 'surprise' and Desai (1939) has shown that, in human subjects, this is associated with an arrest of ongoing movements.

The third way in which protean behaviour may induce conflict is also related to its erratic nature, and arises too from the sheer number of different sequential stimuli to which the predator may be required to react. When a predator is faced with the possibility that any one of a number of alternative stimuli is about to occur and where each of these stimuli requires a different response, the predator is in a state of uncertainty. Berlyne (1960) has described how uncertainty can be measured in terms of information theory. He also points out that these situations of uncertainty essentially generate conflict, since anticipatory arousal of incompatible responses is involved. In the case of the single erratic display, this would mean incompatible orienting responses of the predator. The degree of uncertainty will increase the greater the number of alternatives available in the protean display and the more irregularly they are arranged in sequence.

Uncertainty is involved not only within individual protean displays, but may be effective on a larger scale, owing to the variety of the kinds of display given by the prey on different occasions. Thus if a predator is aware that its prey has a number of behaviours which it can employ in its avoidance technique, anticipatory conflict is likely to automatically increase the predator's reaction time (Driver and Humphries 1966). This point has been noted independently by Robinson (1969) in his studies on the defensive behaviour of Panamanian orthopteroid insects. He remarks that in several species, particularly phasmids, there are a number of escape responses in the repertoire, often deployed without systematic order, and that it seems quite possible that there would be survival value in having escape responses that the predator cannot accurately predict. He argues that this would prevent reduction in efficacy of the escape behaviour due to anticipation by the predator.

An important consequence of the organization of escape behaviours in such a way as to create uncertainty is that anticipatory conflict will be induced in *experienced* predators. This is a significant feature of protean displays since many non-protean escape behaviours, although effective against naïve predators, tend to lose their effectiveness as protective devices once the predator has, by experience, learned what to expect.

A fourth conflict-inducing feature, the novelty found in many protean displays, is however likely to be more effective against naïve opponents. Many of the more intense and bizarre visual and acoustic components of the displays are likely to be novel to inexperienced predators, as in the berserking of sea pansies, cuttlefish transformations, convulsive flashing skipping of lepidopterans, and indeed perhaps the whole gestalt of convulsions in small mammals and others. Berlyne (1960) points out that novel stimuli will resemble to about equal degree a number of familiar stimulus patterns and by generalization will, to some extent, arouse conflicting responses appropriate to all of them. It is relevant that Hebb (1946) shows that avoidance of novel objects is greatest when they also have a number of familiar features. Novel stimuli will also, while they retain their novelty, provoke uncertainty as there is no way of telling what will follow them. Melzack (1952) found that, when quite harmless novel objects with moving parts were shown to dogs, the latter responded by crouching or turning away. Unusual events or objects often have the property of blocking the continued expression of current behaviour patterns, as shown for example by Heiligenberg (1965) in the cichlid fish *Pelmatochromis kribensis*, where novel stimuli — red or brown rectangles — temporarily blocked the fishes' capacity to attack conspecifics and greatly increased responsiveness to frightening stimuli. This kind of blocking of activities is probably due to the elicitation of an orientation reaction which Sokolov (1963) shows is produced by any sudden unexpected onset or modification in any kind of stimulation. The orientation reaction is characterized by a general rise in muscle tonus, the arrest of ongoing actions, and the turning of sense organs toward the stimulus source. If stimulus conditions become more intense, the orientation reaction gives way to a defensive reaction in which the sense organs are turned away from the stimulus.

We have shown previously that protean displays have two important characteristics — their irregularity of pattern and their rapid complex transformations. Both of these characteristics will induce conflict, but in different ways. Irregularity of sequence will generate uncertainty and defeat anticipation, whereas the transformational aspect will arouse conflicting orienting and motivational responses, create novelty, and to a lesser extent reduce the effectiveness of anticipation. In much of protean behaviour the irregular and transformational aspects operate jointly and augment each other's effects.

Factors influencing conflict intensity

In addition to considering the ways in which conflict is generated by protean displays, it is important to know what factors influence the intensity of conflict. Three factors which have been shown to be responsible for determining the degree of conflict between competing response tendencies are: their nearness to equality in strength, their absolute strengths, and their number (Brown and Farber 1951; Berlyne 1957*b*). Berlyne (1957*a*) has obtained evidence that an increase in any of these factors will lengthen the reaction time. For example, he shows that where a subject is required to make a single response, but has a free choice from among several different stimuli displayed simultaneously, the reaction time is always longer than where the choice is forced by only one out of the same number of possible stimuli being displayed at a time. This is because in free choice the response tendencies are more nearly equal. We suggest that, in terms of its effect on an observer, the free-choice situation is a laboratory parallel of a scatter display, whereas the forced-choice situation parallels the single erratic display. Berlyne shows that in both free and forced choice, reaction time increases with the number of alternatives. The individual absolute strengths of competing response tendencies may normally be expected to be high because of the predator's need to make a capture.

Much of what has been said so far regarding conflict has concerned the predator's response tendencies. However, perceptual processes may also become confused and less efficient in a protean stimulus situation. This is well known in human experimental psychology where, for example, a listener may be required to answer rapidly and accurately a series of questions while at the same time pressing a button every time he hears a buzzer. As might be expected, the buzzer task interferes with the ability to answer the questions because it involves a different and conflicting response system. But if the subject is told that as well as responding to the buzzer he must also press if a bell sounds, then a further decrement in verbal performance occurs even if the bell does not in fact ring. The wider field of perceptual readiness, or attention, reduces skill performance. It seems probable that similar decrements in predatory skill may accompany the perceptual need to anticipate, recognize, and react to a great variety of possible changes in the behaviour and appearance of protean prey. In addition, Bruner and Postman (1949) showed that, if expectations of normal associations between stimulus patterns are denied by the presentation of incongruous patterns, perception may become distorted. Brief exposure of a red six-of-clubs among a sequence of normal playing cards will, for example, produce reports of a black six-of-clubs, a red six-of-hearts or diamonds, or purple or brown sixes. Incongruous stimuli also require longer exposure for recognition. The unexpected associations of stimuli in many protean displays may therefore similarly induce perceptual errors, to the prey's advantage.

Protean displays involve an increase in the variety of behaviour, the unexpected juxtaposition of behavioural elements, the possibility that a given element may be followed by any one of a number of elements, and the appearance of novel patterns. Laboratory findings of experimental psychology demonstrate that features of this type will induce conflict in the reator, and that conflict, at the very least, will increase the reaction time. In terms of information theory we may say that the communication channel is not merely flooded with information, but that until the conflict is resolved the channel is effectively blocked. We propose that the basic survival value of the protean display resides in this effect upon the predator's readiness or speed of reaction.

Experimental evidence for conflict induction

Protective displays much simpler than protean devices have been shown to induce conflict and significantly delay the predatory response under semi-natural conditions. It remains only to provide field evidence that protean displays contribute to survival under natural conditions. Little field research has been directed to protean displays recognized as such, but they have been indirectly included in other studies. There is an immense amount of anecdotal evidence relating to the survival value of displays such as the single erratic, distraction, mobbing, united erratic, convulsive deterrent, and scatter displays. The sources of much of this evidence have been given in Chapter 4. There is a great need for further detailed quantitative field observations and experiments specifically designed to analyse protean features of escape displays. Fortunately, other important supporting data are provided, almost accidentally, by certain intensive and quantitative experimental studies which were not deliberately designed to cast light on the problem with which we are concerned.

Roeder and Treat (1960, 1961), for example, were primarily interested in showing that the possession of a tympanic organ is of survival value to moths. Their work, and further research on the acoustic sense of moths by Roeder, provides us with the most detailed available informtion on a predator–prey interaction involving protean display. This interaction will now be briefly outlined.

Bats are the main aerial predators of night-flying moths and find their way and their prey in darkness by using sonar (Griffin 1958). Many moths, such as geometrids, noctuids, and some sphingids, have auditory organs sensitive to the ultrasonic frequencies used by the bat (Roeder and Treat 1957). Schaller and Timm (1950) and Treat (1955) described erratic dives and turns made by moths when exposed to artificial sources of ultrasound. It seemed reasonable to assume that the warning of the bats' approach given by the tympanic organ is of survival value and hence adequate to account for the evolution of the moths' auditory sense (independently in several families) by natural selection. Roeder and Treat set out to discover if this assumption is justified.

The moths' flight was tracked photographically in the field as they responded to a series of artificial ultrasonic pulses simulating bat cries. Some moths showed no change in flight pattern as they encountered the sound; others reacted dramatically, abruptly and repeatedly changing flight direction in a bewildering variety of manœuvres; some dived directly down into the grass; others, flying at a greater distance from the sound source, turned away from it and flew off at high speed. Observation on 402 field encounters between moths and feeding bats showed a similar range of reactions. The behaviour of each moth was recorded and it was noted whether or not the bat captured it. Comparison of the percentage of survivors in moths which gave evasive manœuvres, with the percentage surviving when there was no reaction, showed that for every 100 reacting moths which survived there were only 60 surviving non-reactors, a selective advantage of 40 per cent. Roeder and Treat suggest that this figure is 'more than adequate to account for the evolution of the moth's ear through natural selection'. We suggest that the same may be said for the evolution of the protean nature of the flight paths, for the majority of the manœuvres through which the ear's selective advantage is realized are protean. And they are all protean when compared one with another.

Webster (1963) has shown that under laboratory conditions bats soon learn to track by echo-location and capture food objects tossed in the air in a simple ballistic trajectory. Roeder and Treat (1961) comment that the erratic manœuvres seem to be a natural answer to the predictive ability necessary for the bats to plot an interception. Roeder (1967*a*) analyses further the advantage which the single erratic pattern gives to the moth. A cruising bat (*Myotis* or *Eptesicus*) emits 8 to 10 cries per second, each cry lasting some 10 msec. Thus a moth flying at 10 km/h would travel about 28 cm during the interval between successive cries. Consequently the moth's flight path must appear to the bat as a temporal series of echo 'dots'. This necessarily makes it much more difficult to predict sudden changes in the prey's course than if it were perceived continuously. The bat is able to compensate slightly for this shortcoming by increasing the rate of emission of its pulses, sometimes to as much as 150 per second, so bringing the echoes closer together. Even so, since Roeder (1966*a*) has shown that bats are unable to detect the echoes from moths at distances greater than 3–4 m it is evident that the bat has little more than one second in which to track and capture the moth on the basis of echo dots situated upwards of 3 cm apart. Only some 100–200 echo 'dots' will be available to guide the bat's attack. Further, since the maximal echo is returned when the wings are at right angles to the pulse sources (Roeder 1963), erratic manœuvres by the moth will produce rapidly varying echo intensities and, when the moth's anteroposterior axis is parallel to the direction of sound, may blank out some of the 'dots'.

The single erratic response is used primarily when the bat has detected the moth and is close to it, and the moth is subjected to high-intensity ultrasonic pulses (Roeder and Treat 1961; Roeder 1967*b*). Although moths such as *Caenurgina erechtea* can detect bats at distances over 30 m (ten times the distance at which the bat detects the moth) they react to such attenuated pulses not by erratic manœuvres but by systematically turning away from the source of sound and flying swiftly away (Roeder 1967*a*). The use of the single erratic only at close quarters is further evidence that it is a special adaptation to an emergency situation in which the moth has little chance of escape by direct fleeing.

A quite separate feature of certain moth-bat interactions may be yet another example of a protean protective device. Moths of the families Arctiidae and Ctenuchidae not only have tympanic organs sensitive to the cries of bats, but as described in Chapter 4 are themselves capable of producing bursts of clicks having a strong ultrasonic component. These sounds are generated at a rate of 1000 or more per second by repeated contractions of a leg muscle operating microtymbal organs on the metathorax. Dunning and Roeder (1965) found that arctiids such as *Halysidota tessellaris* and *Pyrrharctia isabella* would emit the clicks in response to bat cries. Since these moths show flights reactions similar to those of noctuids when exposed to bat ultrasound, Dunning and Roeder tested the hypothesis that the clicks might have a protective effect against bat predation. They trained captive little brown bats, *Myotis lucifugus*, to catch mealworms projected into the air from a solenoid-operated 'gun'. The bats soon learned to intercept successfully (though sometimes only striking and dropping the mealworm) on about 95 per cent of their attempts on the nutritious projectiles. When tape-recorded bat ultrasound was presented during the chase, the interception rate fell to about 60 per cent. When the recorded ultrasonic clicks of a hand-held *Halysidota* were played during the chase, interceptions dropped dramatically to about 15 per cent.

Roeder (1967*a*) notes that some noise-making arctiids are distasteful to bats and consequently interprets the acoustic display as aposematic advertising. However, it should be noted that Blest (1964) showed an inverse relationship between click generation and unpalatability in New World arctiids. When unpalatability reaches a certain level, the capacity to produce ultrasound is lost. Further, in the more palatable arctiids like *Melese* and *Ochrodota*, the threshold for sound production is lower than in more distasteful types such as *Automalis* and *Cissura*. The fact that palatable arctiids produce sound more readily argues against an aposematic function for clicking. The finding that bat sounds — and extraneous white noise — are less effective as protection to mealworms than extraneous arctiid clicks is no more than might be expected if the arctiid clicks do indeed function in a protean manner as an acoustic jamming device. For it would then be surprising if the acoustic characteristics of the moth sounds were not much better adapted to confuse the bat than other quite different sounds which happened to be readily available for the experiments. We note too the comment of Dunning and Roeder (1965) that in the presence of moth sounds there were uniformly low contacts 'in spite of ample opportunity for the bats to learn that these sounds did not warn of any noxious target organism'. This finding is more consistent with the hypothesis that the moths' sounds are a jamming device than with the idea that they are aposematic. A similar remark may be made concerning the fact that an evasive single erratic is often given by sound-emitting arctiids chased by bats. As a jamming device the clicks may function simply to disorientate the bat, or they may actively deter it in a manner acoustically parallel to the deterrent colour changes of cuttlefish and berserking of sea pansies.

Moving on now to other work, Blest (1957*a*) obtained evidence relating indirectly to the flashing skipping of the small tortoise-shell butterfly, *Aglais urticae* (p.37). On the approach of a predator this butterfly, like several other species, may suddenly open its wings to reveal the bright colour of their upper surfaces. Blest showed that the display of bright colour when the wings were simply opened and remained open tended to make yellow buntings *Emberiza citrinella*) avoid the butterfly. It is hence reasonable to conclude that the skipping display, where the colour is exposed repeatedly — though irregularly — and in combination with an unusual form of movement, is likely to be at least as effective against predators, if not more so.

An additional important finding in Blest's work was that on repeated exposure not all of the birds learned that butterflies' wing displays are harmless; some, instead of habituating to simple systematic displays, persisted in their avoidance of the butterflies and consequently were unable to learn to overcome the display by experience. Hogan (1965) makes a similar point regarding young chicks of the Burmese jungle fowl, which are unlikely ever to take mealworms if they do not do so on their first encounter with this prey. It would be interesting to discover if protean displays which initially elicit a defensive reaction also retain their effectiveness against some predators due to persistence of the initial response. If the subsequent readiness of a predator to take a particular prey type depends on the intensity of its initial defensive reaction, then the special suitability of protean displays for creating defensive reactions due to their continued unexpected, dramatic, and variable effects becomes particularly significant.

Factors connected with prey density and numbers

The protective effect of numbers, theoretically predictable from the conflict theory work of Berlyne (1960), received little experimental attention until recently when interest in this subject was stimulated by work on optimal foraging theory (reviewed by Pyke *et al.* 1977). There is still disagreement as to the validity of this theory (Krebs 1978), illustrating our own misgivings as to the usefulness of theories which are not based on thorough field observations. A significant problem seems to be lack of information on the confusing effect of numbers or density of prey.

Several observations have come close to the prediction of optimal foraging theory that predators will allocate all their available time to areas of greatest food availability. Such observations include work on predation by great tits, *Parus major* (Smith and Dawkins 1971; Smith and Sweatman 1974), redshank, *Tringa totanus* (Goss-Custard 1977), oystercatchers, *Haematopus ostralegus* (O'Connor and Brown 1977), blackbirds, *Turdus merula*, and song thrushes, *T. philomelos* (Greenwood and Harvey 1978). Also, parasitic wasps, *Nemeritis canescens*, have been found to spend most time in areas of highest host density (Hassell 1971).

However, Milinski (1977) found that sticklebacks, *Gasterosteus aculeatus*, feeding on water fleas, *Daphnia magna*, only attacked highest-density swarms when they were very hungry. The discrepancy here lies in the fact that Milinski was working with mobile rather than static prey; the *Daphnia*, through their numbers and movements, present a highly-confusing target — a phenomenon which we noted in 1970.

This effect has long been suspected. Miller (1922) suggested that the attention of a predator attacking a school or swarm is distracted by the many simultaneously visible potential targets. Allen (1920*a*) had already noted this problem in the case of loons (Gaviidae) attacking sardines. Various other species of fish have been recorded as using schooling behaviour (pp.82–4) as an antipredator device (Eibl-Eibesfeldt 1962; Hobson 1968; Potts 1970; Radakov 1973). A similar function has been suggested for the swarming of the larvae of peacock and small tortoise-shell butterflies, *Vanessa io* and *V. urticae* (Moseback-Pukowski 1937).

Various factors have been suggested as the cause of the confusion. Welty (1934) found that density increase in an experimental *Daphnia* swarm reduced predation by goldfish, and considered the effect to be due to an increase in numbers. So did Neill and Cullen (1974) who, working with various fish and cephalopod predators under experimental conditions, found decreasing capture success with increasing prey numbers. Milinski and Curio (1975), however, thought the effect to be due to swarm density. Müller (1978) and Ohguchi (1978) considered that uniformity of swarm members was the confusing factor.

Hobson (1968) considers that predators upon schooling fishes are confused by the presentation of multiple targets making it difficult for them to fixate consistently on one prey animal. He says that the effect is enhanced by (1) the fact that schooling fishes are all of approximately the same size and appearance ('schooling fishes are noted for lacking external differences between the sexes'); (2) the production of silvery flashes as the prey orientation changes; (3) the increase of prey speed upon detection of the predator; and (4) an increase in the darting movements of fishes within the school changing their relative positions. Hobson suggests that individuals spend similar amounts of time in all positions within the school and doubts explanations of tighter prey bunching upon detection of a predator in terms of greater vulnerability at the edge of the school. The confusion effect is thus paramount. We believe that virtually the same comments can be applied to the predator reactions of flying wader flocks (very little sexual dimorphism; flashing of wing and tail markings; increase of speed; and increase in shuffling) with the added complexification of splitting and crossing-over (Fig. 4.24). We suggest that all these factors will be involved in producing reactor confusion, i.e. increased numbers, increased density, increased uniformity, increased flash-signalling, and increased erratic movements of the prey.

Heller and Milinski (1979) give experimental evidence in support of their thesis that a predator chooses the lowest prey density with which it can achieve a feeding rate to satisfy its hunger because increasing density causes increasing confusion. We suggest that it is neither density nor numbers *per se* which cause the confusion. Neurological conflict will automatically result from the too rapid and too irregular presentation of individual prey animals. In particular, unpredictable changes in the frequency with which, and direction from which, new targets appear will interfere with attack behaviour. Anything which can help a stickleback to concentrate on an individual *Daphnia* — such as different colour or size — would therefore improve capture performance.

Ohguchi (1978) and Müller (1978) found preferences for swarms containing individuals differing in precisely those factors. There are, of course, limits to the degree to which a predator can increase its performance in tackling a swarm of *Daphnia*, even under experimental conditions (Milinski 1979). These limits will ultimately be defined by the information-processing ability of the animal's central nervous system. For example, although the human ear can handle ten thousand bits of information per second and the eye can handle 4 million or more bits per second, the brain can only process all this at around 14 bits per second. We assume that other animals rarely exceed this ability, except for rare specialized instances such as avian visual reactions (p.80). It must be assumed that denser swarms — or higher numbers — of prey will progressively reduce a fish's ability to attend to other stimuli such as approaching predators (Milinski and Heller 1978).

The erratic presentation of prey individuals in swarms of animals such as *Daphnia* must increase the problem for predators to such a degree that natural selection would reinforce any factor contributing to that irregularity. The greatly enlarged antennae of the Daphniidae — by means of which they swim — have long been of interest through the nature of their movement (Baird 1850). This causes the animals to progress in irregular jerks, largely in the vertical plane, and in spirals. The animals also collect in swarms even when there is plenty of open water. It could well be that both the irregularity of movement and the aggregation have been evolutionarily reinforced through the protean principle.

In birds, Kruuk (1964) found much field evidence for the deterrent and distractive value of group displays by black-headed gulls (*Larus ridibundus*) against a variety of predators, with large groups being more effective than small ones. Hoogland and Sherman (1976) show that mobbing of predators by bank swallows (*R. riparia*) is increasingly effective with larger numbers of swallows taking part. We can confirm the effect from our own experiences in gull colonies. It is much more disturbing to be mobbed by several black-headed or herring gulls than by one pair of the greater black-backed gull, *L. marinus* — a much larger bird.

The well-known multiple-erratic/scatter display of impala exploding in all directions when attacked (p.38) must also have a confusing effect upon predators (Jarman 1974). This will be less effective if only a few displaying animals are involved. But we must remember that the prey themselves may be confused, at the conscious level, when they display in numbers. This, according to Schaller (1972) may be one of the reasons why lions are more successful when attacking large herds of ungulates compared with medium-sized (but not small) ones. Large groups may be more likely to contain old and ailing — and therefore more vulnerable — animals than small groups, but the individuals in larger groups are also likely to be more confused by the complexities of the mass display of which they are a part.

The majority of studies of predatory consumption in relation to prey population density have been largely ecological in character (Atwal and Seth 1963; Ivlev 1961; Kruuk 1972; Mech 1970; Schaller 1972; Solomon 1949). Bertram (1979) suggests that the prey of large mammalian predators derive distinct advantage from group living, but he has no hard evidence to support his suggestion. In any case herds are protected by the dilution effect whereby the larger the group, the less is the chance of an individual healthy prey animal being caught in one predatory attack (Hamilton 1971). Experimental investigations of the confusing effect of numbers are urgently needed.

Protean displays as learning blockers

Although we have argued that the main survival value of protean displays arises from their effects on the predator's speed of reaction, they have another important property which, on theoretical grounds, should greatly enhance their protective properties; they include devices for reducing the likelihood that the predator will learn how to deal with the display. Learning depends upon the recognition of regular repeatable relationships. Therefore, to the extent that the relationships between the components involved in protean displays are not repeated regularly, the predator is prevented from learning how to cope with them. Variations in the order and extent of changes during single erratic displays are a good example of this, for the detailed sequential relationships in the display are unlikely ever to be repeated exactly. Berlyne (1960) has shown that, in paired-associate learning experiments on human subjects, response terms receive significantly fewer correct recalls when the stimulus terms — various visual patterns — are more complex or incongruous. Complexity and incongruity are common properties of protean display components.

In addition, differences in the form of displays between various individuals, or in the same individual on different occasions will, as previously mentioned, generate uncertainty due to the difficulty or impossibility of learning to predict the prey's reaction. This is not say that the predator will always fail to improve, by experience, its capacity for dealing with some protean displays. It may be able to learn to deal more effectively with those displays which depend heavily on novelty to create conflict in the predator. It may also learn that certain types of response to an erratic display are unrewarding and therefore may try other, possibly more effective, general strategies. Nevertheless, the irregular structure of a typical protean display makes it impossible, in principle, for the predator to learn useful detailed tactics for, apart from novelty, the other conflict-inducing aspects are fundamentally more resistant to defeat by learning. It is these latter aspects which essentially distinguish the protean display from systematic behaviours such as the eye-spot or flash displays of moths and butterflies. Systematic escape responses such as these are highly susceptible to defeat by experienced predators, as Blest (1957*a*) has shown for flash and eye-spot displays and de Ruiter (1952) has demonstrated in the case of cryptic mimicry in stick caterpillars.

The protean nature of a display is not only likely, by introducing irrelevant events, or 'noise', to retard the rate at which a predator is able to learn its more systematic aspects, but it also sets a fundamental limit on the extent to which such learned countermeasures can be effective. As has been pointed out in earlier chapters, protean displays are usually essentially superimposed on other quite systematic forms of escape behaviour. Further, the elements employed in the display are to some extent systematic; the behaviour is always structured to some extent. For example, a zigzag must of necessity contain changes in direction — this is predictable. In certain instances it may also be predictable that the change of direction will be in two dimensions, not three. But within these very broad limits what is unpredictable, at least to the predator, are the extent and time of each change in direction. This is the detailed information which would be necessary for successful interception.

The protean display is thus not so random as to be formless; it is a structured system within which predictability is reduced to a minimum. In this connection the findings of Game Theory are highly relevant. Rapoport (1962) explains that when two protagonists have a more or less equal chance of success, as in encounters such as the 'which hand is it in?' game, 'It is reasonable to conclude (and it can be proved mathematically) that the best pattern is no pattern'. In other words, the way around the problem is to increase unpredictability to the maximum possible for that particular system; the behaviour of the system becomes protean. A protean system always retains some degree of predictability, though the greater the number of possible arrangements of its components, the lower the probability of predictions being correct. For example, in the game referred to above, a fully random organization of the two possibilities would set a fundamental limit of 50 per cent to predictability; this cannot be overcome by learning processes in the opponent. Since protean displays generally contain many more possibilities than the 'which hand is it in?' game, it is evident that their fundamental maximum predictability will generally be very low indeed. This still holds true even when we allow for the fact that the deployment of some elements is usually imperfectly random. It is worth noting too that the fundamental resistance of protean displays to learned countermeasures also applies to the evolutionary development of set capture mechanisms by the predator. We shall take up this point again later, for it may be connected with the phylogenesis of highly-flexible predictive tracking and choice mechanisms in predators, and consequently may have implications for the emergence of some important properties of advanced central nervous systems.

Protean displays as part of the behavioural mosaic

Before leaving the problem of function, we must remind the reader that erratic behaviours are not always necessarily protean. Occasionally they may fulfil quite different functions. For example, certain leaf-mimicking mantids and butterflies make irregular movements resembling the trembling of leaves in the wind (Hingston 1932; Cott 1940; Edmunds 1974). In this case erratic behaviour is part of the cryptic device of special resemblance.

In other instances, irregularities and transformations in behaviour appear to function primarily to signal to, and draw the attention of, another member of the same species. One example of this is the stooped submissive flight of the purple martin, *Progne subis*, in which a defeated bird assumes a hunched posture and flies in an irregular, apparently uncoordinated manner (Johnston and Hardy 1962). Another example is the epigamic display of lapwings (p.49).

Certain displays which we consider to have a significant protean content have been interpreted as having a primarily social function. In particular, Wynne-Edwards (1962) considered that the pronking of springbok, and various other bouncing and colour displays in fleeing mammals serve primarily as intraspecific signals, possibly being turned secondarily to the diversion of predators. This may be so, but without real data one cannot say which function is primary. The flash displays of fleeing mammals (and fishes) may have both intra- and interspecific functions, one function being more important in some species, and the other in different ones (Guthrie 1971).

The attention-drawing effect of stimulus complexity for human subjects is discussed by Berlyne (1958) and Vitz (1966). This property is obviously suited for use in the visual signalling of social animals when they need to gain the attention of conspecifics whose attention may be directed elsewhere. But it is also relevant to the predator–prey situation. For example, it may operate in a predator's luring antics, in distraction by autotomy where the cast-off parts draw attention by complex movement or luminescence, and in avian distraction displays of the injury-feigning type where the attention-drawing effect of stimulus complexity complements the protean evasive function and maintains the predator's interest in a lure it cannot capture. The same effect accrues from the mass pronking of springbok and impala.

It is clear that, in most species, protean displays are integrated into defensive variety. Also, anarchization necessarily accompanies more regular behaviours. However crazily a stoat may dance, it also runs and jumps normally; however varied and confusing a starling's song may be, it still has territorial and epigamic functions. We eagerly await the detailed elucidation, in specific phenomena, of what is protean and what remains orderly.

Summary

The validity of protean theory rests on three main points: first, the widespread occurrence of unpredictable phenomena in predator–prey encounters. This we have demonstrated to be the case (Chapter 4). Second, protean displays must be of survival value and therefore subject to the forces of natural selection. Third, the action of protean displays must be explicable in the terms of the predator's response and perception systems. The second and third categories, concerned with function, are examined in depth.

We suggest that protean displays act as inducers of conflict and confusion, using information theory for the theoretical framework within which the idea of confusion can be specified. Protean displays induce conflict in the reactors' response systems by increasing the number and variety of stimuli presented and by the appearance of unusual forms of behaviour. The predators' neurological communication channels become effectively flooded with irrelevant information.

Evidence is drawn from the literature demonstrating the occurrence of conflict in the predator when forced to choose between alternative stimulus–response chains in predator–prey situations. Experimental evidence supports field observation and theory.

Protean displays induce neurological conflict in several ways, depending on the timing, number, and novelty of the elements involved. The greater the neurological conflict induced in the predator, the greater its reaction time. This would seem to be a central factor in the natural selection of protean phenomena. Field evidence supporting this is drawn from bat–moth encounters and other predator–prey interactions.

It is shown that protean displays also include devices for reducing the likelihood that predators will learn how to deal with them. Protean displays represent a structured system in which predictability is reduced to a minimum.

7
Evolution and anarchization

> I have often thought, in my long and practical career and at my age (69 years) how much pain and how many researches are necessary to probe to the depths a rather complicated phenomenon. The greatest difficulty comes from the fact that we are too much accustomed to attribute to a single cause that which is the product of several, and the majority of our controversies come from that.
>
> *Liebig*

Biological explanation is obviously incomplete without consideration of all significant aspects of the evolutionary process. Equally clearly, it would be unrealistic to attempt to produce a well-rounded explanation of protean phenomena at this stage. For many years, particularly since 1966 when G.C. Williams seriously questioned the nature of adaptation, biology has been bedevilled by a multiplicity of arcane adventures into the nature of evolution, from philosophical flights of fancy to arid mathematical analyses. And with the technological explosion, most biologists seem to have lost sight of the virtue — clearly demonstrated by Charles Darwin — of painstaking and thorough observation of animals in their natural environment. Without this, it is futile to attempt sweeping generalizations and mischievous to propose grand new theories. With evolutionists stumbling tentatively towards objectivity (Cain 1979) it is not surprising that recent evolutionary symposia display a degree of confusion that is normally only found in protean contexts.

In view of these comments we must indicate the sense in which we ourselves use the term 'adaptive'. Adaptive traits, we believe, cannot be usefully considered isolated in time or in space. Each must be seen in the context of an ever-changing relationship with environment (internal and external) from conception to death — or at least to reproduction. As Dobzhansky (1956) noted, it is the trajectory of this relationship which confers upon the genotype or the individual its fitness to survive and to reproduce. Like Gould and Lewontin (1979) therefore 'we support Darwin's own pluralistic approach to identifying the agents of evolutionary change'. Only thus can meaningful progress be made.

Our central thesis is ecological and behavioural in character and is founded primarily on the demonstration of the existence of the protean pattern and on evidence for its operation taken largely from the field. However, some of the most fascinating, and potentially fruitful, aspects of protean theory would seem to be found in problems of mechanism and causation, and we now consider these facets of the subject.

In Chapter 6 we summarized the following features as being typical of protean displays: increased variety, unexpected juxtaposition, diversity of sequence, and novelty of factors. These features are attained in a variety of ways in animals of different taxonomic groups, and it must remain largely for future investigations to reveal the extent to which common principles of protean mechanism may apply to common features of display patterns in these diverse organisms. While our main concern will be with central nervous mechanisms implicated in the displays, we must note that many other aspects of organic design play important roles. Special anatomical, morphological, and physiological features are often required, particularly in colour displays, just as physical and behavioural features are reciprocally involved in regular adaptations such as crypsis. For example, anatomical modifications occur in cephalopods for muscular control of the chromatophores; in the Nassau grouper for automatic production of sound by the swim bladder; and in a variety of animals which practise autotomy. Morphological characteristics are involved in colour displays of lepidopterans, and special physiological mechanisms are seen in the luminescent display of sea pansies (Nicol 1955).

Again, the presence of more than one animal is an essential feature in producing the protean effect in the united erratic dispay, in scatter, and in multiple-distraction displays, including mobbing. The mechanisms of protean display are thus not restricted to the neurophysiological level but — to varying extents in different groups — include other kinds of components.

In the light of the protean principle we now re-assess some of the explanations advanced in the literature for certain types of display mentioned in Chapters 4 and 6, particularly injury-feigning, mobbing, and immobility displays. We also consider the evolution of single erratic, united erratic, and convulsoid behaviours.

In this chapter, and also in Chapter 8, we will find abundant evidence to support the suggestion of Harvey and Greenwood (1978) that three points should be made clear when considering prey defences in an evolutionary perspective. These are: (1) antipredator strategies co-evolve with those of the predators (Gilbert and Raven 1975; Curio 1976;) (2) genetically-controlled antipredator strategies can evolve in a population even when predation is not a density-dependent limiting factor on that population (Haldane 1953); and (3) prey are not necessarily trying to defend themselves — their reproductive success might be better served by protecting other animals, even other species (Davies and Krebs 1978).

Injury-feigning

Avian distraction displays of the injury-feigning type (p.95) have been attributed, by several authors, to conflict within, or between, a complex of drives. In other words, several behavioural tendencies are expressed synchronously or in succession, so that no one tendency dominates and a complex, changeable pattern is produced. But there is little agreement as to what the underlying components of the neurological conflict may be. Armstrong (1949*b*) refers to the 'self-preservatory' and the 'sexual and parental' behaviour systems. He also suggests that distraction displays often incorporate elements of displacement behaviour originally associated with epigamic and aposematic displays. In contrast, Simmons (1952) and Tinbergen (1952) argue that the conflict is primarily between the aggressive and escape drives. There is also disagreement as to the immediacy of the conflict; Simmons suggests that conflict in the present is involved, whereas Lack (1941) proposes that conflict between drives has contributed to the origin of the display only in an evolutionary sense and no longer actively shapes the behaviour. Rowley (1962) also disagrees with Simmons and considers that although 'rodent run' distraction display may have had its evolutionary origin in drive conflict, the pattern is now emancipated from the original drives in the sense defined by Tinbergen (1952).

We have seen in Chapter 6 (p.142) how the work of Berlyne (1960) and others can help us to clarify the psychological aspects of the problem. But we must ask if the arrival of a predator induces conflict and shapes each distraction display anew on every occasion, or if the general pattern has become evolutionarily fixed by separation from the control of the original complex of drives? In the latter case, the predator would trigger an integrated predetermined sequence, the features of which have been selected through their past successes in protecting both lure and prize against predators. We suggest that both processes may occur, representing stages in the evolution of increasingly effective distraction displays. Even if we suppose the display to be created by immediate conflict we are left with the problem of why the conflict is so finely balanced that it renders the actor highly attention-drawing. We cannot believe that this is fortuitous.

Tinbergen (1951) points out that 'as a rule, the strong activation of instinctive behaviour of one kind prevents the functioning of another pattern' and he cites escape as a reaction which tends to override other activities. The first stage in the evolution of distraction display therefore would seem to require that instead of the escape drive leading to the speedy withdrawal of the actor from the scene, the inhibition of other drives becomes weakened with the result that the actor becomes more conspicuous and the survival of the brood is enhanced. This would lead to the situation proposed by Simmons and by Armstrong — alternate approach and retreat behaviours — which in turn provides the possibility for the evolutionary emancipation of successful patterns. But since it is likely that the effectiveness of protean displays is enhanced by their variation from occasion to occasion, and since such variation is a significant feature of distraction displays, it seems probable to us that, whatever degree of emancipation of overall pattern or of individual components has occurred, some form of conflict in the present is likely to be involved in generating the variability in the details of the displays. The precise balance of conflict is bound to differ according to the unique spatial relationships and other stimulus components in each encounter.

It is difficult to imagine any selective circumstances which might cause a protean display to evolve via emancipation towards regular repetition, though rodent-run distraction does include an element of behavioural mimicry. In the purple sandpiper, *Calidris maritima*, for example, the retreat from the nest is a zigzagging run carried out in a stereotyped hunched posture which gives a striking resemblance to a small mammal. Even here the erratic component is present as a background to the behavioural mimicry and will serve both to protect the actor and to prevent backward extrapolation revealing the position of the nest.

A predator's attention may be maintained, perhaps even strengthened, by the repeated expectation of a capture. This could arise either as a result of previous capture of an injured bird, or through repeated arousal of expectation of capture as the prey approaches from time to time, or is approached more closely. Such a phenomenon seems to operate in experimental rats at least (Sheffield *et al.* 1951). Also, as goal-seeking behaviour again and again seems to be almost successful, the predator will experience something akin to the 'salted nut phenomenon' (Hebb 1949). '*L'appétit vient en mangeant*' (see Flugel 1948). It would be logical for arousal to be even stronger after the predator had *almost* succeeded than it would be after a brief taste of success.

In the light of this approach we now present our model of injury-feigning distraction display in diving ducks (Fig. 7.1). This is based on numerous field observations of waterfowl behaviour supported by cine and still photography and extensive field notes. It can readily be adapted to model the behaviour of other injury-feigning birds including waders, game birds, and rails. There are, of course, major theoretical difficulties involved in constructing such models (see, for example, Hinde 1960) but we stress that our model is only an operational analogue, and may help us to understand a motivational complex which is — for the present at least — otherwise impenetrable.

The model — developed from the work of Tinbergen (1951, 1969) — suggests an explanation of distraction displays in terms of the hierarchical organization of behaviours via structures in the central nervous system (CNS). It co-ordinates our observations of prey reactions to predators with theories of the organization of control centres in the CNS by means of the 'innate releasing mechanism' (IRM) (Lorenz 1935; Tinbergen 1951). We realize that there are differences of opinion as to the nature of the IRM (see, for example, Hinde 1970). The releasing mechanism may not be wholly innate, but our model fits the observable facts. Thus, certain 'action patterns' are shown as responses to appropriate releasing stimuli when the associated 'blocks' are removed. It may be that the disinhibition hypothesis (Andrew 1956; van Iersel and Bol 1958; Rowell 1961; Sevenster 1961; Fentress 1965) is closer to the truth than hypotheses of block removal. Or there might be a preference for some form of synaptic threshold phenomenon. We suspect that the reality involves a complex of many factors, with their balanced activity controlled by external stimuli. However, while the present model may seem naïve, it does show that the classical model can be elaborated to show how the erratic behaviours of injury-feigning may occur as *natural variations of centrally-organized responses to standard stimuli.*

A simpler design of this model was developed over the years as a teaching aid for the illustration of duckling behaviour. Only when we tried to see if it would allow for distraction displays did we find that it suggested an explanation of injury-feigning. This supports our long-held belief in the heuristic and explanatory value of the hierarchical principle; a belief supported by Dawkins (1976*b*).

Our model assumes that there is, in the brain, a 'defence integrator' ('defence centre' in Tinbergen's original work) which sends out impulses controlling defence behaviour via a hierarchy of lower integrators. The integrators may be discrete collections of brain cells, diffuse arrangements of interacting cell aggregations and fibres, or any combination. The defence integrator receives impulses from a generating mechanism of its own, and from various internal and external sensory stimuli. Like other mechanisms it is probably also subject to hormonal influences. Impulses flow out of the defence integrator aimed at six target action patterns, the activation of a particular action

Model of the organization of distraction display in Anatinae

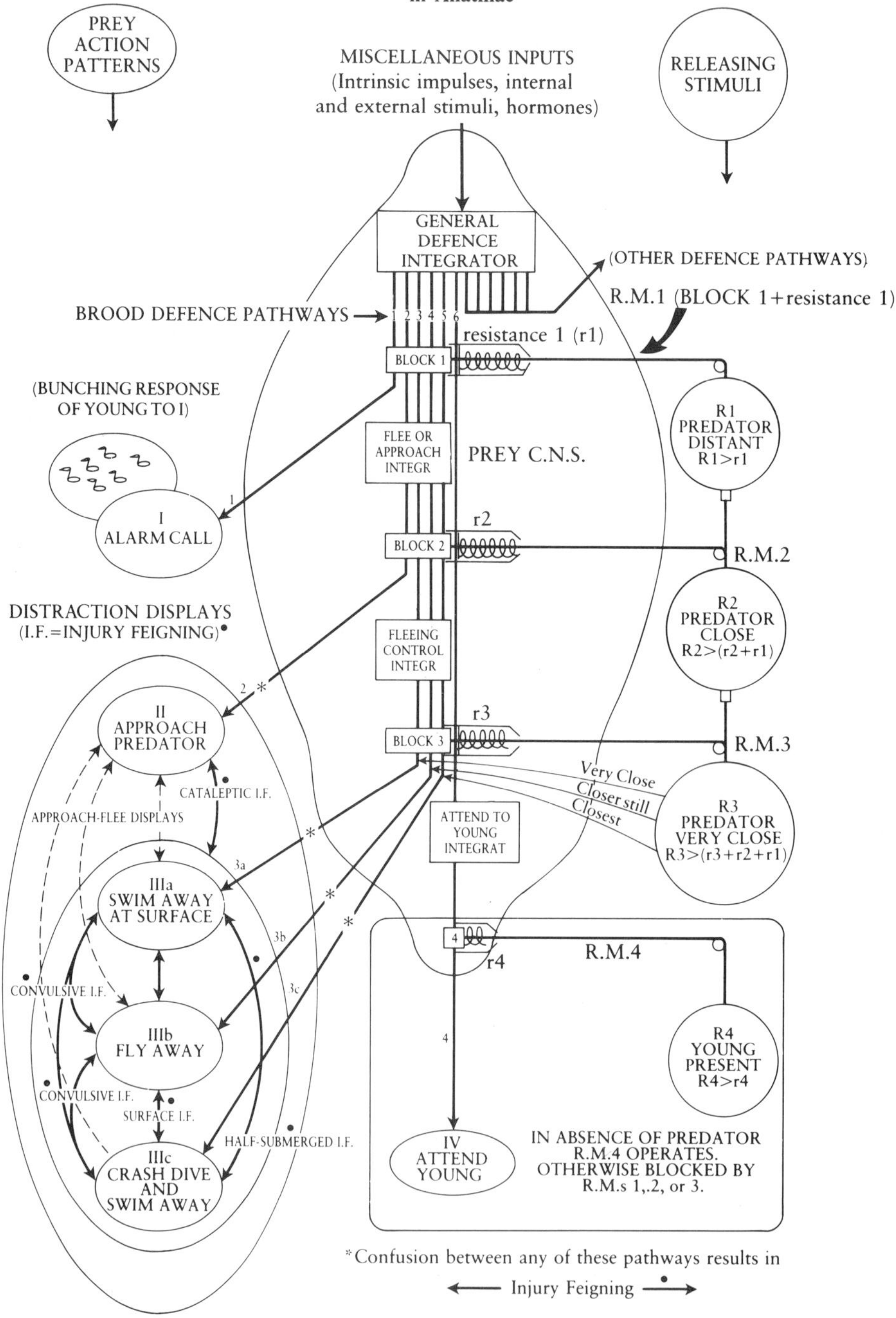

Fig. 7.1. Model of the organization of distraction displays in Anatinae. The behaviour of the prey is indicated in the ellipses to the left. The releasing stimuli from the predator are the circles to the right. In the centre are the neuronal mechanisms in the CNS.

When a predator appears at a distance (R 1) the appropriate releasing mechanism (RM1) unblocks channel one and the alarm call is given. At the same time, impulses flow through the other channels but are not seen as activity of the animal because after charging the mechanism of the next level — the approach integrator — they are in turn blocked (BLOCK 2). Only if AP II (approach predator) is 'required' does RM2 operate (R2 > r2+r1); then the duck approaches the predator. If, however, the predator gets very close, r3 is overcome and Ap III (a–c) operate. The behaviour now becomes protean.

The large ellipses show how distraction displays, including injury-feigning, may come about. AP III (a–c) include the protean zigzags characteristic of the single erratic display. Which AP is activated depends on detailed information, the sequence of AP's in the whole system involving an escalating response to increasing danger. If the predator is 'very close' the duck will swim away at the surface; if the predator is 'closer still' the duck will fly away. If the predator is 'closest', so that the prey is in *immediate* danger, channel 3c operates and the bird crash-dives, swimming away under water.

AP III (a–c) increase predator–prey distance and also alter the distance information in the R3 input pathways. As the prey moves from 'crash-dive' situation to 'fly-away' situation, and as one sensory input changes to another, confusion occurs between channels 3b and 3c. This results in 'surface' injury-feigning in which the bird swims erratically away beating the water with its wings as if unable to take off. Similarly, conflict between channels 3a and 3b, or 3a and 3c will result in 'wounded flight' or 'half-submerged' injury-feigning, respectively. Conflict between all three channels results in convulsive injury-feigning in which the actor thrashes about on the spot in a convulsive manner. This is not especially dangerous for the actor, for when the releasing input changes to 'predator closest', channel 3c alone operates and the convulsive condition changes to 'crash dive and swim away' under water. Conflict between any of channels 3 and channel 2 sometimes occurs and results in cataleptic injury-feigning. The displaying bird, torn between the drives to both flee and approach, remains on the spot in a temporary trance-like state. Again, this condition is replaced quickly by another as the releasing situation changes.

Occasionally, one of APs III (a–c) will take the actor so far away from the predator that R3 no longer operates and BLOCK 3 is replaced. AP II is then re-activated and there will be alternation between AP's II and III resulting in approach–retreat behaviour. In a 'successful' distraction display, stage III, with or without stage II, will eventually result in the predator getting lost or giving up. R4 will then operate and the parent bird will re-join its young. We can see from this that it is theoretically possible for the erratic nature of injury-feigning to arise from conflict between the components of the attack and fleeing drives, as suggested by Tinbergen (1952) and Simmons (1952). Tinbergen also noted (loc. cit.) that the parental drive is always involved too, but that its role is uncertain. Our model may explain this point.

The model also allows for continued injury-feigning when the prey is fleeing and the predator following, as with a duck being chased along a stream or a wader being pursued over the ground. If prey and predator maintain roughly equal distance between them, as often happens, R3 will continue to operate and the detailed stimulus controls will continue to vary between channels 3a, 3b, and 3c.

pattern being mediated by the appropriate releasing mechanism. This involves unblocking the impulse channels or drive pathways when the releasing stimuli are presented. We represent this in the model as a mechanical system of weights (Releasing Stimuli, R) which, when applied, overcome resistances (r) removing the blocks.

In a neutral situation, when a duck with young on water does not observe a predator, all defence behaviour pathways are closed and only the attentiveness of the female to the young is activated. When a predator appears, however, this attentiveness is blocked and there is an escalating series of responses to escalating danger, with a corresponding increase in the protean nature of the response. Detailed operation of the model is described on p.167.

Such models may only be analogies, but if they reflect observable behaviour, one may gain confidence in one's assumptions about the kind of connections that may exist between stimulus and response. For example, the appearance of a predator does result in ducks giving the alarm call. The ducklings' response is to gather in the 'danger shadow' of the female who then *approaches* the predator — presumably for the acquisition of more relevant information (Kruuk 1964). These behaviours are systematic and highly predictable, but thereafter in a full predator–prey interaction sequence, the behaviour becomes increasingly protean. The erratic behaviour of action patterns, AP III may be explained by the activation of a general fleeing drive without clear directional stimuli: the prey zigzags away. Then come the three forms of 'normal' approach–retreat distraction display which can be explained, in our model, as alternation between drives and action patterns AP II and III. Their protean nature is enhanced by the to-and-fro component. If the connections were as we suggest, then one would get the behaviours we describe; these are, in fact, observable and, so far, the model follows classical theory.

The true injury-feigning displays seem to be different from more fixed action patterns in that — as they must be for success — they are less regular and predictable than 'normal' behaviours. They, like the patterns they mimic, seem disorganized, due, we believe, to conflict between the organizing drive pathways. Exactly what causes the apparent conflict between action patterns AP III is a question of conjecture and, we trust, experiment. Candidates for examination include reverberation between circuits, alternation between drives, simultaneous excitation of incompatible drives, inefficiency of directing stimuli, activator neuron/inhibitor neuron switching, and reciprocal inhibition. If the connections we suggest are approximately accurate, apparent injury could result from confusion between any two of action patterns, AP II and AP III. We know that birds are easily capable of the spatial discrimination necessary for them to be able to distinguish between distances R3(a), R3(b), and R3(c). Whatever the cerebral mechanisms involved, the present model would allow injury-feigning to appear as a result of conflict and confusion between channels 3a, 3b, and 3c, or between channel 2 or any of channels 3. Thus, a basically regular hierarchical structuring of innate behaviour can be so organized as to appear disorganized.

The suggestion that injury-feigning is a very ancient mechanism is hardly to be quarrelled with, for it must go back to a time before the earliest ostrich (Struthionidae) at the beginning of the Pliocene about ten million years B.P. Ostriches perform elaborate injury-feigning displays, as is well-known (see for example Van der Post 1961) and commonly include the broken-wing display. This presumably developed at a time when the ancestors of ostriches had the power of flight.

The ancient origin of injury-feigning is also indicated by the general similarities between the displays of so many different families of birds — at least twenty. And it is demonstrated especially by the automatic triggering of approach behaviour — often before the predator has seen the prey. We ourselves have been first alerted to the presence of eider, mallard, moorhen, and ringed plover when the displaying bird approached noisily with injury-feigning. Either this behaviour originated in birds before the various families diverged, or else it has developed by convergent evolution selecting similar pattens of behaviour based on the common avian brain structure. Whichever is the case, injury-feigning has clearly been of great value for a very long time. We believe that without the protection afforded by its protean nature this would not be so.

We can see that the significance of injury-feigning is threefold, summarized as follows: (1) There is an apparently random relationship between the movements of the actor and the position of the prize. It is thus unprofitable for the predator to try to discover that prize by relationship with the actor's movements; (2) the seemingly uncoordinated activity of the actor results in behaviour indistinguishable from that of an injured bird (or, sometimes, mammal). The reactor's attention is thereby maintained; (3) the unpredictable behaviour of an injury-feigning bird significantly assists it in avoiding capture.

Finally, we have to note some disagreement with Wilson (1975) who writes 'there can be little doubt that the adults engaging in the displays endanger their own lives while reducing the risk for their young'. The protean principle protects the displaying birds so effectively that altruistic considerations are unnecessary. We have no record of an injury-feigning bird being captured, though there are many accounts of the escape of such birds. This is not the case with immobility displays of the cataleptic type. Shamming dead can only be of selective value through its protean occurrence, and many cataleptic animals are killed by their captors. Injury-feigning however is fundamentally connected with the protection of the young and it is a selective requirement for the adult to survive.

Mobbing

Mobbing displays (pp.100–6) have widely been considered to involve an approach–escape conflict, but there has been disagreement about the precise nature of this. The approach component has been variously considered to express primarily an aggressive drive (Simmons 1952; Goodwin 1953) or investigation (Hinde 1954*a*; Marler 1956). Marler notes that chaffinches, *Fringilla coelebs*, are likely to react to any unusual object by showing behaviour similar to mobbing and that in this context the behaviour gives the subjective impression of curiosity or inquisitiveness. Kruuk (1964) has shown that a prey animal may approach a predator in order to acquire relevant information.

Labelling the approach tendency as 'aggressive' or 'investigatory' however does not help us to explain the origins of mobbing displays. This is, rather, an example of how the context within which we become aware of behaviour patterns may influence the causal interpretations we put on them. The central problem is why the mobbing pattern should be shown to particular types of object. Thus we may ask why any assumed neurological conflict in the participating individuals should be expressed by a distinctive spatial oscillation of those animals, rather than in any other way. After all, behavioural conflict in other kinds of approach–withdrawal situation, such as those involving food or social partners, may be expressed by quite different patterns.

Kruuk (1964) in his detailed study of the antipredator behaviour of the black-headed gull *Larus ridibundus*, showed clearly the way in which the various defensive behaviours are specifically related to a given type of predator. For example, birds of prey such as kestrel, merlin, peregrine, and short-eared owl all evoke the formation flight (united erratic display) in the gulls, whereas swooping mobbing attacks are made against crows, herring-gulls, and certain mammalian predators. Kruuk demonstrates convincingly that the display evoked by a particular type of predator is that which is most likely to be effective against it, either by protecting the displaying adult or by disrupting the predator's search for the brood. He also notes daily and seasonal fluctuations in the readiness with which the various antipredator reactions can be elicited and relates these fluctuations to the changing relative vulnerability of brood and adults. Significantly, he goes on to list as components of the displays 'attack, avoidance, attraction towards the predator which we will call "exploration", attraction to the nest site and reactions to con-specifics', and continues 'we are therefore forced to conclude that all antipredator behaviour is the result of the simultaneous elicitation of the basic patterns listed above, and that the differences between reactions to one predator and the other are quantitative — differences of degree rather than of kind.'

Kruuk's findings and conclusions underline our previous point — it is not possible to explain the beautifully adaptive differences between the various forms of antipredator behaviour merely by describing their relationship to inferred conflicting components. We have to ask a further question about this conflict: how has it become organized to produce for particular classes of stimuli distinctively different behaviour patterns, each neatly countering the specific hazard? The answer must involve the differential survival values of the various possible patterns of behaviour which might be shown in a given situation.

Again, we must view the mechanism of protean display in an evolutionary light. Published explanations of injury-feigning and mobbing displays have stressed that the display mechanisms involve conflict between major drive systems. The general approach has been to show that behavioural elements normally classifiable in other contexts (i.e. in other drive systems) are present, and to conclude that the display is thus due to an underlying conflict between these drives. However, we agree with Andrew (1961*b*) who, after analysing the motivation underlying mobbing calls, concluded that any treatment of approach and retreat in mobbing in terms of 'drive conflict' will be inadequate at any but a superficial level.

Whether the conflict involved in these displays is in the present or is only a relic of the evolutionary origin of the display components, the fundamental point is to describe the pattern of behavioural elements shown. To us the central problem is not whether aggressive and flight drives are in conflict during predator reactions, but why the resulting pattern of behaviour has certain characteristics rather than others. In particular, why do the situations we are discussing activate a set of mechanisms — whether definable as drive conflict or not — which produce a dramatic increase in the variability of behaviour with a consequent decrease in its predictability? Why, in these situations, should there not be simple, regular, highly predictable behaviour?

Conflict *per se* does not necessarily produce irregular, unpredictable behaviour. In many displays involving conflict, for example courtship displays, a systematic patterned sequence of behavioural elements is a central feature. And in higher mammals, at least, conflict — in the sense of simultaneous arousal of incompatible responses — 'must accompany virtually every moment of normal waking life' (Berlyne 1960). The irregularity of injury-feigning and mobbing cannot be explained by the mere existence of conflict, but only by the particular way in which that conflict is organized. This in turn poses the problem of the forces which have shaped the evolutionary development of this organization.

Anarchization

It can be argued that the predictable and regular way in which the pattern of courtship movements in birds arises from conflict in the present has been determined during the course of evolution. As is well known, ritualization of behaviour has made various signalling functions systematic and clear (Tinbergen 1952; Blest 1961); consequently, any traces of ambiguity and irregularity of sequence have tended to be eliminated.

The converse of this has not, as far as we are aware, been suggested. In forms of behaviour shown in situations where it is of survival value to deny useful information to reactors, irregularity of sequence, or ambiguity, or component irrelevance are all likely to be favoured by natural selection, provided that the alternative means of blocking information have failed, or are inappropriate.

We suggest that the irregular nature of injury-feigning, mobbing, and the single erratic display and its derivatives have evolved — at least in part — by a process rather like ritualization in reverse — by a process of complexification and apparent disorganization which obscures signals of real value to the reactor. In the single erratic and injury-feigning displays this is manifested in the variability of the individual's behaviour, whereas in mobbing and the united erratic display much of the necessary variability is secured as a group effect. We shall see in the next chapter that the same process of confusing complexification seems to have operated in the production of certain polymorphisms and behavioural variations, and in patterns — or, rather, non-patterns — of dispersion in space and in time. We hereby name this development of apparently random irregularity and confusing complexity *The Principle of Anarchization.*

We note here that there are interesting parallels between patterns of behaviour and patterns of surface colouration. Just as the ritualization of behaviour is paralleled by the development of colourful, conspicuous releasers providing clear, unambiguous signals, so disruptive colouration in its chaotic complexity resists perceptual analysis of body outline by the predator and therefore parallels the behavioural anarchy of protean displays. Indeed, one might regard disruptive colouration as a protean phenomenon at the structural level, tending to defeat the predator before rather than after the detection of prey. The irregularity of marginal form which obscures those characteristics of body outline — such as regularity, repetition of detail, and symmetry — which facilitate its recognition, is another manifestation of the same principle. It is seen, for example, in the hind wings of butterflies such as the comma, *Polygonia c-album*. Such considerations underline our opinion that a new analysis of animal defence mechanisms is much overdue.

We have already seen how the principle of anarchization might have operated in the evolution of injury-feigning. Now let us see how it applies to mobbing displays. We take as our example hornbills mobbing a snake (see p.101). The actors, individually weak, but jointly powerful, surround the reactor. The erstwhile predator is thus on the defensive.

Let us suppose the existence in early hornbills of an autochthonous drive to flee when attacked by a large snake or other predator. Also assume the drive to attack small and/or fleeing animals. The diet of ground hornbills includes many species of poisonous animals, including snakes and scorpions, which they have evolved special means of catching and killing and from which they are partially protected by the long, bulky bill. Now add the development of a social organization — as seen in many hornbill species — and we have the prerequisites for overwhelming large snakes.

If, in early hornbills, a small foraging party encountered a small snake, the reptile would soon be killed and eaten. A medium-sized snake might flee, or else it would threaten and strike, in which case the birds would defend themselves. But the snake can only strike at one bird at a time. While it is involved with one bird, therefore, another bird has the opportunity to attack it from behind; only the bird being struck at will have to defend or flee. Attack by the second bird will automatically result in the snake facing that bird, which makes it vulnerable to attack from the first or other birds. Such would be the success of this behaviour — based upon the cohesive elements of social organization and the alternation of attack and defence — that the joint attack would be favoured by natural selection. This would operate by the selection of those individuals showing the behaviour, particularly when operating in small groups.

As a result of natural irregularities in the physical environment — uneven ground, occasional stones, logs, herbage, etc. — the sequence of attackers would be effectively unpredictable. All the snake could learn would be that it would suffer attack from other birds when it was attacking any particular individual. Such learning would result in increased confusion in the snake (see pp. 101–3).

The success of this behaviour could easily result in hornbills attacking larger and larger snakes until one arrives at the situation we see now, in which a small group of birds will overcome a snake so large that one individual, or even two, could not possibly kill it. Thus a relatively simple degree of anarchization resulting from environmental irregularities could be reinforced by natural selection. This selection would operate specifically upon social cohesion and the reinforcement of attacking or fleeing in appropriate circumstances. The result would be an automatic increase in the anarchization of the display, and hence in its effectiveness.

We see no reason to involve considerations of altruism or group selection in this kind of behaviour. It could result simply from selection operating upon appropriate behavioural mechanisms in the normal way. As Charles Darwin put it (1859) 'In social animals [natural selection] will adapt the structure of each individual for the benefit of the community; *if each in consequence profits by the selected change.*' (Authors' italics). Or, as it is stated elsewhere, 'So whatever you wish that men would do to you, do so to them, for this is the law' (Matthew 7:12). It is interesting that this point made in two of the greatest publications is reflected in some of the latest. The Tit for Tat strategy, developed as a computer program for dealing with the problem of 'the prisoner's dilemma' (Axelrod 1984), would seem to reflect organic solutions to conflict already developed through natural selection. Our explanation for the efficacy of mobbing — arrived at several years ago — proposes one mechanism whereby such a solution can be achieved in an actual living example. Doubtless there are others, but with only slight modifications our scenario for the evolution of mobbing behaviour in ground hornbills would serve to explain all kinds of mobbing. As long as there are appropriate built-in attack and defence behaviour control mechanisms in the mobbers, protective reciprocity will be the automatic result. This may have the appearance of the 'altruism' seen in man but, far from involving any sacrifice, mobbing behaviour is more or less equally beneficial to all mobbers, and does not put the actors in any unusual danger. We are pleased to note that Wynne-Edwards' most recent publication (1986) reveals that cooperation within a group often increases the fitness of its members.

The united erratic display

Species which show the united erratic display are generally characterized by high speed and great manœuvrability, characteristics which demand an associated increase in co-ordinational ability between individuals. Flocks of manœuvring ducks, starlings, and waders, or shoals of fish, move as one, and we have not seen nor found records of collisions between individuals in the tightly-packed groups. The rapidity and complexity of manœuvring would generally seem to demand a very short reaction time between the onset of a turn in one individual and its onset in a neighbour (see p.176). For this reason, it would be efficient for relatively simple signals to be used, for it is a general principle that the simpler the signal, the shorter the reaction time. There is some evidence that surface features on many of the appropriate species are so organized as to provide such simple signals.

In fishes it seems likely that visual communication is important in maintaining school integrity and co-ordination (Magnuson and Prescott 1966; Cushing and Harden-Jones 1968; Pilcher 1973). The distracting silvery flashes which we see when the fishes' vigorous swerving moves them out of their normal orientation to the incident light may well be involved in co-ordinating the direction and timing of school movements as a whole.

In the case of birds which use the united erratic display we must assume that a visual mechanism is involved, the eye of the bird being so much superior to that of other animals in receiving information on small angles and movements (Pumphrey 1961; Welty 1964). As we have already suggested (p.80), the wader wing-bar may have evolved in this context, and it is significant that the light-coloured, conspicuous wing-bar is ideally placed to be seen by a bird slightly behind and to one side, as occurs in tight flocks (Major and Dill 1978). Further, the wing bars of waders show very little divergence from a basic simple pattern (Fig. 4.25), a feature which accords well with the fact that waders often fly in flocks of more than one species, and with their relatively minor use of the wings in epigamic display.

In the Anatidae however, particularly in the ducks (Anatinae), which also commonly use the united erratic display, we see the development of wing patterns in the same place but in a more diversified form. This greater diversification can be related to the fact that ducks show a wide use of wing specula in epigamic displays (Lorenz 1941; Johnsgard 1966). Also, they fly in mixed-species flocks less frequently than do the waders. Nevertheless, the specula of ducks show as little complication as may be consistent with functional efficiency, and we believe that they are used in making directional changes immediately obvious to adjacent members of the flock. We suggest therefore that wing patterns in birds have sometimes evolved in connection with flock co-ordination and manœuvrability as well as for their more generally recognized functions of species and sexual recognition. This suggestion is entirely compatible with recent work on wader flock movements. Potts (1984) proposes the 'chorus-line' hypothesis of manœuvre co-ordination in bird flocks based on analyses of cinefilm of flocks of dunlin, *Calidris alpina*. He shows that manœuvre waves — initiated by single birds turning back in towards the main body of the flock — begin relatively slowly but reach 'mean speeds three times higher than would be possible if birds were simply reacting to their immediate neighbours'. (Mean reaction time initiator to neighbour 67 ± 24 msec; mean propagation time within manœuvring flock 14.6 ± 6.7 msec; mean species reaction time, laboratory measured, 38.3 ± 3.1 msec.) The high speed of wave propagation seems to be achieved in a similar manner to that of a human chorus line; individuals react to an approaching wave rather than to their immediate neighbours. In bird flocks the wave accelerates, and we note that in large wader flocks at least, one can observe this with the naked eye.

Neurological as well as morphological features must be involved in the complex phenomena of anarchization. As much of the irregularity in mobbing and united erratic displays results from the fact that several individuals display together, it follows that selection has probably favoured the development of intraspecific responsiveness and co-ordination mechanisms at the neurological level. For example in birds, particularly colonial-nesters, mobbing by one individual is likely to precipitate mobbing by others. Such responsiveness ensures that the predator is presented with as complex a display as possible. Again, a loose flock of ducks, waders, or starling, rapidly becomes a tight flock on the appearance of a falcon, the individuals responding in the necessary manner to make the united erratic display effective. This is clearly a fixed response which has been developed through natural selection.

Single erratic displays

Although the single erratic pattern is so common and is found in such a diversity of form and in such a range of animals, few attempts to explain this type of behaviour have appeared in the literature except for certain limited examples. It is in fact unlikely that any single explanation will fit all cases. Thus, drive conflict explanations of the single erratic aspect of distraction displays, even if we accept them as adequate, cannot be extended to cover other types of single erratic display. Interpretations of the normal undulating flight of many passerine birds — finches, for example — in terms of the efficient utilization of muscular power in relation to aerodynamics and body weight (Schaeffer pers. comm., 1968) cannot take in zigzagging manœuvres provoked by the arrival of a predator, and probably cannot account for much of the variation in the rhythms of undulation which occur in normal flight. Again, habitually erratic flight in small insects cannot in general be attributed to the demands of any overriding aerodynamic principle (Pringle pers. comm., 1968). Nor can irregularity be generated aerodynamically. As in erratic displays on land or in water, these protean irregularities in flight must have their source in irregular, though organized, patterns of nerve impulses.

We must point here to a widely-held misconception, which is that random behaviour indicates a more primitive underlying mechanism than is necessary for orderly behaviour. It is naïve to assume that the simplest and most easily-achieved state of a mechanical system is the random one. It is much easier to design a mobile toy to travel in a straight line than to make it randomly erratic, always allowing of course for environmental influences such as variations in the surface over which the toy is travelling. To get it to follow a curved path on a flat surface, or add variations of speed, irregular alterations of direction, or variation of rate of change, becomes progressively more difficult and requires more intricate and unusual mechanisms. This also applies to biological systems. Organized — though overtly erratic — variations of direction and speed in prey animals, i.e. protean flight, necessitate the evolutionary development of special control mechanisms.

In most cases where a single erratic display is released by the approach of a predator we suggest that the condition for the generation of irregularity is the arousal of the escape drive in the absence of a dominant directing stimulus. As is the case with many other instinctive acts, the stimuli releasing or arousing escape activities are not necessarily those which direct the activities in relation to the environment. The functional distinction between directing and releasing stimuli is stressed by Tinbergen (1942, 1951). The response systems for these two types of stimuli may however be neurally connected. Escape is essentially a directed response so that normally both types of stimuli must be present. Thus, for many passerine birds, escape responses released by the sight of a hawk will be directed by other visual stimuli arising from nearby vegetation or other potential cover.

Where an adequate directing stimulus is lacking we suggest that the arousing stimulus, in producing a high level of escape drive also lowers the threshold for the triggering of other possibilities of direction of escape. Thus, eventually, escape movements occur in the absence of adequate directing stimuli. The animal then takes a number of directions in succession, since there is nothing to select one direction rather than another.

The hypothesis predicts that the higher the escape tendency, the greater is the likelihood that fleeing will be of the single erratic type. It is thus relevant that Grant and Mackintosh (1963), dealing with intraspecific agonistic behaviour of small mammals, find that high-intensity retreat 'is characterised in the rat, the mouse, and the hamster by undirected bouncing movements, interspersed with very rapid retreating'. Lower-intensity escape movements do not have this erratic nature.

Frequently, the predator will act as a partial directing stimulus. However, it is then acting essentially in a negative way, merely inhibiting movement towards itself, and still leaving a number of largely equipotential possibilities leading away. Thus may the typical zigzag fleeing be produced, as in waders and plains mammals.

In cases where even the releasing source cannot act as a directing stimulus, a series of erratic responses in all directions is to be expected. This situation occurs in highly mobile prey animals which detect a predator via olfactory, auditory or tactile channels in such a way that the precise location of the predator is uncertain; only its general proximity is indicated. Thus, for many small insects living on tree trunks the first sign of danger from insectivorous birds may be a violent tactile stimulus or a sudden shadow, and the exact direction from which the beak will descend is not likely to be susceptible to rapid analysis by the insect's visual and central nervous system — which is very limited in its powers of integrating information (Vowles 1961). Immediate response gives the best chance of survival, so again escape must be initiated in the absence of directing stimuli, producing the highly contorted rapid zigzagging and whirling so characteristic of, for example, chironomid and psychodid flies. Similarly, a rodent flea receiving a sharp tactile stimulus as a result of antiparasite grooming by the host will scuttle and leap in any direction, even back towards the host, for it is incapable of analysing the host's position in less than several seconds.

An essentially similar situation occurs on the sea shore when sandhoppers and shore-hoppers have their drift-line debris cover suddenly removed by a predator. The crustaceans' immediate response is simply to leap, the direction of leap being immaterial. As the substrate is irregularly uneven, and the animals may be facing in any direction, their take-off is influenced accordingly. There is no way a predator can tell which way a hopper is going to jump. Our preliminary experiments support this assumption. Further, as these animals are highly gregarious, the disturbed group explodes in all directions, producing the combined scatter/multiple erratic effect.

This general situation also applies in higher animals and on substrates which are more uniform. Hanby (1983), for example, notes that on one occasion hunting lions 'had managed to catch two wildebeest who had stupidly run right at the lions as their brethren fled the other way'. And we have noted (Driver 1960) how crêches of eider ducklings will scatter in all directions under water, particularly if one approaches quitely under cover of darkness. Sometimes some of them surface closer to the intruder than they were originally.

Some of the most instructive data in this area are supplied by Roeder in relation to moths; his studies constituting the most detailed analysis of the mechanism of any protean display to date. Many noctuid moths can react to their bat predators through the possession of a pair of tympanic organs. Roeder (1966*a*) recorded the impulse pattern in the moth's auditory nerve in the field and showed that for the moth, *Caenurgina erechtea*, the tympanic organs can detect ultrasonic cries emitted by cruising bats 30–40m away. The bat's ear, in contrast, can detect the moth at only 3–4 m because it must depend on the much fainter echo of its own cry reflected from the moth. The countersonar system of the moth thus has at least ten times the range of the bat's sonar.

Roeder (1962) noted two basic kinds of noctuid moth reaction to bat sonar: if they are a considerable distance from the ultrasonic source they turn and fly away from it in a direct path, but close to the source they react with erratic, often diving, responses (Fig. 4.12). Laboratory studies by Roeder (1966*b*, 1967*b*,) showed that low-intensity ultrasonics emitted by a loudspeaker situated to one side of a tethered flying moth would consistently induce it to attempt to turn away from the sound, whereas high-intensity ultrasound evoked erratic turning tendencies and periodic interruption of flight.

Each tympanic organ of the moth has only two receptor cells, and only one is activated by low-intensity ultrasonics. Sound intensity giving the approximate distance of the bat is coded by the shorter intervals between consecutive nerve activity spikes, and its bearing can be derived from the shorter latency of response to louder sounds. Comparison of the latency on the two sides will thus reveal which receptor is facing the bat, and Payne *et al.* (1966) have shown that the moth's wings act as auditory baffles giving further information as to the vertical bearing of the bat.

High-intensity ultrasound saturates both tympanic organs, in the sense that a maximum response is given by both. Since directional cues depend on a differential response by the two ears, directing stimuli cannot be received at all when the bat is at close quarters — that is, less than 4 m away. And this is the very distance from which the bat itself is first enabled to home on the moth. In straight flight the bat can quickly overhaul the moth, so that directed turning away would in any event be useless. The fact that the auditory organs cease to give directional information we interpret not as an indication of any limitation in their information-processing ability due to their simple construction, but rather as an adaptive cut-out of a mechanism which has ceased to be of survival value.

Deprived of directing stimuli, but not of the escape-arousing stimulus of continuing bursts of ultrasound, the moth's flight becomes wildly erratic. Here we have clear evidence of the occurrence of a single erratic display resulting from the strong activation of an escape tendency in the absence of specific directing stimuli. That the erratic manœuvres do not primarily function to obtain auditory information, but are basically escape tactics is shown not only by Roeder's neurophysiological work but also by the occurrence of these manœuvres in other stress situations, as during release from captivity or when subjected to strong light.

While we have shown the causal circumstances in which erratic display occurs, we have not been able to show what mechanisms generate the irregularity and why undirected escape should be erratic. In this connection we re-emphasize our original point that erratic behaviour is likely to require a more complex mechanism than orderly behaviour. So, given that erratic display has survival value as attested for noctuids by Roeder and Treat (1961), we can accept that natural selection may have produced such a mechanism. The attempt to describe the neurophysiological mechanism involved would seem to be the next step in research.

As an example of one possible line of investigation let us consider anticipatory arousal and the part it might play in bounding and pronking displays. Anyone who has worked much with sheep will be aware of their habit of bounding erratically as they are driven through a gateway, particularly if it is an opening in an otherwise visually impervious wall or fence. It is as if they are reacting to an imagined attack. They bound thus when released from restraint, as after shearing or marking. Deer and antelope jump in a similar manner when crossing a road. The behaviour is essentially similar to the bouncing component of play-fleeing in fawns and lambs, and even one-year-old sheep will bounce in this way when running across a field to join companions. Clearly, this behaviour is readily evoked.

It may be that the anticipation of danger as the animal emerges from cover is enough for the reaction to be evoked by the view of the open space beyond the gate or up and down the road, even though there is no danger there. Any soldier or policeman, bursting into a room containing unknown but expected dangers will know how high such arousal can be. Undoubtedly, this has resulted in many unnecessary deaths in human agonistic encounters when a lethal act was evoked too readily, one recent, sad, example being the child shot by a policeman bursting into his room in pursuit of a suspect. In animals in general this reaction will sometimes be evoked by actual, rather than potential, danger and this would facilitate any subsequent arousal. Also the protean insurance element inherent in the behaviour could increase its selective value and potential for genetic reinforcement.

A similar situation seems likely to exist in stotting and pronking (pp.45–6). Blanford (1888–91) notes that the Indian antelope *Antilope cervicapra* 'has the habit of occasionally springing into the air, all the members of a herd bounding, one after the other. This is done . . . before they are much frightened, when the herd is first moving off'. L van der Post (1961) records similar behaviour in the springbok. Kruuk (1972) notes how Thomson's gazelles stott at the beginning and end of a chase when the tendency to flee is high but not at the optimum. Stotting is more common and more pronounced in fawns, as is play-pronking in springbok, which suggests an ancient, deeply-rooted mechanism. Several television and big-screen wildlife firms have shown this.

Two known neural phenomena could be involved in the evolution of these bouncing displays. First, they could be allied to the reflex jump when startled — that experienced by humans when they exclaim 'ooh, you made me jump!' Second, the anticipatory arousal could apply when tommies stott at the beginning of a chase when they see the approaching predator some time before they begin to flee. And it could apply at the end of a chase as they concentrate upon the predator to see if it is in fact safe to slow down. In between they have to rely on speed or, when very hard-pressed, on the standard two-dimensional single-erratic display.

Bounding displays must increase the overt complexity of the prey's behaviour, which means increased attention on the part of the predator, but also an increased reaction time. When several animals bound in view simultaneously the predator may also become confused — as happens in sticklebacks preying upon water fleas (Milinski and Heller 1978; Heller and Milinski 1979). A predator may even have learned that only strong animals bound and that they are less worth pursuing than lower bounders (Zahavi 1977) — though we do not support this interpretation ourselves. Whatever the reason, it would seem that the establishment of the neural pathways involved in bounding displays depended upon selective predation on the less erratic prey. Bounding thus became autochthonous in primitive ungulates.

In some other groups, e.g. the Lagomorpha (including rabbits and hares), we see convergent behaviours. The black-tailed jack-rabbit, *Lepus californicus*, occasionally jumps higher than usual when running

(Anthony 1928). The white-tailed jack-rabbit *L. townsendi*; bounds very high when disturbed and continues thus in flight, bouncing on stiff legs like a pronking antelope (Coues 1875).

In the various animal groups, rump patches and dorsal or lateral stripes have evolved according to the raw materials which were being moulded by the pressures of natural selection. The flashes have the common function of reinforcing the predator's attention to the displaying animal.

In rodents the original startle response seems to have been followed by the development of erratic switching from one flight direction to another, doubtless because, through the operation of the protean principle, this was more effective than directed fleeing. This switching has become emancipated from its original neural connections over time and now occurs automatically in an imminent capture situation.

Convulsoid behaviour

Some species in which escape is usually direct show erratic behaviour when the escape route is blocked, or when a refuge is not immediately to be found. Gray and Tarbit (pers. comm., 1964) describe this behaviour in lemmings when approached by a human observer. The slower the approach the greater the likelihood that the lemming would quietly run away if on fairly bare ground, or crouch cryptically if in grass or ling. A sudden close appearance of the intruder on bare ground, and blocking the only available escape route, produced squeaking, showing of teeth, facing the intruder, and a vigorous erratic bouncing on all four legs over an area of about a square metre. Chance (1957) suggested that the wild undirected running, aggression, and bouncing convulsion which constitute the audiogenic seizure in deer-mice (*Peromyscus* spp.) when subjected to the sound of jingling keys in the laboratory, might be related to the relative vulnerability of deer-mice to predation in the wild, for they nest on the ground, not in burrows, and may well be attacked while foraging in open spaces where erratic behaviour may be the only defence tactic with any possibility of success. Chance had no direct evidence for this belief, but demonstrated (1954) that in laboratory cages individuals of *Peromyscus* normally susceptible to jingling keys would fail to show seizures if provided with a small 'hut' as a refuge. Such individuals would give the usual startle reaction, then run to the refuge. In the absence of the refuge the running would become progressively more wild and erratic, then, passing through a leaping aggressive phase, would subside into a cataleptic phase. In other words, the seizure and refuge escape behaviour were interchangeable, suggesting that the seizure might itself be a form of escape reaction. Griffiths (1942) had made somewhat similar findings. We would add that Dice *et al.* (1963) note that 'In nature, all *Peromyscus* probably depend heavily upon hearing to warn them of potential danger'.

The importance of Chance's work was that it brought audiogenic seizures and some other forms of convulsoid behaviour into the broader perspective of biology from the narrow confines of medical pathology. We have shown in Chapter 4 that convulsoid behaviours are common in many prey animals and often serve adaptive functions in confrontation with predators. They are even used in some predators' tactics also. We cannot agree with Chance's interpretation of the convulsoid behaviour of *Peromyscus* as a 'displacement activity due to the frustration of the appetitive flight drive', for it seems to us that his refuge experiments demonstrate once again that erratic behaviour develops when the escape tendency is powerfully aroused in the absence of adequate directing stimuli. In the bare laboratory cage there is no possibility of directed escape. The situations which have been used to evoke seizures in various small mammals (Bevan 1955) may all be expected to induce a high escape tendency and at the same time to block any possibility of performing a genuine directed escape response. The controversy as to whether the seizures have a primarily audiogenic basis, or are basically due to insoluble conflict (Maier 1939; Morgan and Waldman 1941) has thus bypassed the basic point — that the animal is predominantly escape-motivated.

Most researchers, for exampler Maier (1948), have classed seizure-type behaviour as abnormal because animals may damage themselves or even die in laboratory seizure situations. We consider that this behaviour is not necessarily abnormal and, in natural circumstances, is often adaptive. It is the laboratory situation itself which is abnormal, for the behaviour of the animal is adapted not to the laboratory environment, but to the field; and it is hardly surprising that it does not always match requirements for survival in unique and novel situations. With respect to the subject's sensory attunements and natural response modes the laboratory situation may be totally alien (Driver and Corning 1968).

In field conditions a prey animal either escapes rapidly from the escape-arousing stimulus, or is killed, so situations of prolonged blockage of escape are rare, and even then temporary catalepsy may be of survival value, as we have indicated previously.

In addition, it would seem on theoretical grounds (Vitz 1966) that the visual and auditory complexity of the display would be likely to be strongly attention-drawing and induce a marked 'orientation reaction' in the avian or mammalian reactor. This suggestion is supported by experimental work with humans (Berlyne 1957*b*, 1960). A well-known feature of the 'orientation reaction' is an arrest of ongoing movements (Sokolov 1963). Seen in this light the convulsoid displays of vertebrate predators, even wolves (Mowat 1963), become more credible.

We have seen on p.166 that injury-feigning can be tentatively explained by a model based on the hierarchical organization of behaviours by means of releasing mechanisms (RMs). We now present a model to show how this approach can also suggest an explanation for the fox convulsoid display or 'crazy dance' (Fig. 7.2). In this model, which can also be applied to stoat and weasel crazy dances, the portrayal of the RMs is simplified. Their detailed structure would be as in Fig. 7.1.

Our crazy-dance model is based on the lability and interchangeability of killing, grooming and play behaviours. Some action patterns (APs) are common to two, or even all three of these primary activities. We assume that the organizing centres — or some equivalent motivational complex — which we call the integrators', can be operationally interconnected. This is a reasonable assumption considering, for example, how readily play behaviour in Canidae can change to fighting, and how, in a disturbed bitch, the licking and nibbling of grooming of newly-born pups can become biting and killing. Furthermore, grooming, play, and fighting behaviours are commonly seen in Canidae and other groups as displacement or vacuum activities.

An unchecked hunt in foxes will proceed smoothly through the sequence of action patterns indicated, each specific releasing stimulus unblocking the appropriate AP and loading the integrator of the next level until the consummatory act of ingestion is achieved. If a hungry fox is presented with food ready for ingestion, the intermediate APs are bypassed and the fox eats.

However, if the sequence is halted by Block 3, displacement from the Attack Integrator to the Grooming Integrator may occur. This involves motivation of Roll, Shake, Scratch, and Lick APs which, in the case of a fox on the edge of a small pool on which swim a brood of ducklings and their mother (Fig. 4.32) results in the duck approaching the fox until it is *almost* within range of a pounce. Such a situation could result in the Pounce occurring as vacuum activity. The Pounce of feeding could thus be redirected to any appropriate object in the vicinity, as in the redirection of aggressive pecking in herring gulls (Tinbergen 1959). We are aware of the theoretical and practical difficulties of distinguishing between displacement, redirection, and response to suboptimal stimuli (Ficken and Dilger 1960; Kruijt 1964; Hinde 1970). For the present we are simply concerned to demonstrate that it is possible, at least theoretically, that convulsoid behaviours can be systematically organized, yet have an unsystematic appearance.

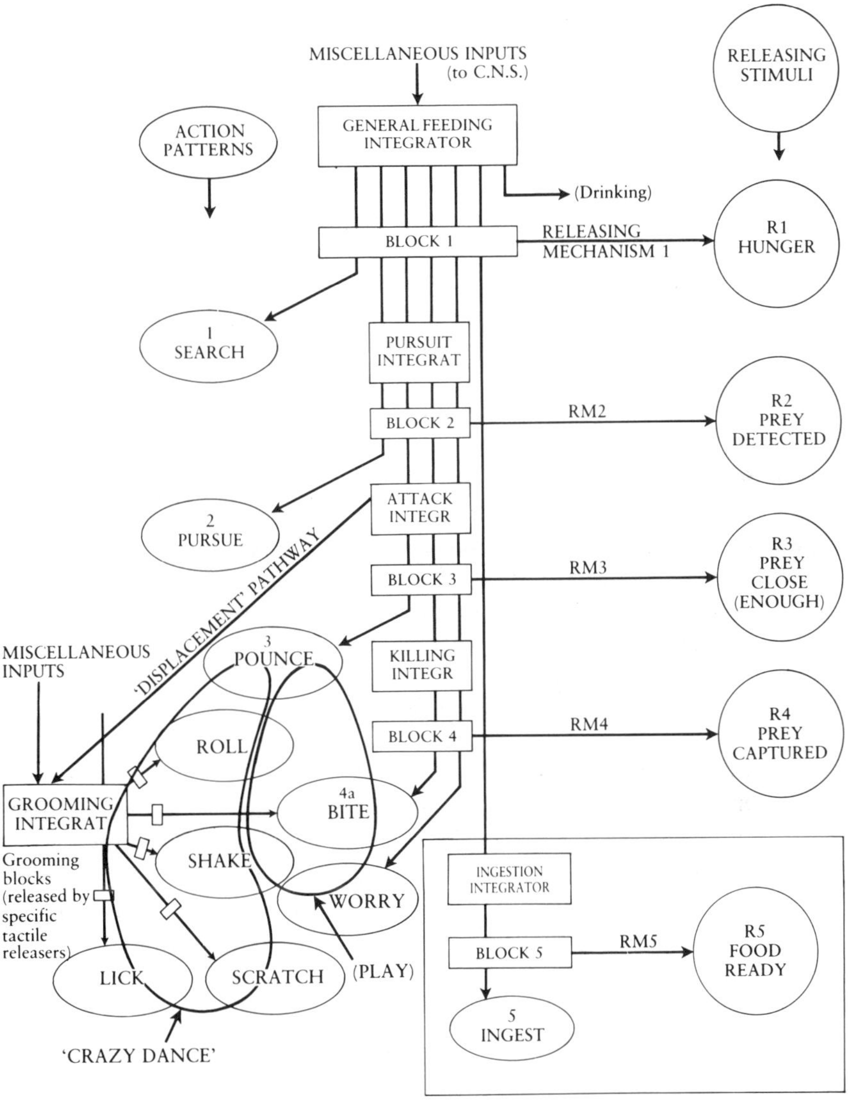

Fig. 7.2. Model of the organization of fox feeding behaviour, including the 'crazy dance'. (Details of releasing mechanisms as in Model 1 — here simplified.) In a quickly successful hunt Action Patterns 1 to 5 will proceed smoothly. When the sequence is stopped at Block 3, a 'crazy dance' may occur. This is likely to happen, for example, when there is a high attack drive but the prey is not quite close enough for a pounce. If the fox is hungry and food is ready, this system allows the approriate response: RMs 1 and 5 operate, and the fox eats.

So, if we combine Pounce with the APs of grooming, we have the makings of a convulsoid display. A computer-produced series of Grooming APs associated randomly with equal loading, and including Pounce as occasional vacuum activity gives the sequence of behaviours in Table 7.1. It can be seen that the Pounce and Roll are occasionally activated together. One may assume that this would result in somersault-like behaviour. Other associations, such as Roll-Bite-Shake, and Pounce-Shake-Lick, would be almost as bizarre. In a natural situation the sequence of behaviours would presumably be modified by variations in the inputs from the various sensory pathways which stimulate grooming. For example, a wet or dusty fox would be more likely to Shake than a dry and clean one. In fact, autonomic activity engendered by the conflict situation may be central to the structuring of the crazy dance. Such a situation may promote vasomotor or pilomotor activities which provide skin stimuli which, in turn, may elicit grooming behaviour, just as dryness of the throat may elicit drinking (Andrew 1956; Morris 1956). But, whatever the mechanism involved, the fox provides an unsystematic and attention-drawing signal and the prey approaches, or at least does not flee. Thus may the crazy dances of predators result in the capture of prey.

As with our previous model based on the hierarchical principle (p.166) we find that the exercise emphasizes the heuristic value of that principle. For, coincidentally, the present model also suggests a model for the motivational structure of play behaviour.

It seems probable that the convulsoid behaviours of rodents, which seem to be more nearly akin to classical convulsive seizures, may normally depend upon a progressive withdrawal of inhibition (Chance 1957; Krushinskii 1962) and that the appearance of a progression of miscellaneous activities as the seizure proceeds reflects the spread of this disinhibition. The initial disorientated running which often involves violent collisions with obstacles, sometimes to the extent of inflicting physical injury, is therefore understandable as a breakdown in the discrimination of environmental information due to loss of inhibition. It may well be that the non-convulsive single erratic display itself may be generated by a similar but more limited reduction in the inhibition which is so essential for effective discrimination. Without discrimination, directed escape becomes impossible and the actor will provide very little systematic information to which a reactor can respond because — in part at least — it is not reacting systematically itself.

We may expect the extent of zigzagging in the single erratic display and the readiness with which it is evoked at a given level of escape drive to be adjusted by natural selection against disadvantageous ex-

Table 7.1. Computer-generated simulation of combination and succession of postures seen in crazy dance.

Lick	0	Shake–Lick
Pounce–Lick	Roll	Roll–Bite–Shake
Shake–Scratch	Pounce–Roll	0
Pounce–Lick	Pounce–Shake–Lick	0
Scratch–Lick	0	Roll
0	Bite	Roll
Roll–Bite–Shake	0	Roll–Scratch
Shake	0	Roll
0	0	Roll–Scratch
0	Roll	0
0	Lick	0
0	Bite–Shake	Pounce–Roll–Scratch
Shake	0	Bite
0	Bite	Roll–Scratch
0	Pounce	Bite
0	0	Scratch
Shake	Roll	Scratch
Pounce–Roll	0	Scratch
Bite–Scratch	Roll	Roll–Scratch
Roll–Bite–Lick	Shake	Shake–Scratch
0	0	Pounce–Roll–Scratch
0	Roll	Pounce
0	0	Bite–Lick
Shake–Scratch–Lick	0	Shake
0	0	Bite
Bite–Scratch	0	0
Lick	Roll–Lick	Roll
0	Shake	Pounce
Scratch	0	0
Shake	Pounce	Scratch
Lick	Scratch	0
Bite–Lick	Roll	Roll–Scratch
0	0	Bite–Shake–Lick
Pounce–Shake	Lick	0
0	Bite	Roll–Bite–Lick
0	Roll	Shake
0	0	Pounce–Lick
Bite–Scratch	Bite-Shake	Roll
Roll	Lick	Shake
0	0	Scratch–Lick
Roll	0	Pounce
0	Roll	Bite
Roll	Lick	Bite
Pounce–Bite–Shake	Pounce–Lick	Shake

tremes. The influence of genetic factors upon the single erratic mechanism and its role in convulsoid displays is shown by the inheritance of susceptibility to waltzing and seizure behaviour in small mammals (Dice *et al.* 1963) and by the artifical selection of the 'tumbling' characteristic in pigeons (Darwin 1859).

In cases where single erratic behaviour occurs habitually in the absence of predators there is obviously no need to suppose the existence of an escape drive, but here again locomotion may be subjected to directing influences. Variations in movement through the successive activation of a number of different directing mechanisms could be selected for through influences other than those already mentioned. For example, among the insects of the aerial plankton the more predictable fliers are more likely to be captured by insectivorous birds, and such differential predation would provide a selective force favouring the evolution of erratic flight. The same point would apply in dancing midges and displaying lapwings (p.48–55), for many species of animals become more conspicuous to predators during their reproductive displays. The resulting dangers may be minimized by erratic movements or by erratic vocalizations such as alternating songs in crickets and grasshoppers, duetting in various birds, and mimicry in bower birds and starlings (pp.241–42). These are examples of what we call protean insurance.

Immobility displays

In immobility displays where the prey has been detected, or even captured, the survival value of the behaviour depends entirely upon the protean principle. The predator must not be able to predict the likelihood of the prey 'coming to life'. To this end it would seem that in many species the captured animal often experiences neither pain nor fear (see below). We believe that this condition has been selected for in evolution, for an animal frightened or in pain will struggle or scream. Absence of these reactions may deny the predator adequate stimuli for the elicitation of the consummatory act of killing and eating, thus enabling the occasional prey animal to escape.

This would seem to have been demonstrated when Livingstone, the explorer, was caught by a lion (Livingstone 1865).

> The shock produced a stupor similar to that which seems to be felt by a mouse after the first shake of the cat. It caused a sort of dreaminess in which there was no sense of pain nor feeling of terror, though quite conscious of all that was happening. It was like what patients partially under the influence of chloroform describe, who see all the operation, but feel not the knife.

As with certain avian instances (p.109) the prey was able to escape when the predator's attention relaxed.

Such a trance-like state will often, of course, be unsuccessful as an escape device, for the predator may be so hungry that it devours the prey immediately it is caught. But the condition could still be said to have a systematic function, for it blocks fear and pain in billions of the prey animals which die every year. Without this regular aspect of the phenomenon the protean element could not operate, for a predator will not readily release a struggling animal. A pre-condition for trance-like immobility displays in captured animals therefore is the reduction of fear and pain. This is a matter of considerable import for those interested in animal welfare as well as for pure scientists. We call the mechanism involved *auto-analgesia.*

Supporting evidence for our theory comes from many biologists who have worked on predator–prey relationships in the Serengeti and elsewhere, and from current work on naturally-occurring pain-killers. It is clear that, in general, ungulates caught by dogs, big cats, or hyenas struggle very little, even when being dismembered or disembowelled. Schaller (1972) for example writes:

> After prey has been pulled down, especially if this has been done quickly, it struggles surprisingly little, sometimes even failing to thrash its legs. An extreme instance of such behaviour was an uninjured buffalo that lay on its side while a lioness chewed on his tail. Animals in such a situation seem to be in a state of shock.

Published illustrations of this phenomenon include photographs of hyenas and their prey (Kruuk 1975) showing a captured wildebest both standing and sitting without struggling. And among recent TV wildlife films illustrating the phenomenon fortuitously, one showed a wildebeest, captured but only slightly damaged by a pride of lions, lying or standing almost quiescent for at least 20 min while the young, inexperienced lions 'practised' their pre-ingestion techniques. Another showed a Thompson's gazelle, captured by a female cheetah and carried limply in its mouth to the cubs, 'coming to life' after being put down, and giving the cubs excellent 'practice' in the chase. Such situations, which must be replicated innumerable times every year, suggest that prey auto-analgesia has advantages for the predator also. Prey animals which always struggled as vigorously as they were able would deny to juvenile predators the necessary experience for the maturation of innately-based capture and kill behaviour. There could thus be a reciprocal advantage and a double selection pressure inherent in the blocking of fear and pain through auto-analgesia.

This condition seems to be akin to that experienced by many people in road accidents and similar mishaps: a state of shock, with considerable amnesia. It is a state readily appreciated by those, including the present authors, who are accustomed to vigorous contact sports during which one may receive, unnoticed, injuries which would cause considerable distress if received off the 'field of battle'. Such situations are operationally similar to real battles in which serious, even fatal wounds may pass, for a while, unnoticed. Indeed, it is well established that a bullet may be cut out painlessly if it is done almost immediately after impact (W.G. Hayes, pers. comm. 1981).

Auto-analgesia would seem to be closely allied with, if not identical to, the neurophysiological state of horses when 'the twitch' is applied.

> The twitch is a 20 to 100 cm long wooden stick with a loop of rope at the end. This loop is twisted rather tightly round the upper lip of the horse. After application of the twitch the horse becomes quieter, appears somewhat sedated, the eyelids drop, and its hostile attitude decreases. The horse's interest in its surroundings diminishes and it becomes difficult to stimulate it to walk. The tolerance and acceptance of pain increases' (Lagerweij *et al.* 1984).

We remember a farmer effectively applying the twitch bare-handed, the horse's upper lip simply being grabbed and twisted. One wonders if this action, or a similar application in man, could be the origin of the term 'being led by the nose'? It has been use by farriers and grooms for centuries — possibly from the time the horse was first domesticated.

Lagerweij *et al.* suggest that the twitch, like acupuncture, brings about the activation of a pain-decreasing mechanism, and that endorphins may be the mediators in both. Endorphins and enkephalins — the endogenous opiates — were first recognized as recently as the mid-1970s, but their general biological significance is considerable. They have a most extensive distribution in the body, 'linked to the endocrines, associated with behaviour, influential in mental breakdown, critical in pain control, most relevant to sex . . . ' (Smith 1984).

It seems likely that endorphins are the chemical mediators in auto-analgesia, and that acupuncture and the twitch are cultural derivatives of an evolved mechanism. We suggest that auto-analgesia developed originally through its role in blocking sensations of fear and pain in prey animals when attacked, so that they could perform their escape behaviours with optimum efficiency. This mechanism has been found to be secondarily useful as a protean device promoting escape in a proportion of captured animals. Natural variations in external and internal environmental conditions will automatically result in any sequence of prey captures being effectively protean to a predator with respect to the likelihood of the captured animal suddenly making a real escape attempt. Auto-analgesia thus helps to make possible the escape of some of those captured animals which happen to 'come round' when the predator's attention is relaxed. And it is likely that this mechanism operates more effectively in prime, healthy prey animals than in old, very young, or sickly ones, the latter group already being more stressed in the broad biological sense (Selye 1956). Auto-analgesia will thus have a higher selective value in those individuals which have a potentially greater contribution to make to the population gene pool.

The trance state, with the blocking of pain and fear, would seem to have evolved separately in birds and mammals. Many lower vertebrates, and invertebrates, also show immobility reactions but there is considerable variation in the incidence and form of the displays. Some insects become immobile when attacked (see, for example, Edmunds 1972; Robinson 1969*c*). However, most insects caught by mantids or spiders struggle until immobilized by wrappings, dismemberment, or anaesthetization.

In many cases, particularly in amphibians and reptiles, immobility behaviour may be truly cataleptic, basically similar to catalepsy in man — 'a kind of hysterical or self-induced trance, in which the patient remains fixed in even the unlikeliest position, as in the children's game "statues"' (Wingate 1972). But the well-known instances of certain death-feigning amphibia 'unrighting' themselves when turned the right way up suggest the operation of a 'fixed' action pattern — the basic overt unit of instinctive behaviour whose form has been developed by natural selection.

Support is given to the 'fixed' action pattern suggestion by work on the best known immobility display of all: the 'playing possum' behaviour of the Virginia opossum, *Didelphis virginiana*. When sleeping normally this animal has its eyes and mouth closed, and its feet are usually concealed. When attacked however, it collapses with its eyes open, lying with the feet visible and the claws clutching the ground. In this condition the opossum can be handled for long periods without moving — as can the African ground squirrel *Xerus erythropus* (Ewer 1966). There is no apparent difference between the body temperature, oxygen consumption, blood chemistry, and electroencephalogram of a normal opossum and one that is shamming dead (Francq 1968). And the display is even seen spontaneously at 120 days of age — the stage of development, around weaning, when it would begin to be needed (Norton *et al.* 1964).

C.W. Nicol informs us (pers. comm. 1985) that the racoon dog or tanuki (*Nyctereutes procyonoides*) is well known in the Nagano prefecture of Japan for its habit of shamming dead, particularly when cornered or grabbed suddenly by the scruff of the neck. 'Recovery' is usually a very cautious procedure. This behaviour is so well known that the term 'tanuki jiji' (old man) is used for one who masks his true feelings.

Norton *et al.* 1964 suggest that death-feigning opossums are in fact quite aware of everything that is happening around them. This is an interesting concurrence with old reports that a death-feigning European fox, *Canis vulpes*, would 'often cautiously open its eyes, raise its head, look around and finally scamper off' if its attackers had withdrawn (Carrington and Meader 1911). The striped hyena, *H. hyena*, behaves in a similar way (Conway 1972*b*).

The variety of apparent mechanisms for the immobility response is of interest, for such variety would increase the protean effect upon polyphagic predators through the associated variety of form of the response. The response may be truly cataleptic, particularly in the case of animals which die unaccountably while in the immobile state. Or it might be akin to the human faint or syncope, a stress reaction which overstimulates the vagus nerve, resulting in slowing of the heart, relaxing of abdominal blood vessels, and a consequent lowering of blood pressure in the brain arteries. It may be a 'fixed' action pattern as indicated by the opossum studies. Or — as seems most likely — these responses to attack are behaviours in the process of evolutionary test and development, subject to a variety of operational factors, and only reaching a characteristic and stable species-specific state in a few species such as the opossum. But, whatever the mechanisms involved, they seem to have the common function of confusing the opponent, firstly by giving an inappropriate signal to the predator, and secondly by their unpredictable occurrence.

The grotesqueness of some immobility displays is no hindrance to the efficacy of the behaviour, for the central operational requirement is confusion of the opponent. It could therefore be better not only to appear unpredictably and unaccountably dead, but also unusually dead. This might explain the curious immobility postures in insects studied by Darwin (quoted in Carrington and Meader 1911). Darwin compared the natural death postures of seventeen species with their feigned death postures and found that 'in no instance was the attitude exactly the same, and in several instances the attitudes of the feigners and of the really dead were as unlike as they could possibly be'.

Anarchization and the diversity maturation hypothesis

We trust that the reader is no longer in any doubt that anarchization has been a significant factor in the general evolutionary process, particularly in the more advanced system of mature biomes. On this last point we are pleased to acknowledge recent work on biome maturation, not only for its intrinsic interest but also because it blends well with our own ideas.

A.H. Clarke (1978 and pers. comm. 1985) is developing an hypothesis which links the extent of intraspecific and interspecific diversity with the amount of progress towards biome maturity. He postulates a continuum of biome development with three principal phases. Phase I, *initial occupancy*, involves relatively little interspecific competition, selection pressure being largely imposed by the physical environment. Intraspecific diversity is considerable and there is noticeable morphological overlap between certain species. Phase II, *early biological selection*, sees a progressive development of intra- and interspecific competition, and selection by predators. Overall diversity of colour and form decrease because some morphs are much better adapted than others. Phase III, *biome maturation*, is characterized by species saturation and the evolution of highly specialized prey strategies and tactics. Predation by searching image is increasingly employed. In multi-species communities, non-cryptic prey species develop highly contrasting aspect morphologies, while in communities with few species intraspecific diversity is promoted. Both of these prey responses seem to be promoted by apostatic selection. We comment on these Phase III matters in some detail in our next chapter.

We have seen that, when the armaments between protagonists are equal, it always pays one of them to develop a protean tactic. It would seem that regular, non-protean tactics are developed first. Only when such tactics have been optimally developed on both sides does anarchization come into play. For example, certain big cats developed speed and stealth to enable them to catch plains ungulates such as antelope and zebra. Antelopes in turn developed jinking abilities for avoiding capture. This was then met by the special adaptations of cheetahs for following the jinks. Hyenas and hunting dogs tackle the same problem by hunting in groups. Flocks of swifts in the air or of cormorants in the water benefit similarly from their temporal and spatial association. The erratic movements of a prey animal escaping from one predator may place it in the path of another. Thus the usual sequence is, first, the optimal development of regular armaments, followed by a protean tactic by one contestant, which may, in turn, be countered by an antiprotean device in the other.

Clarke's Phase III, biome maturation, provides by far the greatest incidence of protean strategies and tatics. Protean displays of all types, as catalogued in Chapter 4, are seen, and biome maturation also allows for the development of the complex intra- and interspecific variations in appearance and distribution outlined in our Chapter 8. When all the regular avenues of adaptation have been fully explored, protean developments are necessary if further advances are to be made. In this respect protean theory and the diversity maturation hypothesis seem to be mutually supportive.

Before moving on to outline some evolutionary counters to anarchization we emphasize our view that the behaviours we describe are structural complexes which have a meaningful connection with environmental circumstances, even though the initial events which open the way to evolution are fortuitous (Monod 1972). In this context we cannot resist quoting one of Watson's analogies (1974): '...in the course of evolution it is the predators who have determined the ... posture by coming along like drama critics and picking out and eating all the bad actors'.

Antiprotean devices

Protean weaponry is very difficult to counter but, just as protean devices of prey animals have developed through the continual struggle for existence with predators, so do some predators evolve means of combating the protean armaments of their prey. If there is a solution to be found, natural selection may well find it. The development of widespread and well-established devices to counter erratic behaviour is of course further evidence, if any were needed, for the existence and efficacy of protean displays.

We now survey the principal countermethods which predators have developed in their struggle for existence with protean prey. We classify these methods according to type of predation.

Aerial plankton feeders

Nightjars (Caprimulgidae), swallows and martins (Hirundinidae), swifts (Apidae), and some others, have become specialized for capturing large numbers of flying insects in their evolutionarily enlarged gapes fringed with special bristles. This adaptation, we believe, is not just a development for scooping in as many small prey as possible. It is also a counter to the protean movements of the prey. Flying insects with low manœuvrability, such as ants in their nuptial flights, are taken readily by many species of birds which do not normally take insects in flight. Thus various species of gulls, crows, starlings, and sparrows, and many others will readily feed upon slow-flying insects.

Many bats have an enlarged insect-catching surface in the interfemoral membrane stretching between the bases of the hindlegs and incorporating part of the tail. They may also use the wings to sweep an insect to the mouth. In view of the great efficiency of bats' flight and ultrasonic detection mechanisms we assume that they would be able to counter any predictable avoidance manœuvres without the need for an increased area of capture organ. The wing and tail membranes enable them to catch some of the prey animals which employ a protean jink at the last moment (Fig. 7.3).

Terrestrial insectivores

Ostrom (1972) suggests that the wing feathers of *Archaeopteryx* may have provided an increased catching area for trapping small prey such as insects, i.e. an antiprotean device. Harrison (1976) however suggests that the protean element was in the ability to take flight unpredictably, which it could not do, of course, without wings. We suggest that both factors may have operated. Harrison's reservations concerning the wings of *Archaeopteryx* being used for prey capture were based on the dangers resulting from consequent abrasion. We note however that several species of birds use the wings in fighting. Male jungle fowl, *Gallus* spp., and their domesticated derivatives trail the wings on the ground during display. And many raptors hood their captured prey with the wings in a manner that may cause abrasion. It can be argued that these species are able to compensate for abrasion by virtue of the annual moult which automatically renews the flight feathers. But we may also argue that, although *Archaeopteryx* may not in fact have had an annual moult, a certain amount of abrasion of the wing feathers would have been no hindrance, for it is generally agreed that the bird flew poorly. Use of the wings for prey capture therefore would not necessarily hinder flight, though it might make that flight less predictable.

Analogous countermeasures are seen in the outflung forelimbs of certain predators when they leap upon their prey. This is seen in cats when they jump at birds taking flight (Leyhausen 1965), and in many of the jumping spiders (Salticidae) when they make their capture leap. If the predator could predict which way the prey was going to move, it could be more accurate in its leap and the increased catching area would be less important. The same principle applies when shrews leap upon arthropods such as grasshoppers and shore-hoppers landing after a flight or jump (J.C. Pernetta pers. comm., 1970). The shrew can hear the touch-down of both flying and jumping prey, and leaps at the sound, attempting to cover the prey with its body until its captive can be bitten.

Aquatic predators

In the case of whales which feed largely on krill (*Euphausia* spp.), even the enormous gape is an insufficient counter to the prey's movements,

Fig. 7.3. Bat countering single erratic display of moth. Bat follows moth's general flight path and collects its prey with the aid of wings and interfemoral membrane.

Fig. 7.4. The bubble net of the humpback whale — an essentially antiprotean device. Prey are hindered in their scatter and single erratic behaviours by a cylinder of bubbles blown by the whale (drawn from Earle 1979).

for the single erratic display of krill is outstandingly effective in avoiding capture. Numerous workers have noted this, including Earle (1979), on whose account we have based the following outline of one whale species' countermeasure.

In order to keep the krill and other prey such as small fishes in a limited volume of water so that they may be more easily ingested, humpback whales *Megaptera novae-angliae* have evolved a means of constructing a temporary cage (Fig. 7.4). Beginning at depths of perhaps 15 m, the whales emit bubbles from the blow-hole while swimming upwards in a spiral. The rising bubbles form the wall of a cylinder inside which any prey animals are retained. The whale then simply swims up through the cylinder and ingests its prey.

It seems that two, maybe more whales may co-operate in bubble-netting, but B. Dolphin (quoted in Giddings 1984) believes that when a number of whales are involved, they are more likely to be in competition. Giddings (1984) has excellent photographs of this activity.

We suggest that there may be a degree of co-operation, depending on familial relationships and other circumstances.

This strategy of containing large numbers of prey animals within a small volume of water as a counter to the single erratic display is used by other aquatic predators too, but the other recorded instances involve a number of predators encircling or herding the prey. Martinez and Klinghammer (1970) describe this tactic in killer whales, *Orcinus orca*, which surround porpoises. Accounts of the similar restriction of fish movements by a number of species of cormorants, pelicans, and anhingas (Order: Pelecaniformes) have occurred in the literature over many years. Heinroth for example, writing in Stresemann (1927–1934), reports Indian darters, *Anhinga anhinga*, fishing co-operatively in a group several hundred metres long. Recent reports have tended to confirm the co-operative aspect of this behaviour, discounting the suspicions that the birds simply congregate at prey concentrations — as gannets and boobies (Sulidae) do when diving upon fish shoals. Cramp (1969) describes white pelicans (*Pelecanus onocrotalus*) fishing socially, 'their beaks dipping in perfect synchronisation. They often form a great arc driving the fish before them into shallow water by beating furiously with their wings and feet'. Batten (1969) gives further detail for the Dalmatian pelican, *P. crispus*:

> Sometimes over a hundred birds form a closely-set line parallel to but some distance from the shore. They proceed then to swim towards the shore splashing and paddling rapidly, driving any fish before them. As they approach nearer the bank, a crescent is formed by the end birds increasing their speed. By the time the depth of the water has decreased to about 18 inches the crescent rapidly becomes a circle due to the birds on both ends closing in and joining up. Inside this circle the water may be alive with fish, on which the pelicans gorge themselves.

Curio (1976) also gives examples. The double-crested cormorant, *Phalacrocorax auritus*, is well known for its communal fishing. Bartholomew (1942) describes the antumn fishing behaviour of this species in San Francisco Bay. From a dozen to several hundred birds form long closely-packed lines on the water at right angles to their direction of movement, moving steadily forwards. Birds throughout the flock make short feeding dives, up to half the birds being under water at a time. The birds drive the fish before them and, apparently, limit their dodging.

Over a period of six years we have frequently seen shags, *Phalacrocorax aristotelis*, fishing communally in Orkney. Typically, up to fifty birds (occasionally 200 or more) swim steadily along in loose formation some 50 m from the shore, diving as they go, with birds at the rear of the group occasionally flying ahead to pitch at the front. The underwater conditions here — as we have established by diving — involve a variable bottom, with numerous rocks interspersed with sandy areas, and luxuriant kelp (*Laminaria* spp.), at a depth of approximately 2 fathoms at low water. The conditions are thus typical of the lowest region of the littoral zone on British rocky shores. Fishes of various species abound. The free-swimming species, such as pollack (*P. pollachius*), grey mullet (*Crenimugil labrosus*), whitebait (*Clupea harengus*), and stickleback (*Spinachia vulgaris*), feed in and out of the kelp forests. Bottom-living species such as flounder (*Platichthys flesus*), eel (*C. conger*), rockling (*Ciliata mustela*), and butterfish (*Centronotus gunnellus*) forage in the substrate. When these animals are disturbed they take cover, the free-swimming species in particular performing single or multiple erratic displays. When successful, these displays end in concealment. However, when a number of shags are fishing simultaneously, any fish disturbed by one bird may be captured by another, in reciprocal fashion. The loose fishing formation may thus be regarded as a means of increasing the capture rate when faced with this combination of protean displays and concealment. In the absence of cover the close formation is more effective, as in the San Fransisco Bay example.

Kinnear (1978) commenting on the synchronous diving of great northern divers *Gavia immer* suggests that they 'actively co-operate under water to obtain food'. Taylor (1978) suggests that small groups of puffins (*Fratercula arctica*) co-operate in attacking schools of fish such as sprats (*Sprattus sprattus*) and sand eels (*Ammodytidae*), breaking down the schools' cohesion so that individuals may be captured.

In assessing these instances of communal feeding we must be careful not to jump to conclusions. Krebs (1974) for example has shown that, though great blue herons (*Ardea herodias*) feeding in flocks take in food at a faster rate than herons feeding singly, this is not due to the number of birds *per se*, or to co-operation, but is the result of the herons congregating only when feeding conditions are good. We suggest that this is not the explanation for flock feeding in Pelecaniformes, for these birds may aggregate in close ranks — through which the fishes have difficulty in escaping — and actually drive the prey before them, often finally encircling them or hemming them in against the shore. Herons, not being specially adapated for swimming, overcome the fishes' single erratic display by surprise and speed of capture. In the Pelecaniformes we have the desiderata of a social organization (they are usually colonial nesters) with the ability to swim well at the surface and better beneath it, and adaptations to feeding in shallow water. Thus pelicans, anhingas, and cormorants, by virtue of their numbers and feeding behaviour, can form traps that are functionally equivalent to whales' bubble nets. It will be interesting to see if other examples come to light. We suggest that they are most likely to be found in predators which are capable of social aggregation and pursue fish in shallow water. The cormorants of Lake Malawi seem to provide just such an example (Anglia TV *Survival* Programme, 29 Nov. 1982).

It is gratifying to note that both line-abreast and mass techniques are used by man for fishing in shallow waters. The line-abreast method is used by many littoral peoples. We quote just one — perhaps unexpected — example, that of the 'Meggies' (Cleethorpes-born) of north-east England (P.J. Little, *pers. comm.*, 1986). During the periods of big tides, particularly in the autumn, parties of up to twenty people walk, line-abreast, through the shallow stretches of water left on the sandy shore by the receding tides. As they go they stab randomly, or with aim, for flatfish, with metal-tipped wooden spears up to twelve feet long.

The mass technique is used by a number of African tribesmen such as the Nuer of the Upper Nile (Evans-Pritchard 1940) and the Dinka of Southern Sudan (Ryle *et al.* 1982). At the end of the long dry season the fish have retreated to the few remaining shallow pools and lakes, and tribes such as the Azonde Dinka gather in thousands for communal fishing and festival. Here too the technique is to stab, with or without aim, with long slender spears, although the men are often crowded together and wading quite deeply.

Another counter to the single erratic and scatter displays of aquatic prey is the charging of certain predatory fish into and through prey fish schools, striking or snapping as they go, then returning to pick up the bits. In fresh waters this may be seen in pike (*Esox lucius*) and trout (*Salmo trutta*), among others. Trout sometimes use this tactic in numbers, charging into a shoal of fish fry or small young from different directions (B. Webster, pers. comm., 1974). Marine predators such as swordfish (*Xiphias gladius*) and marlin (*Makaira indica*) charge into shoals of mackerel, tuna, herring, and other fishes from below. Thresher sharks may swim through such shoals at the surface, thrashing about with the specially developed caudal fin, in addition to sometimes striking at individual prey or encircling a shoal in numbers (Gordon 1977).

Quadrupedal hunters

On land, the single erratic display does not usually have the advanced three-dimensional aspect it has in water or air but it is nevertheless very difficult to counter. As we would expect, tactics have evolved to cope with this. Cheetahs in the Serengeti depend for 90 per cent of their food on gazelles, which they run down. The single erratic display of the gazelles is the healthy cheetah's main problem. Schaller (1972) notes that a cheetah can use its full speed only to draw close to a gazelle; after that it has to follow at a reduced rate in order to follow each twist in the unpredictable path traced by its quarry. Describing unsuccessful hunts he writes:

> Once a cheetah slipped and fell when making a sharp turn after a gazelle, on another occasion the hunt failed because the cheetah was unable to select an individual from the compact herd, and on the third occasion the cheetah lost sight of its prey after it had mingled with others. Twenty three other unsuccessful runs terminated when a cheetah gave up the chase after 200–300 m, particularly after it failed to follow a sharp turn by a gazelle. It seemed exhausted, panting at a rate of some 150 breaths a minute, and seldom attempted another rush for at least half an hour . . .

Another instance:

> '. . . she sees about 15 gazelles at 250 m, approaches slowly to within 150 m, and suddenly runs towards them, the first 130 m moderately fast as she chooses a victim, and then 160 m at full speed after a yearling. She slaps but misses as the gazelle dodges and the hunt is a failure.

All the reasons for gazelles escaping cheetah noted here — and catalogued Chapter 4 — depend on the protean principle, the single erratic display being dominant. This is not to say that the prey always escape, of course, but it does force the cheetah to choose younger animals or ailing ones. For example, another cheetah, chasing about twelve gazelles at moderate speed, selects the smallest and then 'sprints after it, follows three zigzags closely, and after 130 m, catches it in a cloud of dust'.

The cheetah's structural countermeasures against such single erratic displays involve special adaptations to enable it to turn faster and more accurately than can such a large animal normally. In addition to changes in overall body configuration we see strong non-retractile claws for extra grip when turning at speed, and a long, club-shaped tail (0.8 m out of 2.1 m overall) which can be swung as a counterbalance and is a very effective aid in turning (Conway 1972*a*; D. Bygott pers. comm., 1979).

Wild dogs also have difficulty in coping with the defences of gazelles, even though they hunt in numbers. Schaller (1972) notes 'about half of the hunts after adult gazelle ended in failure, sometimes because the dogs lost sight of the prey but more often because they simply gave up the chase after 0.5 to 3.5 km'. Obviously it is not an economic proposition for several animals to expend large amounts of energy trying to capture prey so small that it cannot satisfy their needs. The lion — a stalk-and-charge predator — has only 15 per cent success when a single animal hunts gazelles (Schaller 1972). This increases to 31–33 per cent when two to six lions are involved, but the prey then has to be shared, and Caraco and Wolf (1975) point out that energetic requirements tend to restrict the group size of lions hunting such relatively small prey. Therefore, the use of large numbers to overcome both protean and regular aspects of gazelle escape behaviour is not an energetically efficient practice, and the majority of gazelle hunts are undertaken by lions in pairs or alone.

A single dog finds it almost impossible to cope with the single erratic display of a smaller animal. However, man has used his knowledge of artificial selection to breed a type of dog which can cope with the speed, and to a certain extent the manœuvrability, of rabbits and hares. He has crossed greyhounds and whippets (sometimes salukis, borzois, Afghans, or deerhounds) with various types of sheepdog to produce 'lurchers' which have both speed and stamina. These are the hunting dogs of gypsies and other folk who live close to nature.

'Spring', owned by one of us, was a king among lurchers. An outstanding hare catcher at a year old, he caught and brought to hand nearly a thousand rabbits in his second year, and quicky learned to adapt his hunting techniques to prevailing conditions. His acceleration was so great that any rabbit he decided to pursue — any that he had a good chance of catching before it reached cover — had only one worthwhile weapon at its disposal: the single erratic display. Time and time again — occasionally several times in the same pursuit sequence — a rabbit would jink, Spring would overshoot, and the chase would re-start with the rabbit that much nearer a refuge. Sometimes a rabbit jinked too soon and Spring simply cut the corner. At other times it jinked too late, or not at all, and was seized without difficulty. But a large percentage of encounters involved a single erratic display of one or more jinks, to one side or the other or both, when the rabbit was out of reach by one to six dog-lengths. The advantage then was with the rabbit. But Spring even had a technique for dealing with some of these occurrences. If the rabbit was close enough, Spring would throw his head sideways at the jinking prey, reaching out with his long muzzle on extended neck, but allowing his body to continue on its previous path. Frequently, the result was a spectacular somersault, but more often than not the rabbit would be clamped in Spring's jaws (Fig. 7.5. Thousands of greyhounds and lurchers have this ability, but only man's techniques of artificial selection have made it possible for the dog to overcome the rabbit's single erratic display so successfully.

It seems hardly necessary to add that when the protean principle is

Fig. 7.5. Lurcher dog catching rabbit. A good lurcher has the speed, build, and agility to deal with the single erratic display of rabbits and hares.

consciously employed by man, it is almost impossible to counter. In the civil context, where the authorities are restricted in the escalation of their countermeasures, the protean principle dominates. Contemporary examples include air hijacks, armed civil strife in Northern Ireland and the Middle East, and terrorism generally. During the Persian Gulf crisis of later 1987, the 'unpredictability' of Iranian tactics was noted repeatedly in the media as a significantly problematical factor. In wartime, with regular weaponry becoming ever more destructive and intolerable, military strategy must increasingly involve the utilization of the protean principle.

Summary

Biological explanation is incomplete without broad evolutionary consideration. In addition to the central nervous mechanisms involved in protean displays, special anatomical, morphological, and physiological features often play a part.

Injury-feigning, mobbing, united erratic and single erratic displays, convulsoid behaviours, and immobility displays are considered in the light of the protean thesis, particularly with respect to their evolution.

Various drive conflict theories as to the origins and mechanisms of distraction displays are noted. We suggest that the first stage in the evolution of distraction displays must have been a weakening of the escape drive and a reduction of inhibition of other drives. The actor could thereby remain conspicuous to the predator, and the possibility of evolutionary emancipation of behaviour patterns is provided for. Some conflict in the present probably remains for the generation of variability within and between various displays. A model is presented to suggest how injury-feigning behaviour may be organized neurologically to appear erratic yet remain controlled.

Mobbing displays have also been widely considered in the light of approach–escape conflict, but this can only lead to superficial explanation. The fundamental point is the description of the pattern of behavioural elements employed; then one can ask why this pattern has certain characteristics rather than others. In particular, one needs to ask why these conflict situations produce an increase in the variability of behaviour and a decrease in its predictability.

We suggest that an evolutionary process has occurred which is, in effect, the opposite of ritualization: a process of complexifiction and apparent disorganization which denies to the reactor relevant information. In injury-feigning and single erratic displays this is seen in the variations in the behaviour of individual animals; in mobbing and united erratic displays the variability is a group property. We name this process *anarchization*. With this in mind, an explanation is suggested for the efficacy of mobbing, as seen in snake-mobbing by ground hornbills.

United erratic displays involve remarkable co-ordination within groups. It appears that visual signals have been developed, or enhanced, to assist this co-ordination. In fishes, the flashes as they turn with respect to incident light may operate in this way. In birds it is suggested that the wing bars of waders and the specula of ducks play an important part in flock co-ordination.

Irregular behaviour does not imply a more primitive organization than regular behaviour. It is easier to design a machine to travel in a straight line than to make it behave erratically. This principle is also likely to apply to biological systems and we suggest the development of irregularity-generating mechanisms in the evolution of protean displays.

Where a single erratic display is released by the approach of a predator we suggest that the generation of irregular behaviour is the result of the arousal of the escape drive in the absence of adequate directing stimuli. When such a stimulus is lacking, a high level of escape drive might lower the threshold for the onset of all possible escape movements. The higher the escape tendency the greater the likelihood that fleeing will be erratic.

Consideration is given to the reaction of moths to bats' ultrasonics, as shown particularly in the work of Roeder. As a result of the construction of the moth's tympanic organs, the close approach of a bat using its sonar deprives the moth of any directing stimuli but strongly arouses the escape drive, and the moth's flight becomes wildly erratic.

Erratic bouncing and pronking displays are considered in a broad evolutionary context. It is suggested that anticipatory arousal may have been involved in the formalization of these displays.

Audiogenic seizures in *Peromyscus* can be interpreted in the light of the escape tendency being strongly aroused in the absence of adequate directing stimuli. Laboratory experiments with rats which involve seizures during which the animals may suffer injury or death may be readily understood in this light.

A model of the fox convulsoid display, or 'crazy dance', is presented, showing how it is theoretically possible for such behaviours to be organized systematically while appearing unsystematic.

It is suggested that, neurophysiologically, classical seizure behaviour depends on a progressive reduction of inhibition with a concomitant breakdown in the discrimination of environmental information. The basic single erratic display may even be generated by a similar, though limited, reduction in the inhibition necessary for effective environmental discrimination. Thus could the process of anarchization develop.

Immobility displays function by denying to a predator adequate stimuli for the release of the consummatory act of killing. In higher vertebrates, at least, they must also involve trance-like states which do not permit sensations of pain or fear. Thus, occasionally, a captured immobile prey animal is enabled to escape when the predator's attention is relaxed. It would seem that captured prey animals suffer much less than is generally thought. We name the process *auto-analgesia.*

We emphasize our belief that protean display patterns have a definite relationship to certain neurophysiological mechanisms, this relationship being shaped by the processes of natural selection.

The importance of protean displays is reflected in the evolution of certain categories of antiprotean devices. We consider such developments in aerial plankton feeders, terrestrial insectivores, aquatic predators, and quadrupedal hunters. In the human context, protean tactics are almost impossible to counter.

8
Dispersion and diversity

> 'There is one point connected with individual differences, which is extremely perplexing: I refer to those genera which have been called "protean" or "polymorphic", in which the species present an inordinate amount of variation.'
>
> *Charles Darwin* 1859

> 'The irregular organisation and diversity of the components of a protean display in the individual seems to be paralleled by an analogous phenomenon of irregularity of occurrence and diversity of attributes at the population level.'
>
> *Humphries and Driver* 1970

The success of the usual scientific strategy of concentrating research on those events and processes which are regular and predictable has led to very little attention being paid to irregular events. We ourselves only developed an interest in this field when chance observations initiated the idea that irregular behaviours might occur in a systematic context. Subsequently, the focusing of our attention on phenomena that are erratic, variable, disorderly, and unexpected has enabled us to pick up anarchistic properties arising from, but not immediately apparent in, confrontations of a predator with its prey. We now refer particularly to the behavioural effects of the way animals arrange their distribution within their habitats, especially their pattern of spacing (dispersion), and their intra- and interspecific diversity of appearance and behaviour. We will see later that such considerations will take us from the free-living modes of life into such unlikely topics as endoparasitism, and the intermittent production of offspring.

In our first review of protean phenomena (Driver and Humphries 1966) we pointed out that variations in individuals' repertoires of antipredator behaviour within a species contribute to the generation of uncertainty in the predator, since experience with one prey individual will not make anticipation of the next individual's escape tactics any easier. We commented that a parallel argument might be made to classify certain colour polymorphisms as protean, in the sense that the predator's anticipation of prey appearance on the basis of one encounter would be confounded by experience with a different morph, so hampering the overall rate of predation. In this chapter we shall develop these ideas more extensively, drawing together a number of studies which have been published on certain systematic aspects in this area. In viewing the matter against the background of our study of protean displays we believe that the information takes on additional significance. We also include a consideration of dispersion because, as we shall demonstrate, dispersion patterns are in some ways significantly related to irregular diversity and other protean systems. We stress that in these further descriptions, as in our previous examples, we do not necessarily claim that the phenomena involve anarchization before anything else. Rather, they should be seen as parts of the adaptive mosaic, the protean significance varying from instance to instance.

Searching image and area-concentrated search

A predator's success, or otherwise, in capturing cryptic prey is not solely dependent on the balance of armament between attacking predator and alerted prey — the situation we have primarily considered so far. It depends also on the general strategies of the prey for avoiding detection, on the strategies of the predator for increasing detection frequency, and on the previous experience of each protagonist in detecting and capturing, or escaping, the other. Optimal foraging theories have been surveyed by Krebs (1978).

There is abundant evidence that predators of many types will, after encountering a suitable prey individual, show behaviour patterns which increase the probability that the next prey attacked will be of similar appearance. This is brought about partly by the predator's perceptual discrimination processes temporarily favouring one type rather than another — the phenomenon which gave rise to the term 'searching image' (Tinbergen 1960) — and partly by area-concentrated searching after finding the first prey. (We believe the term 'area-concentrated search' to be more accurate than the 'area-restricted search' of Smith (1971); see also Curio (1976).

Morris (1954) describes how thrushes, *Turdus philomelos* and *T. merula*, after carrying a snail to be broken open on their 'anvils', will often return to resume hunting at the place where the snail was captured. This behaviour would seem to be common in thrushes, at least in *T. philomelos*, and is similar to our observations of oystercatchers, *Haematopus ostralegus* feeding on limpets *Patella* spp., where one bird, or a pair, will repeatedly fly between a partly submerged rock where they obtain the molluscs, and a feeding rock where the shell contents are devoured or fed to the young. Banks (1957) and Dixon (1959) record area-concentrated searching in larvae of coccinellid beetles hunting aphids, as does Chandler (1969) for syrphid fly larvae hunting aphids. The same general phenomenon is recorded for herring larvae by Rosenthal and Hempel (1970) and for three-spined sticklebacks by Thomas (1974).

Croze (1970) noted the importance of area-concentrated searching in his experimental studies of predation by carrion crows, *Corvus corone*. Using artificial 'larvae' made of flour and lard tinted with food colouring, Croze showed, for example, that dense prey populations with an interprey distance of 1 m only had one-tenth as many survivors as well-spaced populations of equivalent numbers in which the interprey distance was 6 m. It follows that area-concentrated searching might be expected to favour the evolution in the prey species of behaviour which leads to a scattered distribution, against which such a strategy would be minimally effective. It would seem that this, in fact, has often occurred, as later examples will indicate.

Nevertheless, prey distribution is not governed solely by predatory pressures; food supply and other circumstances must produce local patches of prey abundance in relation to which the predator's area-concentrated searching is likely to remain effective. An extreme example is the clumping of aphids. This renders them vulnerable to syrphid and coccinellid predators (hover fly larvae, ladybird larvae and adults), but it has the compensating advantages of optimal food supply (easier access to better food at the growing tips of plants) and defence against insect parasitoids which can effectively attack only the outside individuals. Thus, while for many prey species the success of the predator strategy can be reduced by the evolution of behaviour producing a scattered distribution, in some species clumped dispersion will be the norm because of other requirements.

Scattered distribution and the confusion of predators

Area-concentrated searching remains an effective strategy to the extent that many prey populations do not maintain an ideal scatter. Furthermore, evidence is accumulating that predators alter their behaviour to suit the particular prey distribution they have encountered. For example, Driver (1960) noted that young eider ducklings leave tundra pools when their prey (insect and crustacean larvae) fall below a certain density, and walk to the next pool. Rosenthal and Hempel (1970) found that when prey become scarce, herring larvae swim faster and in a much less meandering path than that employed in the 'search swimming' associated with denser prey populations. And in a detailed study of the searching behaviour of blackbirds, *Turdus merula*, Smith (1971) has shown that the birds' searching pattern changes in relation to the distribution of the prey. The searching behaviour is different for sparse populations of prey compared with clumped, regularly spaced, or randomly scattered ones. For clumped and randomly scattered prey populations, the birds showed changes in turning behaviour after a capture, which resulted in them remaining in the vicinity. This is an adaptive strategy for clumped dispersions but not particularly advantageous for a perfectly random dispersion. Adoption of the strategy may, however, be adaptive in natural field situations where exactly random distributions do not occur and a degree of clumping is likely. Even small departures from randomness would give advantage to changed turning behaviour.

The effectiveness of scattered distribution as a countermeasure to predators' searching behaviour also depends on the predator having a short 'direct detection distance' relative to the interprey distance; in other words, the average distance at which the predator's perceptual systems can just recognize the prey object against its natural background must be much less than the distance between prey individuals. Hence the vital importance of crypticity in this context. Indeed, it is usually those species which rely on camouflage to hinder perception by predators which show the special behaviour patterns that ensure that individuals will be widely spaced in the available habitat. For example, young spiders, peppered moth larvae, and many other arthropods have developed mechanisms for being scattered by the wind. Spiders, for example, may climb to an eminence, spin a thread into the wind, and release their hold.

Another example, the Camberwell beauty butterfly, *Vanessa antiopa*, lays its eggs in large groups, the caterpillars later living in protective clusters with an aposematic colour scheme of black with white and reddish markings. However, just before the last moult the nature of the protection against predation changes. The fully grown caterpillars disperse from the willow, birch, or elm on which they have been feeding and scatter over half an acre or more before metamorphosis to the cryptic pupal stage. Here, at a set point in the life history, crypticity and scattering behaviour appear together in functional harmony. This is not uncommon in butterflies.

In many other Lepidoptera the larvae are cryptic and their scattered distribution arises in the first place from the pattern of egg laying. This is seen in the poplar hawk moth, *Laothoë populi*, the females of which fly some distance after laying each egg, even when kept in captivity. The scattering of the brood in this way is found in several other sphingid moths and in other lepidopterous families. Neotropical nymphalids of the Genus *Euptychia* provide good examples. *E. renata* and *E. hermes* glue their eggs to dead leaves or twigs or some other plant some distance from the larval food plant. *E. penelope* allows its eggs to drop to the ground beside the larval food plant (Singer *et al.* 1971). Such scatterings will tend to avoid the competition for food which may occur when many larvae capable of growing to a large size all arise from a single clump of eggs. They will also act as protection against predators, and against parasitoids such as chalcid and ichneumon wasps which oviposit in the prey eggs or larvae. A similar range of egg distribution behaviour is seen in the highly cryptic stick insects and leaf insects of the Order Phasmida. The wide dispersion of eggs in some of these relatively immobile species is achieved by hurling them to a distance as they are laid.

Brower (1958) has suggested that polyphagy in herbivorous insects may develop and be maintained by the less effective predation resulting from the prey's scattered distribution on several plant species. This is an attractive idea. Herbivorous insects which specialize on a single plant species — often on a particular structure of that plant — are particularly vulnerable. Because the host species is likely to have a more clumped and predictable distribution than would an assortment of different plant species, predators such as insectivorous birds may concentrate their search on the easily recognized structures bearing the whole of the prey population.

In a study of the composition of the diet of nestling great tits, *Parus major*, in mixed woodland Royama (1970) showed that the birds appeared to specialize in searching those patches of the habitat which give the best yield. The earlier work of Gibb (1966) had shown that a habitat patch might be as small as a pine cone. If the density of the eucosmid moth, *Ernarmonia conicolana*, in the cones was high enough, blue tits, *Parus caeruleus*, concentrated their foraging upon the cones. At lower densities — below 10 per 100 cones — the tits sampled the cones occasionally but did not specialize on them.

Spacing-out of cryptic species is also seen in vertebrates. The fish, *Vimba vimba*, spawns in compact masses if there are no predators, but scatters the spawn when predators are present (Tinbergen *et al.* 1967; Croze 1970). Tinbergen (1953) suggested that the spacing-out of nests by herring gulls, *Larus argentatus*, may reduce chances of discovery of the camouflaged young by predators, and has demonstrated the validity of this suggestion experimentally (Tinbergen *et al.* 1967).

Crook (1965) describes a relationship between exposure to predation and spacing-out of nests in certain euplectine weaver birds. He notes

> These pack their territories into areas of grassland outside which they forage in flocks for grass seeds and insects. Species with the largest territories nest in

sparse vegetation, those in smaller "neighbourhoods" in denser vegetation and the most congregated in crowded reed clumps surrounded by water.

Lack (1968) illustrates the striking convergence between African Ploceinae (which includes *Euplectes*) and New World Icteridae (blackbirds etc.) in this relationship between predation and nest dispersion.

Clearly, there is a widespread connection between territoriality and reduction of predation in cryptic species of birds. We suspect that further work may also establish this connection in arthropods, for example butterflies such as the speckled wood, *Pararge aegeria* (Davies 1978), and mantids (Mackinnon 1970).

Central to all this is the point that a successful predator, by its discovery of one prey object, is enabled to adopt patterns of behaviour which increase the probability of further kills. In effect these behaviour patterns are equivalent to predictions: first, that suitable prey is close at hand, and second, that it has a particular appearance. The predator acts as if suitable prey densities will be associated with habitat patches of a certain type; it acts as if suitable prey will be found close to the first capture point; and, as we describe later, it seems to seek prey of similar appearance to early captures.

Any device reducing the predictability of the placement of prey individuals within the habitat is by definition a protean device. It should be noted that under natural conditions dispersion can rarely be regular and orderly and, even in non-random patterns as seen in the dispersions produced by territoriality, the known position of one individual gives relatively little information about the position of others.

A scattered distribution might be thought to become somewhat less effective when completely regular (an eventuality which would almost certainly arise experimentally rather than naturally) but it is important to recognize that it will only become less effective if the predator is able to discover a large enough proportion of prey individuals to reveal the regularities behind their spacing — and this is impossible in widely dispersed populations with good individual camouflage.

In Croze's experiments (1970) artificial prey objects were laid out in consistent geometrical patterns, the main reason for not investigating irregular dispersions being the difficulty of finding 'surviving' prey objects even in regular patterns! Similarly, the effect on the predator, which had no information as to the overall pattern, was the defeat of its predictive abilities. As in other protean devices the essential point is that the common features potentially assisting prediction of successive events — in this case the causal relationships between positions of vulnerable prey — are hidden from the predator, which is given nothing systematic upon which to base its actions.

The role of spacing out as a protective device is seen too in the scatter hoarding which has been noted in several animal species, including the thick-billed nutcracker, *Nucifraga caryocatactes* (Swanberg 1951). For two or three months of autumn these birds store nuts of hazel (*Corylus avellana*) or pine (*Pinus cembra*) in the ground in the spruce forests where they have their territories. The nuts are buried in the ground in small heaps of varying sizes. Not only do the birds find the nuts again in winter with snow on the ground, but there is often enough for the feeding of the young next spring: 80 to 90 per cent of the stores are successfully excavated by the owners, which apparently have an excellent memory of their location. Yet there is no systematic pattern in the positioning of the stores. It is as if the nutcracker 'as the farmer says, takes the clouds as a mark' when it hides its nuts (von Linné, known as Linnaeus 1748). Recent work with Clarke's nutcracker *N. columbiana* suggests that the birds do utilize local landmarks. Even so, if there is effective randomness in the positioning of the stores, a robber, finding one, would have no clues as to the location of others.

Certain cryptically-coloured species have a clumped dispersion which, on superficial consideration, might seem to be maladaptive if matched against predators employing searching image behaviour and area-concentrated search. But closer examination of these clumped cryptic species shows that there are adaptive reasons for the clumping, and that protean defences, notably scatter displays, often help to mitigate the effect of a predator discovering a prey group.

In some instances, such as family parties of nidifugous birds and many primate social groups, the survival value of grouping arises particularly from the fact that inexperienced and more vulnerable members may benefit from older individuals' superior foraging skills and greater powers of detecting and avoiding danger. The very formation of a group may permit the employment of defences which would otherwise be impossible, and these group defences are often protean — as in the united erratic display of waders and ducks, the scatter display of young nidifugous birds, mobbing by colonial sea birds, and distraction displays by bustard groups or patas monkeys — to give but a few of many examples. Whether the creation of such defensive displays is itself ever sufficient to account for the development of clumped dispersions is uncertain; it seems more probable that in most cases the displays have developed as defensive consequences of groupings which have arisen as a result of other causes.

Clumped dispersions in aquatic prey

In many species, particularly aquatic forms, clumped dispersions appear to have the primary function of reducing predation by increasing the effective interprey distance. The marine copepod, *Calanus*, for example, one of the most abundant genera of animals in the world and an important food of many fishes, typically occurs in a very patchy distribution. Frogman observations show that there are often series of widely separated groups, each consisting of a dozen or so individuals maintaining close proximity (Marshall and Orr 1955). In such a markedly clumped dispersion a predatory fish after encountering one group must search for much further than the average interprey distance before finding another group. This is of particular advantage if the average intergroup distance is considerably greater than the visual range of the predator but the average interprey distance is less than that visual range. The advantage of grouping by the prey might be thought to be cancelled out by the fact that the predator, although discovering prey less frequently, now discovers not one individual but several. However, the scatter display overcomes this objection by taking most individuals out of visual range by the time that one or two have been eaten.

Calanus individuals when disturbed scatter vigorously from the group by horizontal or oblique leaps of several feet, changing direction at each leap in a very similar escape movement to that of the littoral prawns. In *Calanus* the speed and length of these leaps are very much greater than in the small, jerky movements of normal locomotion. This is achieved by means of the synchronous activation of the dorsal longitudinal muscles and those of the antennules and eight swimming legs. Synchrony is brought about by common fast-conducting innervation by branches of the same giant nerve fibre. This arrangement promotes a rapid scattering of the group and also produces an effective single erratic display for each individual. It is so specialized that, clearly, there must have been considerable selection pressure involved in its shaping.

Brock and Riffenburgh (1960) reach similar conclusions regarding the schooling of fishes, though their arguments apply only to the advantage of the clumped dispersion as a device against visually searching predators; they do not discuss the value of scatter displays in enhancing this advantage. Olson (1964) has made similar suggestions on the basis of a theory of search developed in relation to antisubmarine warfare.

The advantage of clumped dispersion increases as the probability of detection of uniformly distributed prey increases; thus at night, as the possibilities of visual detection fall, schools often disperse, the dispersal itself presumably being brought about by the breakdown of visual co-ordination between the schooling individuals.

Clumping may be an effective antipredator strategy in most pelagic animals but not in palatable terrestrial species because the factors determining whether or not the predator detects a prey individual in open water are different from those pertaining to terrestrial, or bottom-living aquatic, cryptic prey. This point does not appear to have been noted before. The direct detection distance for an individual predator and its prey species is an average value for those distances over which a capture is initiated. Sometimes on land the predator will detect a prey at a relatively great distance; at other times it may almost fall over the prey before discovering it. On other occasions it may not detect the prey at all, even at close quarters.

Two factors are involved in determining the likelihood of prey detection. First, the nearer the predator is, the more detailed is its visual analysis of any given area of the habitat and consequently the greater its chances of success. Second, the more effective the prey's immediate camouflage, the more difficult the predator's task of visual analysis and the greater the prey's chance of avoiding detection. Now, in most terrestrial situations the habitat is structurally and perceptually extremely complex; therefore crypsis is a highly effective device. In these circumstances a substantial proportion of the prey will be likely to escape detection even when the predator is much closer than the direct detection distance. Area-restricted search would therefore militate against clumped dispersion and favour over-dispersions. In open water however there are serious limits to the effectiveness of cryptic devices such as transparency in smaller invertebrates (Hardy 1956), or the special ways in which fishes reflect incident light (Denton and Nicol 1966).

Visually hunting aquatic predators are consequently likely to detect almost any prey individual which they approach closely enough in open water. If prey density is such that a random dispersion would tend to place at least one individual within the predator's visual direct detection range from the point at which it made its previous kill, then random dispersion would enable the predator to move more directly and efficiently from one prey individual to the next. In such circumstances clumped dispersion, with groups separated by many times the direct detection distance, will lead to waste of time and effort as the predator cruises 'out of touch' with the prey population. Schooling or grouping is thus adaptive as such in open-water prey, and the scatter display which it makes possible is an additional defensive bonus. It is worth noting that although this clumping strategy cannot be overcome by significantly increasing the visual range of the individual predator, some predatory fish such as tuna indirectly increase the area of visual sweep by forming a widely-spaced school within which individuals detecting prey indicate the fact to others by producing bar-like patterns on their flanks (Magnuson and Prescott 1966).

Some processes mediating searching image formation

So far we have primarily discussed the relationship between area-concentrated search and the changing effectiveness of crypticity with distance on the one hand, and prey dispersion and defences on the other. Against this background we now examine searching image behaviour in more detail and then suggest some of the defences which the prey may develop against it, particularly those which possess protean properties.

It is important to realize that 'searching image' is a concept that arose from ecological studies and, consequently, is defined ecologically. Some authors, notably Hinde (1970) and Dawkins (1971*a*) have questioned the usefulness of the term on the grounds that it has been used in slightly different senses by various authors. No alternative terms are proposed, however, and as the term has been consistently applied ecologically to a distinctive and important phenomenon, we see no grounds for abandoning it, though other phenomena should be carefully excluded from its terms of reference. There is, of course, no currently available technique for discovering whether or not a non-human animal has anything resembling an 'image' in its 'mind' at all, but this is not the problem.

The term 'searching image' gained its current usage from the work of L. Tinbergen (1960) who noted that the proportions of different species of insect larvae in food brought to nestlings by tits (*Parus* spp.) were not consistently related to the abundance of each species in the surrounding pine wood. When an insect species first appears in the habitat its risk is unexpectedly low; then, as its numbers build, risk quickly increases beyond what would be expected if the birds were collecting larvae at random. The timing of increased risk differed for individual tits. To explain these results Tinbergen assumed that as prey density increases beyond a certain point, each individual tit undergoes learning processes reinforced by repeated captures of palatable prey, and he termed this the formation of 'specific searching images'. In sparse populations, contacts with the prey are too infrequent for reinforcement to maintain the response. Searching image is thus defineable as an increase in risk to a specific category of prey from an individual predator.

Several subsequent studies have contributed evidence for searching image behaviour in various vertebrate predators, and we note Allen (1972*a*, *b*, 1974, 1975), Allen and Clarke (1968), Beukema (1968), Cook and Miller (1977), Croze (1970), Gibb (1962), Horsley (1978), Manley *et al.* (1972), Maskell *et al.* (1977), Mook *et al.* (1960), and Owen (1965, 1969). Holling's investigations (1965, 1966) suggest that learning processes related to prey density will affect prey risk in vertebrates generally, and are borne out by Horsley's studies (1978) in which wild passerines predate artificial baits. Beukema's work (1968) on stickleback predation is useful in that it shows that searching image formation enables the predator to detect the prey at increasingly greater distances up to a limit at which the average direct detection distance stabilizes. Searching image behaviour is consequently not merely a switch from one type of prey preference to another — it is accompanied by a striking increase in the effectiveness with which the new type of prey is detected. And it is this latter property which may well be the main advantage the predator derives from the behaviour. For instance, de Ruiter (1952) showed that for the jay, *Garrulus glandarius*, a single chance discovery of the very effectively camouflaged stick caterpillars of the moth, *Hyloicus pinastri*, dramatically increases discovery rate. Dawkins (1971*a*, b) provides evidence that, in domestic chicks feeding on cryptic versus conspicuous grain, central perceptual changes underly this searching image behaviour.

At the perceptual level a process similar to that experienced by the collector 'getting his eye in' may be operating, so that perceptual isolation of the camouflaged object from its surroundings begins to occur more quickly or at greater distances. Alternatively, or additionally, responsiveness to certain learned key features of the prey may be raised, so that attack is elicited not only to the prey type itself but also to objects which closely resemble it. There is experimental and field evidence for this in birds, fishes, and primates (Meesters 1940–41; de Ruiter 1953; Morgan 1956; Beukema 1968; Robinson 1973). But whatever the perceptual or neurophysiological explanations of searching image behaviour may be, the ecological effects are not in question, and we will therefore now move on to discuss some of the ways in which prey animals may counter this predatory strategy.

Adaptations which counter searching image behaviour

Change of colour and form

A common protean device against searching image behaviour is the apparent change of colour and form. This tends not only to protect the displaying individual but also may be expected to hinder the learning of the prey's appearance by the predator. We have documented this behaviour in some detail in Chapters 4 and 5 and therefore pass on to other protective devices which possess protean properties.

Dispersions

The wide dispersions which act so effectively against area-concentrated search are also likely to hinder searching image formation. The data of

L. Tinbergen (1960) on capture rates of *Panolis* and *Acantholyda* larvae indicate that occasional capture of a prey type does not lead to searching image formation until population density increases to a level such that captures can occur in sufficiently rapid succession. In the context of area-concentrated search, clumped dispersions would therefore be expected to lead to searching image behaviour appearing at overall prey population densities much lower than those at which scattered or over-dispersion would enable the behaviour to develop.

Cryptic fragility

Another device which hinders the development of efficient searching image hunting is cryptic fragility. Blest (1963) pointed out that, whereas aposematic insects are generally tough and heavily sclerotized, cryptic insects are more delicately built so that, for example, upon capture a cryptic moth's wings rapidly lose their scales and thereby their colouration. Blest suggests that the function of this fragility may be to shorten the time available for the predator to learn exactly what the normal prey looks like. Conversely, in aposematic distasteful species it is to the advantage of the prey population for the predator to have extended opportunities to learn the appearance of the prey. Thus we see that the tough construction of such moths not only assists the predator's learning, but also gives the individual being sampled a better chance of survival, particularly once its distasteful nature has been recognized. Individuals of cryptic species would also stand a better chance of survival if they were toughly constructed. The fact that so many are not suggests a strong advantage for the role of fragility in hindering learning by predators. Such fragility has presumably evolved through kin selection.

The cryptic fragility of moths is very reminiscent of the fragility of many brittlestars and scale worms of the littoral zone in which, we suggested, a similar process is operating. The individual distractive distintegration of the more specialized species of both these groups, where inessential organs are cast off conspicuously — in some cases luminescing, as well as moving independently — but where enough of the body is retained for survival and regeneration, may perhaps have evolved on the basis of the initial adaptation of cryptic fragility which is so common in these groups. The distraction aspect has presumably evolved through its enhancement of individual survival while still hindering the predator's learning of prey appearance.

Our suggestions here seem to be supported by the work of Lindahl (1979) who studied the mosquitoes of tree boles. As a general principle he suggests that 'selection for increased handling time could occur if there is a chance for escape without undue injury after the initial capture.'

Polymorphism

A further countermeasure against the prediction implicit in the searching image strategy is provided by certain colour polymorphisms which render the appearance of various prey individuals sharply different within a species. We hope to show that there are good reasons for believing that the anarchistic changes of aspect presented by differently-coloured individuals — a protean phenomenon — cannot be coped with effectively by a predator's searching image behaviour.

There are many examples of intraspecific polymorphisms in which a biological function of a quite regular nature can operate. The work on these is reviewed by Ford (1964, 1965), Wickler (1968), and Owen (1980). Examples include the mimicry of various distasteful species by certain butterflies and other insects, and seasonally varying green or brown cryptic adaptations of certain caterpillars living on deciduous trees. The ramifications and complexities of the various kinds of polymorphic mimicry are much too extensive, confused, and confusing for us to deal with here. But there are other colour polymorphisms which we would like to consider.

Many of the caterpillars and leaf-hoppers which live on the foliage of bushes and trees have a number of different colour forms which can be found on the same part of the tree at the same time of the year. The caterpillars of the sawflies, *Arge ustulata* and *A. fuscipes* (Hymenoptera) feed at the edges of leaves and, in Britain at least, occur in two colour forms, both feeding on birch. One form is symmetrically pale green, but the other is asymmetrically pale green/dark green, the left side being darker and always being uppermost as the caterpillar sits on the leaf edge (Claridge and Edington 1965). This provides the animal with highly effective crypsis, the birch leaves being much darker above than below. Searching image behaviour will protect whichever morph the predator is not seeking. Doubtless there are many similar species which have escaped the attention of biologists because their attention-escaping adaptations are so effective!

There are numerous instances in the Lepidoptera. The caterpillars of the mottled umber moth, *Erannis defolaria*, for example, may be basically black, yellow, or brown; they are also protected by their occurrence on a variety of trees.

The hawkmoths (Sphingidae) provide many examples of species with larvae which are polymorphic for colour. In the convolvulus hawk moth, *Herse convolvuli*, there are several such morphs (Poulton 1888; Bell and Scott 1937; Pinhey 1960; Edmunds 1974). In Africa one food plant is *Ipomaea aquatica* which has yellow-green stems tinged with purple. Third and fourth instar moth larvae are green, or yellow with purple stripes. The green caterpillars rest on the undersides of the green leaves, while the yellow and purple caterpillars rest on the stems. The final instar larvae are all brown and hide in the earth during the day. Thus, while colour pattern and behaviour combine to give crypsis, the variety of form will hinder searching image development in predators.

A similarly complex example in which polymorphism is probably functional against predators is the New World sphingid, *Errinyis ello*. In the Galapagos the young larvae are green and rest on green leaves. Late instars are green, brown or grey, the last two normally resting on brown twigs (Curio 1965). In Jamaica the last instar has at least four morphs: green, green/grey, blue, and brown. The first three feed and rest on the leaves of *Poinsettia pulcherrima*, where they are cryptic. The brown morph rests during the day on the trunk, where it is cryptic, only feeding at night. Although larvae on the leaves are at a great selective disadvantage (ca. 88 per cent) against predation by the wasp, *Polistes crinitus*, the larvae on the trunk have to cope with lizards, for example, *Anolis lineatopus* (Curio 1970*a*–*c*). Insectivorous birds are potential predators of all. In such a situation, polymorphism could well function as a hindrance to searching image development, even though *Polistes* do not hunt visually, for other senses could mediate searching image behaviour. For example, Soane and Clarke (1973) trained house mice, *Mus musculus*, to take either peppermint- or vanilla-scented bait. As Edmunds (1974) concludes, 'If prey whose predators hunt by smell are common, then selection should favour the evolution of scent polymorphisms, just as it favours colour polymorphisms in species whose predators hunt by sight'. Here is another field with considerable research potential.

Intraspecific diversity is well known in a number of insects and molluscs living on or among herbaceous flowering plants. The meadow spittlebug, *Philaenus spumarius*, exists in about twelve colour forms (Owen and Wiegert 1962; Hutchinson 1963) which often occur together on the same food plant (Whittaker 1968). Most of the colour morphs are cryptic, and the polymorphism may be explained by predators developing a searching image for the commoner morphs, giving the rarer morphs a selective advantage. The form *marginella*, found in females only, is conspicuously black and cream. Thompson (1973) suggests that predators learn to avoid it through association of its colour pattern with lack of reward (because of its leaping ability). This could effectively protect the population as long as other morphs are present as potential prey.

Several species of land snails show striking colour polymorphisms. *Cepaea nemoralis* and *C. hortensis* provide well known examples (Ford 1964, 1965, and further referenes in this section). Both are common in western Europe, and have been since Neolithic times at least (Diver 1929).

> They are easily recognised since the shell is yellow, pink or brown, rarely lilac or other colours, and is striped with any number of bands up to five (very rarely more) any of which may fuse with adjacent ones, and any (including all) may be absent. It is very rare indeed for a colony of either species to include only one banding-pattern; usually a large number are present, that is, each colony is highly polymorphic. Colonies containing both species are not uncommon . . . (Cain 1963).

Sheppard (1967) illustrates some of the more distinctive morphs of *C. nemoralis*, summarizes the genetics of the polymorphism, and demonstrates a correlation between shell colour and background features of the habitat. It is interesting that, in Britain at least — where insolation is not a critical factor — the polymorphism is maintained against all backgrounds, although some morphs appear much more cryptic than others (Fig. 8.1).

We are loath to add to the reasons suggested for *Cepaea* polymorphism (see Jones, *et al.* 1977) but there is so much anarchization in the appearance of *Cepaea* populations, that predator searching image formation is bound to be affected. We hope that this suggestion will simplify rather than complicate the matter. It does, in fact, support the early suggestions of Cain and Sheppard (1954) and Haldane (1955) that *Cepaea* polymorphism could be maintained by the uncommon forms being at an advantage because of their rarity, and the common ones being at a disadvantage because of their commonness. Searching image confusion could be the mechanism whereby this comes about.

Owen (1965, 1969) studied the colour polymorphisms of the African land snail, *Limicolaria martensiana*. Typically, the shell is streaked, but there are also three different 'pallid' forms. Owen found that the population density increased significantly with increasing numbers of morphs present. Thus, when the whole population was streaked, the population density was less than $1/m^2$. When the population contained all four morphs the total density was greater than $100/m^2$.

Heterosematism

Cain (1983) studied the differing colour patterns of the upper and under sides of snails. He points out that 'In many species of land snails with globular, trochoid or depressed shells, the colour pattern is markedly different above and below the periphery of the shell'. He gives many examples, particularly from von Martens (1867), and notes that in Europe at least, plain-coloured species are mainly noctur-

Fig. 8.1. Some morphs of *Cepaea*. Four left specimens *C. hortensis*. Five right specimens *C. nemoralis*. Shells are yellow, pink, or brown, with 0 to 5 dark bands in almost any combination. Note that the harvestmen (centre) and the carpet moth (top centre) are well camouflaged, but the different snail morphs vary in this respect.

nal or leaf-litter dwellers, while the patterned ones sit out, at least occasionally, in the daytime (Cain 1977). Furthermore, species exposed to strong sunlight and protected against overheating by being largely or wholly white above are frequently banded below, sometimes polymorphically. In *Cepaea nemoralis* plain yellow, pink, and brown morphs are well known, as are morphs with a single peripheral dark band on one of these three backgrounds. All of these morphs appear plain from above, but from below the contrasting lip, callus, and mantle are conspicuous. In banded forms all combinations of the five possible pigment bands are known, and morphs with bands below the periphery only are common, but morphs with bands above and not below the periphery are uncommon. Thus forms with bands 00345, 00(345), 00:45, 00:(45), and 00045 (parentheses indicate fusion) are common, while 12300 is rare.

Cain supposed that such differential colouration could be an adapation evolved to counter a predator's searching image behaviour when a snail has been dislodged and falls to the ground. He carried out tests to establish which side comes to rest uppermost after the animal is dropped and found equal or subequal tendencies to show upper and under sides. This result supports his supposition. He notes that 'a random fall producing one of two markedly different images in an overall ratio of 1:1 may be the most effective strategy as in the game "which hand is it in?" (Driver and Humphries 1970).' He calls this phenomenon *heterosematic* (signalling different things). Here we have an anarchistic visual phenomenon which will have a protean effect upon searching image formation whether the prey population is polymorphic, in the usual sense, or not. All will be protean in terms of the incidence of any particular colour pattern in any particular encounter with a predator.

An effectively similar phenomenon had already been suggested by Arthur H. Clarke Jr. (1965) after studying the scallop superspecies, *Aequipecten irradians*, from the eastern American seaboard. These bivalves vary considerably in the incidence of white right (lower) valves in different populations (from 17 to 100 per cent incidence recorded). This could have survival value in the animals' escape behaviour which consists of an erratic 'flight' through the water, propelled by the snapping shell valves. Clarke (op. cit.) notes that the scallops 'would then present a strikingly different appearance than when at rest. A phenomenon similar to apostatic selection . . . might occur but on an individual basis and to a predator with functional vision the escaping scallop would not look like the same desirable prey which had been at rest'.

Frequently, effects other than the disruption of predators' searching image behaviour can be adduced as explanations for polymorphisms. To refer only to *Cepaea*: in the Pyrenees, unbanded snails may be more common on the hot hillsides than in the valleys because they are more resistant to heat (Arnold 1969). In parts of Wales, brown snails are more preyed upon by glow-worm beetles *Lampyris noctiluca* than are yellow ones (O'Donald 1968). And there are considerable behavioural differences between the different morphs (Wolda 1963, 1965, 1967) which do not seem to be directly associated with degree of predator success. Even so, whatever the factor or factors causally associated with polymorphism in *Cepaea*, they could all assist in maintaining that polymorphism and thereby help to protect the animal against visually-hunting predators. Mosaic evolution, we must remember, does not only mean that 'Each organ and each system of organs has its own rate and pattern of evolution' (Mayr 1950). This variation must also apply to units and patterns of behaviour and to the agents which influence them.

Apostatic polymorphism and reflexive selection

On the sea shore the flat-topped periwinkle, *Littorina littoralis*, which occurs in large numbers on brown seaweeds such as the *Fucus* spp. and *Ascophyllum nodosum*, shows about a dozen co-existing colour varieties (Fretter and Graham 1962). During the past twenty years or so several biologists, including ourselves, have arrived at a possible explanation for this intraspecific diversity, an explanation which depends on the action of the polymorphism against the searching behaviour of the animal's predators. We give here some detail of the background against which the explanation has arisen, for the idea is still not widely understood — even among biologists — and we believe that in the context of protean theory it takes on new significance.

As with so many other ideas in biology the first appearance of the idea in print was ignored because of the immaturity of theory relating to the ecological, genetical, and behavioural context. The beginnings go back to 1890 when Poulton pointed out that species which show colour polymorphisms may be at an advantage relative to monomorphic species, for 'the foes have a wider range of objects for which they may mistake the prey and the search must occupy more time for equivalent results than in the case of other species which are not dimorphic'.

The next important development was in population genetics. Cain and Sheppard (1954) and Haldane (1955) suggested the possibility that polymorphism in the colour and banding of the shell of *Cepaea* might be maintained by predators taking disproportionately larger numbers of the commoner morphs, so progressively decreasing percentage mortality among the least common morphs until their representation in the population as a whole ceases to fall. These authors did not develop the idea, or even consider it very likely because, at that time, little direct evidence was available. But B.C. Clarke (1962*a*,*b*, 1964, 1969) considerably developed the idea, having noted the work of Reighard (1908) who had published clear evidence of searching image behaviour in the grey snapper fish, *Lutianus griseus*. Clark re-analysed Popham's data (1941) concerning the role of crypsis in reducing predation by fish (rudd, *Leucisus erophthalmus*) on the various colour morphs of the water boatman *Sigara distincta* (Insecta; Order: Hemiptera), and showed that Popham's figures indeed revealed that uncommon morphs were at an advantage. And as Fisher (1930) showed, polymorphism will be maintained in any population where rare genotypes have a selective advantage over common ones simply because of their rarity.

B.C. Clarke (1962*a*) coined the term 'apostatic polymophism' to cover such cases because 'the crucial thing about them is the selective advantage of phenotypes that stand out from the norm', in other words 'the selective advantage of apostates.' As Paulson (1973) put it, apostatic selection is 'selection for variation for its own sake in which a given phenotype is favored in direct proportion to its rarity through frequency-dependent predator pressure'. An interesting mathematical model of apostatic selection has been developed by Elton and Greenwood (1970), though it seems to us that much more field data are required before meaningful quantitative predictions can be made.

At the same time that the theory of apostatic selection was being developed genetically, other biologists were coming to similar conclusions using different examples and different approaches. For many years we ourselves had been studying the general ecology of the littoral zone of Ynys Enlli (Bardsey Island), N. Wales, and had been struck by the fact that several littoral molluscs — as well as other invertebrates — are polymorphic, and that some morphs stand out much more than others, at least to human eyes. This is particularly noticeable in *Littorina littoralis* (=*L. obtusata*) which lives largely in the mid-shore zone on brown seaweeds, particularly *Fucus* and *Ascophyllum* spp. Smith (1976), studying this species in Great Britain and Iceland, has found that the various morphs are cryptically favoured on different algae in different seasons. He did not, however, find that apostatic selection was important in promoting the colour diversity. A.H. Clarke (1978) suggests that the mechanisms responsible for maintaining colour morph variability impact more strongly on tropical than on temperate marine molluscs, and gives numerous examples of intra- and interspecific polymorphism in tropical shallow-water marine molluscs. In the wider context Dobzhanski (1950) noted the greater evolutionary challenge posed by the tropical environment, resulting particularly from 'the intricate mutual relationships among the inhabitants'. We should therefore perhaps expect a greater incidence, and variability of both form and function, in tropical polymorphisms. It may be that the anarchization of appearance in temperate littorinids does not always have an apostatic function, but we suspect that it often does. Certainly, in many colour polymorphisms, including *L. littoralis*, the species is basically cryptic and lives against a background of highly varied colouration and conformation — a situation ripe for the operation of apostatic selection.

The brittlestar, *Ophiopholis aculeata*, found in the north Atlantic, shows variations between individuals in pattern and colour on the arms and central disc. The ground colour may be red, purple, brown, nearly black, light or deep yellow, orange, or pink. The disc may be uniformly coloured or may have a contrasting border or a series of bold marginal or central spots. Other individuals may be finely speckled. The variations and combinations leave no two individuals alike out of hundreds examined (Moment 1962). The same author describes similar variations in the butterfly clam, *Donax variabilis*, which lives in enormous numbers on beaches from Maryland to Texas, moving with the surf up and down the beach and consequently being exposed to predation by wading birds, particularly various sandpiper species (Scolopacidae). The clam shells may be striped or uniformly coloured — white, yellow, orange, green, pale blue, lavender, deep purple, red, pink, brown, or intermediates. Moment comments that massive diversity of this kind 'is usually dismissed as the result of the free play of meaningless mutation in the absence of selection' but 'the possibility presents itself that it is the variation as such which is adaptive, giving a measure of protection against predators'. He says 'the virtually endless variation makes it impossible for a sandpiper to learn that all "pebbles" of any particular appearance are probably little clams for the very good reason that no two clams have the same appearance'. Moment coined the term 'reflexive selection' to cover this situation for, where variation *per se* is adaptive, the frequency of any particular type is determined by a feedback relationship with all the other types (see also Li 1962).

In support of Moment's ideas, Croze (1970) reports an experimental study of predation by wild carrion crows, *Corvus corone*, in which he examined the possibility that polymorphism *per se* increases prey survival. He placed small pieces of meat beneath mussel shells coloured yellow, black, or red-yellow, and found that, with an interprey distance of 12 m, trimorphic populations suffered less heavy predation than monomorphic ones (survival for each morph ca. 39 per cent in trimorphic populations, and ca. 13 per cent in monomorphic ones). These findings led Croze to comment on the polymorphism of *Cepaea* (see p.220):

> If we admit that in the eyes of a thrush, vole, rabbit or any other predators *Cepaea* is likely to have, a downland could absorb with greater or less efficiency several patterns of *Cepaea* camouflage, then the door is open for a positive visual selection hypothesis of the function of polymorphism.

We note that in the bug *Philaenus* (Order: Hemiptera) where colour varieties are controlled via six major genes, some of these are only expressed in the female (Halkka *et al.* 1966), which is consequently more polymorphic than the male. The same is true for a number of butterfly species (Wickler 1968). If polymorphism *per se* assists survival, it is good biological sense for the reproductively more important female to receive the greater protection.

We may conclude that it is against a highly varied background, with many camouflage possibilities, that colour polymorphisms can best be employed without seriously upsetting crypsis. It is important to recognize, of course, that the advantage of polymorphism, both for the overall population and for each morph, is obtained against a population of predators and over a period of time. It depends on individual predators employing searching image behaviour which in any one sortie against the prey population will produce very high mortality of one morph — usually the most common one.

It is because of searching image behaviour that predators concentrate disproportionately on the more common varieties of their prey. With respect to prey colour variants, at least, this frequency-dependent predation is now well-documented for a wide variety of animals, particularly fishes and birds (see B.C. Clarke 1979). Uncommon morphs are consequently at an advantage; their sparse distribution protects them against both area-concentrated search and searching image behaviour. It follows that ineffectively cryptic patterns which could not survive in monomorphic populations may, by virtue of their wide spacing out, be enabled to survive in the same habitat if they occur as rare morphs in a polymorphic species. We note that in some polymorphic species such as *Cepaea*, and more strikingly in the littoral gastropod, *Littorina littoralis*, some colour varieties are, at least to human eyes, much less effectively camouflaged than others.

A further possibility suggested by B.C. Clarke (1962*a*) is that *Cepaea nemoralis* and *C. hortensis*, which often occupy the same habitat and show a parallel set of colour and pattern polymorphisms, may act as a single unit on the predator. This would certainly seem to be the case with some butterfly models and their mimics (C.A. Clarke and Sheppard 1971) and we believe this principle may be widely applicable. Arthur H. Clarke Jr. (1965) had also suspected as much from his work on scallops. 'Apostatic selection' he wrote 'appears to hold promise as a productive new approach in our attempt to understand the selective value of chromatic polyphasy at the species level, especially within certain brightly colored tropical genera of marine and terrestrial molluscs, and perhaps even at the community level, e.g. within coral reef communities composed of brilliant hued fishes, mollusks etc.'.

Croze's experimental results, like the work of Owen on the African snail, *Limicolaria*, mentioned earlier, suggest that polymorphic populations have a considerable advantage over monomorphic ones, and we may therefore ask why colour polymorphisms — indeed, polymorphisms in general — are not more common as an antipredator strategy. The answer seems to be that they are but we haven't recognized them, basically because of the mosaic nature of adaptation and the consequent difficulty of unravelling the complexities of interaction between organisms and environment. Could it be, for example, that the great range of colour pattern in young shore crabs, *Carcinus maenas*, is a polymorphism effective against searching image behaviour of waders and other avian predators? This variability is much less evident in the less vulnerable adults. Again, B.C. Clarke (1979) makes the point that diversity of scents may go undetected because of our insensitive noses. We suspect that innumerable polymorphisms go undetected because of our insensitive biology!

Also, the advantage may not be as dramatic as Croze's work indicates, for his experiments compared polymorphism and monomorphism in groups of relatively conspicuous prey objects. In natural populations, crypticity of palatable prey, whether monomorphic or polymorphic, will tend to be as high as possible, simply because, other factors being equal, the genotypes of the least cryptic individuals will tend to be eliminated. Polymorphism necessarily implies some loss of overall crypticity, and it may be that in most circumstances the consequences of this loss more than outweigh the advantages of being polymorphic. However, against highly variable backgrounds, the balance may be tipped the other way.

In habitats of highly variable background we suggest that an additional factor may operate in giving advantage to polymorphic crypticity. This is the added behavioural freedom which, theoretically at least, it makes possible. Cott (1940) documented in detail the fact that camouflage is generally accompanied by special behaviour patterns which restrict distribution precisely to matching backgrounds. Individuals of a polymorphic species in a highly variable habitat where their food or other resources occur on various backgrounds have a measure of freedom from this restriction, enabling exploitation of a greater range of resources. If this occurs at the expense of some overall loss of crypticity, the loss is likely to be compensated for by the advantages accruing from the polymorphism. This suggestion is supported by the work of Sargent (1978) who studied the North American deciduous woodland moths of the genus *Catocala*. There may be thirty or forty sympatric species of this genus, all with cryptic, bark-like forewings. The hind wings are brightly coloured but concealed when

Fig. 8.2. Ruffs with reeve, showing four of the male morphs.

the animal is at rest. Sargent notes that the fore wings vary considerably within a species and that this presumably reduces the efficacy of utilization of searching images by predators.

Ruffs, anarchization, and evolutionarily stable strategies

We now turn to an extreme example of polymorphism which has puzzled naturalists for centuries: that of the ruff, *Philomachus pugnax*, a medium-sized wading bird of the sandpiper family, which breeds in marshes, meadows, and tundra of the Old World. The ruff is the most polymorphic of birds. It is the only species with marked polymorphism in the male breeding plumage, and all male morphs are so different from the female that she has the unusual distinction of a separate name — the reeve (Fig. 8.2).

The ruff is one of the species of birds which display at a 'lek' — an annual gathering in which the males perform together with distinctive plumages and movements, both of which are particularly well-developed in the ruff. Armstrong (1947), Gilliard (1962), Snow (1963), Lack (1968), Selander (1972), and Wiley (1974) are informative on this subject. The lek displays of ruffs have been described by a number of observers including Turner (1920) and van Rhijn (1973). Like that of other lek species the displays seem to have been developed for conspicuousness. The birds display in the open, sometimes in large numbers, at any time of the day, and may perform at the lek over a three-month period. The resident birds carry out vigorous mock fights, each from its own stand or 'hill' where the ground is worn bare by the repeated trampling. They vibrate like toys on wires or rush around like clockwork mice, with interposed immobile posturings and single or communal flights. All of the twenty-two recorded action patterns are performed in silence except for the sounds of collisions between jousting birds and feather rustling.

Current thinking on ruff polymorphism and lek behaviour follows recent developments in game theory and the concept of the evolutionarily stable strategy (ESS). An ESS is a strategy or mixture of strategies which cannot be replaced by any alternative; as long as it is adopted by most members of a population no other strategy will give greater reproductive fitness (Maynard Smith 1974; Dawkins and Krebs 1978; Parker 1978). Ruff polymorphism is thought by some to be a frequency-dependent mixed ESS with the reproductive pay-offs to central and satellite males being effectively equal over their lifetimes (van Rhijn 1983). This may be so, but it should be noted that the ESS concept is still in its infancy and is very largely theoretical. As Maynard Smith — the principal authority in the subject — puts it (1979) 'There are many cases which it is tempting to interpret as mixed ESSs, but in which crucial data are missing'. And he includes the ruff among his examples. (Avery 1984) notes 'The two processes which are thought to shape lekking behaviour, female choice and male competition, are not understood for a single lekking species'.

On the other hand Gadgil (1972) argued that some such cases could be stable genetic polymorphisms. And van Rhijn (1983) has developed a theory for the evolution of ruff polymorphism involving reciprocal altruism (Trivers 1971). Van Rhijn suggests that satellite males benefit by sneaking copulations from resident males, which in turn benefit from the satellites attracting females to the lek. He notes that this system could only work if the individual residents were clearly recognizable, and postulates that the polymorphism serves this end. (We are reminded of Wynne-Edwards' comment (1962) 'the possibility suggests itself that, like the banner of the jousting knight or the jockey's colours, [the polymorphism] is an aid to individual recognition among a group of competitors.') However, we doubt very much if such a high degree of differentiation is necessary for this recognition. And so, in the absence of hard evidence — experimental or observational — for the ESS theory of ruff polymorphism, we present our alternative, or complementary, suggestion based upon the confusion of predators.

Ruffs at the lek appear to be extremely vulnerable to attack by raptors. Potential predators include the peregrine, the gyrfalcon (*Falco rusticolus*), and, to a lesser degree, the merlin (*F. columbarius*), the goshawk (*Accipiter gentilis*), the sparrowhawk (*A. nisus*), the short-eared owl (*Asio flammeus*), and the harriers (*Circus* spp.). The two most dangerous species for the ruff are the peregrine and the gyrfalcon whose ranges in the palaearctic overlap that of the ruff completely and who would certainly take ruff given the opportunity. They are quite capable of taking birds from or near the ground, the full 'stoop' being reserved for such as ducks or waders in full flight well above the ground. Yet, although many other waders of similar size and ecology are known to form a significant part of the diet of these falcons (Witherby *et al.* 1947), we have as yet found no such records for the ruff. This is in spite of the fact that ruffs are highly conspicuous in both plumage and behaviour and, like the rest of the family, are highly palatable. Indeed, the ruff's toothsomeness was thought to be one of the principal reasons for its extermination in Britain in the nineteenth century (Coward 1950, Series II).

Protection through polymorphism may function as follows. Assuming that birds of prey do hunt by searching image, or any operationally similar phenomenon, each encounter with male ruff in the breeding season will be an essentially new experience. Croze's work (1970), among others, showed that increased prey polymorphism reduces the hunting success of some avian predators. Because of the multiplicity of displaying birds at a ruff lek — from a handful to fifty or more, with many distinct plumages — and with such a variety of postures, a raptor will not be able to develop a searching image for ruffs. Even if the predator is simply attuned to a number of generalized prey images, including 'medium-sized wader', it is likely to be distracted by the number and variety of targets. And if it takes one which stands out from the rest, as predatory birds may (Mueller 1971), that simply benefits the other morphs and could help in maintaining the polymorphism.

A raptor faced with a lek of ruffs is thus in a complex multiple choice situation. We have seen (p.148) that this greatly increases reaction time. And, in man at least, the wider the field of perceptual readiness the less the skill performance (p.148). Previous experience with the protean ruff lek situation will therefore render a raptor less efficient rather than more so. And, finally, the single erratic/multiple erratic/scatter reactions of the prey will have the usual confusing effect upon the predator. This could all be quite enough to prevent birds of prey taking ruffs at the lek in any significant numbers.

Whatever the present mechanism and function of ruff polymorphism it seems to us that selection for that polymorphism through its protective value would have been a simpler matter than the highly complex evolutionary processes suggested by van Rhijn. The foundation would then be provided for the development, if necessary, of an ESS for central and satellite individuals.

Our approach may also contribute to the current debate on the evolution of bird colouration (references in Baker and Bibby 1987). We suggest that the relative significance of sexual selection (Darwin 1871; Hingston 1933) and the Unprofitable Prey Hypothesis (Baker and Parker 1979; Baker 1985) will be influenced by protean factors, particularly in polymorphic species.

Some further avian polymorphisms

An example of avian polymorphism which also may operate against searching image formation in predators is seen in the work of Tuck (1960), who made extensive studies of the murres, or guillemots (Alcidae), which normally lay just one relatively enormous egg on exposed cliff ledges and stacks. Murres' eggs are more variably coloured than those of any other bird, both in background and overlying markings. As Tuck notes 'The ground colour of individual eggs varies from almost pure white to various shades of blue, green and brown. The markings vary from mere small spots to fantastic scrawls and blotches of brown, lilac or black.' In the British form of the common murre, or guillemot, *Uria aalge albionis*, we have ourselves seen almost white and almost black specimens; green, blue, cream, buff, and brown backgrounds; scribbles, streaks, spots, smears, and blotches; and striking zones of black blotching at the blunt end. Such variations have long been known and collected by oologists.

In a limited investigation Tuck (1960) experimented with egg selection by marauding greater black-backed gulls by exposing a murre's egg and a white puffin's egg side by side on nine occasions. In seven cases the gull took the puffin's egg, even though the murre's egg is much larger and must surely represent an obvious target for optimal foraging techniques (see Krebs 1978).

Murres need to incubate closely for obvious reasons of embryonic development (the eggs are laid on the bare rock), and also to avoid egg loss on crowded ledges (Tuck 1960; Uspenski 1956). Older, more experienced birds are much more proficient in egg care than younger adults, which lose almost all of their eggs. But all will, from time to time, be open to danger from marauding gulls and crows. These predators will certainly take eggs whenever they are accessible, and Tuck's limited experiment indicates a high mortality of prey objects which are easily perceived and not protected.

This suggests to us that the polymorphism in murres eggs — the variations of which seem to be genetically controlled — may have been selected for through its confusing effect upon searching image formation in predators. There is heavy egg mortality in murre colonies before taking predation into account. Tuck noted that of 632 eggs marked to study egg replacement, 389 were lost: '19 per cent were smashed by rock falls, eight per cent rolled into crevices, one per cent were taken by glaucous gulls and the remainder either fell off the ledges or were suspected, because of their precarious position, to have done so'. Clearly, if these figures are accurate and representative, the murres require, and achieve, extra protection of eggs against predators. The polymorphism may help to provide that protection.

Payne (1967) noted that polymorphism is widespread in *Cuculus canorus* and many other parasitic cuckoos, and suggests that it has arisen by apostatic selection favouring deviant phenotypes because they are less likely to be recognized by their hosts. Significantly, the polymorphism is restricted to the female — which has to approach the host nest. Payne cites white-breasted and black-breasted phases of *Clamator jacobinus* occurring together in South Africa where, in open areas, *Cuculus clamosus*, also has two forms, one black beneath, the other barred. In equatorial forest, where the female's approach is less likely to be noticed, *C. clamosus* is not polymorphic. Payne also records a very high degree of individual variation in plumage of the female in any one locality in several parasitic cuckoo species, with a striking absence of such variation in non-parasitic cuckoos.

Remarkable colour polymorphisms also occur in the African bush-shrikes of the sub-family Malaconotinae (Hall *et al.* 1966; Owen 1967). In *Chlorophoneus multicolor*, for example, there are five morphs, with underparts black, crimson, pale red, yellow, or buff, and up to three morphs may coexist. Bush-shrikes hunt largely in wooded country, taking invertebrates and small vertebrates, such as frogs, lizards, snakes, small birds, and mice, on or near the ground. It seems reasonable to suppose that their polymorphism hinders recognition of the shrikes by their prey.

Colour polymorphisms in birds are in fact most frequent in predators or nest parasites, particularly in the Procellariidae (petrels, etc.), Ardeidae (herons), Strigidae (owls) and Tytonidae (barn owls), Accipitridae (hawks), Falconidae (falcons), and Cuculidae (cuckoos). Paulson (1973) has independently noted this phenomenon in birds of prey, commenting that small birds and mammals may, through experience, learn the colour pattern and outline of common predatory birds and thus react more quickly and appropriately. He evidences the fact that most open-country predators of mammals, particularly buzzards *Buteo* spp., and predators upon small birds, particularly broad-winged hawks *Accipiter* spp., are polymorphic.

Anomaly and aspect diversity

Diversity of appearance may have survival value not only within a species but also within a group of species. It would seem to be disadvantageous for similarly coloured palatable prey species to share the same habitat, since a predator's experience with one species will help it to recognize others. Reduction in prey risk in such a situation can be achieved in several ways. For example, Brower (1958) has pointed out that mutually exclusive oligophagy (specialization on different food sources) will tend to reduce searching image behaviour where predators search only one type of plant structure at a time.

In our present context we are also interested in the possibility that just as polymorphisms may be produced to counter predation within a single species, so interspecific diversification of appearance may arise in habitats containing several structurally similar cryptic palatable species. Any one of these species would seem to gain advantage from developing cryptic devices which differ from those employed by the others: the predator's experience with the other prey does not expose a prey animal to additional risk. Interspecific diversity of appearance will thus be promoted.

This suggestion is strongly supported by the work of Sargent (1969, 1973, 1978) with the north American species of the cryptic moth *Catocala*. Among the 30 or 40 sympatric species in deciduous woodland, the hindwings, which function as startle devices when crypsis has failed, show marked interspecific diversity. Sargent (1969, 1973) had already suggested that *Catocala* hind wing diversity enhances the startle function of these structures. He later suggests (1978) that

> the degree of startle exhibited by a bird in any encounter with a *Catocala* hindwing pattern is not so closely related to that pattern *per se*, as to the degree of expectation of that pattern which the bird brings to the encounter on the basis of its past experience. Thus, anomaly, defined in terms of departure from expectation, is regarded as a critical factor in determining whether a bird will be effectively startled by a given *Catocala* hindwing pattern.

He comments further that 'an anomalous stimulus, like a novel or protean stimulus, presumably produces a physiological state (e.g. excessive arousal or conflict — Sokolov 1960; Coppinger 1970; Driver and Humphries 1970) that interferes in some way with the completion of normal attack'. Anomalies, in fact, provide yet another example of anarchization acting unpredictably, by virtue of their anomalous nature, on reactors' perception mechanisms so that capture is often prevented. But we support Sargent's comment (1978) that 'This view of anomaly as a deterrent to predation suggests that two species with different hindwing patterns would benefit one another by occurring together and so might come to share an essentially identical niche'.

It may be that the diversity of hindwing patterns in the Genus *Catocala* — certainty in the north American species and probably in the European species also — has, by its effect upon predators, contributed to the multiplicity of sympatric forms. The principle of anarchization may thus be a contributory factor in speciation as well as in individual survival.

Such diversity of appearance presented by prey to their predators, particularly between two closely related species occupying the same habitat, has been termed 'aspect diversity' (B. Clarke 1962*a*; Rand 1967). Selection presumably favours such diversity in the same way that intraspecific polymorphism is promoted by predator searching-image formation (A.H. Clarke 1965, 1978; Rand 1967). When the predators are migratory birds the principle may even affect their prey in different latitudes (Ricklefs and O'Rourke 1975).

It is relevant to all this that Beukema (1968) in his experimental study of a polyphagic predator — the stickleback — showed that, once a new prey type was well known, its presence negatively affected the risk of other prey. It is significant that different species often achieve their match to a particular background by different arrangements of the same basic camouflage, the differences almost certainly requiring different perceptual analysis to recognize the individual from its background. Similar examples to the *Catocala* spp. are not hard to find. For example, Herrebout *et al.* (1963), describing colour patterns and behaviour of caterpillars living in Scots pine foliage, note half a dozen species which are camouflaged with green and white stripes, but no two species having the stripes arranged in the same pattern. Again, this diversity seems to be directly adaptive. In a sense it is the reverse of the Müllerian mimicry which occurs in aposematic insects.

A different kind of advantage deriving from the association of differently coloured species is suggested by Croze (1970) who notes that sandwich terns, *Sterna sandvicensis*, typically nest in small groups in the middle of the nesting grounds of other tern and gull species. The tern eggs and young are a pale off-white colour with black markings and differ sharply in appearance from eggs and young of the main colony. Predators entering the breeding area are likely to develop searching behaviour directed at the 'host' species and ignore the very different sandwich tern broods which are also protected locally by the parents failing to remove empty eggshells as would other gulls and terns.

This effect may also operate in the case of eider ducks, *Somateria mollissima*, which not uncommonly nest in colonies of terns, particularly the arctic tern, *Sterna paradisaea* (Summerhayes and Elton 1928; Lovenskiod 1954; Burton and Thurston 1959). The direction of the predator's searching behaviour to the 'host' species may be as important as the protection afforded by the mobbing of predators common to most gulls and terns — the reason usually given for this association.

Aspect diversity in web-building spiders

We believe that the effectiveness of aspect diversity extends much more widely than between groups of closely related species. If a number of species occupy similar niches in the same habitat, aspect diversity will protect them against shared predators, whether they are closely related or not. Harvey and Greenwood (1978) have already pointed out that animals may maximize their reproductive success by protecting animals other than themselves or their kin. As a distinct example let us consider the defence mechanisms of web-building spiders, bearing in mind that the various species will act as a single unit upon their principal predators: the bats and certain birds.

In such a widespread and successful group as the spiders, found — except for Antarctica — wherever there are insects, one may confidently expect to find a great variety of defensive strategies and tactics, both regular (Bristowe 1972; Edmunds 1974) and protean. Even restricting ourselves to the web-builders, we see many different devices for avoiding capture, as well as for capturing prey. Many species, like the European garden spider *Araneus diadematus*, retreat to a refuge adjacent to the web during times of danger. *A. alsine* hides in a leaf placed above the centre of the web. Some species of *Theridion* and *Tetragnatha* stay in the centre of the web, concealed beneath a leaf or twig. The European *Floronia bucculenta* and the Australian *Phonognatha wagneri* fall to the ground when alarmed and instantly change colour, by muscular contraction, to match the background. There is a protean element here in the unpredictable change of appearance. Many species fall to the ground and perform a standard convulsoid display, leaping around in all directions in a totally unpredictable manner. Others drop on a thread — to varying distances — and perform a suspended convulsoid display.

Species of tropical and subtropical *Argyrodes* sit in numbers in the webs of the larger argiopid species, avoiding searching-image formation by their resemblance to captured insects. Many of the Argiopidae and Uloboridae with silvery bodies make silvery patterned structures in their webs, against which they are less likely to be seen. *Argiope pulchella* builds a large cross in its web and often sits thereon with its arms aligned with the arms of the cross. Marson (1947) found that this device was most highly developed when the web was in bright light, was rarely a complete cross in dull places, and in the dimmest light was often absent. It would seem that this is another antisearching image device, particularly as the spiders do not always sit in alignment on the cross.

Ewer (1972) considers that the variety of arrangements of the silvery device in *Argiope flavipalpis* reduced the likelihood of a predator learning to associate the appearance of the device and the presence of prey. Studies of the various species which build these devices (Ewer 1972; Robinson and Robinson 1970; Marples 1969) suggest that there is an inverse correlation between density of spider population and likelihood of a device being built. This again supports the antisearching image theory and the protean principle.

Further combinations of camouflage and anarchization are seen in species of *Cyclosa* and *Uloborus*. They add to the web, aggregations of silk, some of which contain egg-sacs or remains of prey and are of similar size and shape to the spider itself. The Asian *Cyclosa mulmeinensis* constructs a number of false hubs in its web with collections of debris at their centres, each looking like the web-owner. In this case the egg-sacs are strung together and look like a twig in the web.

Edmunds (1974) records a cyclosid web in Ghana with two dummy spiders made of prey remains in the web, and the owner resting at the hub. A week later this web had a very different arrangement of devices: a vertical row of nine equally spaced dummy spiders above the hub. Again the principle of anarchization adds to the effectiveness of crypsis.

Cyclosa tremula seems at first to be maladapted, for the dummies it makes in its web are grey, while the spider itself is black and white. However, when the spider is disturbed it trembles violently, and the oscillations result in the black and white merging to grey. This spider thus achieves with movement what zebras obtain with distance: in each case the black and white colouration unexpectedly changes to grey.

Some web-builders show a high degree of intraspecific behavioural variability in territorial disputes (Reichert 1978). It would seem that the structure of spiders' behavioural control mechanisms predisposes them for diversity. In addition, many species are sympatric. The resulting variety of both intra- and interspecific defence mechanisms must make it very difficult for a predator to (a) learn the appearance of the prey, and (b) deal with it once perceived.

Anarchization of appearance in insects

Certain insects, particularly in the Order Hemiptera, have developed the most weird and bizarre shapes and structures which have hitherto defied explanation. Some of the tropical membracid bugs have sculptural extensions of the exoskeleton giving them an appearance which reminds one of some of the more extreme flights of fancy of science fiction illustrators (Fig. 8.3). *Cyphonia hirta*, *C. clavata*, *Heteronotus tridens*, *H. abbreviatus*, and *Bocydium rufiglabrum*, for example, have the thorax extended into grotesque configurations of spines and bristles, hooks, and spheres, which bear no resemblance to other earthly structures. This certainly seems to be a protean phenomenon, for what could be more anarchistic to a predator seeking insects, and normally responding to cryptic, mimetic, aposematic, startle, or fleeing displays, than an object that appears to resemble nothing else on earth? The predator's sensory attunements would not enable it to recognize the creature at all.

The anarchistic principle may also apply to the incredible south American fulgorid bug, *Laternaria servillei*, which, although only 3 cm long, has its head skeleton prolonged into an excellent 'imitation' of that of an alligator, complete with eyes, nostrils, and gaping tooth row (Fig. 8.4). The similarity is so remarkable that the Indians call the bug 'alligator snake' (Cott 1940). The supposed terrifying function of this adornment has been discussed by Poulton (1924). But an insectivorous predator is hardly likely to be startled by a 3 cm alligator, unless it had attacked baby alligators and suffered the protective anger of the mother, which is unlikely. But if the alligator-like head of the bug does remind a potential attacker of a real alligator, or any lizard, it is likely to be confused and its reaction time will be increased. If a predator does not recognize an alligator 'mimic' as such its bizarre nature may still hinder or prevent the predator's recognition of the animal as potential prey.

Behavioural polymorphism

We have already proposed (Humphries and Driver 1967, 1970) that intra- and interspecific diversity of adaptations for escape behaviour function as protean devices against predators. The first suggestion that individual behavioural differences might subserve defensive functions was made by Chance (1959) in the somewhat special case of the seizure behaviour of the deermouse *Peromyscus*. He noted that only about 5 per cent of mice collected from wild populations would show seizure behaviour when subjected to the sound of jingling keys, and he termed this apparently genetically-based difference in behavioural repertoire 'polyethism', by analogy with polymorphism. Arguing that the seizure behaviour is a form of escape reaction, he concludes 'lack of uniformity in the escape behaviour of different individuals will have the effect of preventing the predator from learning what to expect. The

Fig. 8.3. Some examples of bizarre Hemiptera. 1, *Bocydium rufiglabrum*; 2, *Ennya chrysura*; 3, *Alchisme grossa* (immature); 4, *Heteronotus abbreviatus*; 5, *Emphusis bakeri*; 6, *Sponogophorus hoffmanni*; 7, *Cyphonia hirta*; 8, *Cyphonia clavata*; 9, *Heteronotus tridens*. 2, 4 and 6 are cryptic against appropriate backgrounds. The spines of 3,7,8 and 9 may be deterrent. All are unreal and anarchistic in their effect upon potential predators.

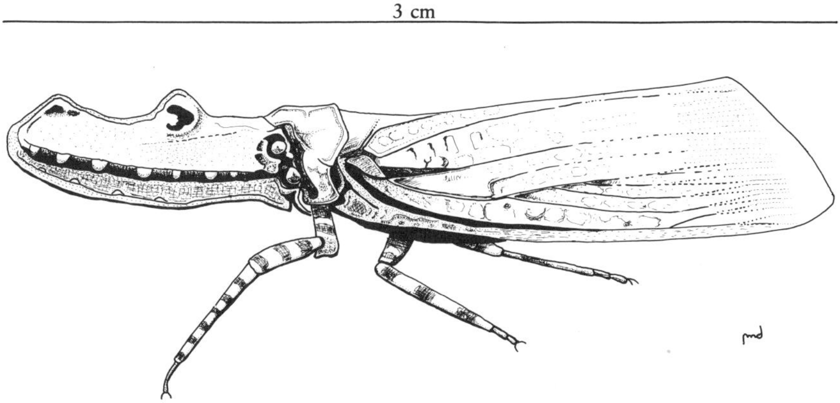

Fig. 8.4. The South American alligator bug *Fulgora*. Potential predators will find its appearance confusing, at least.

situation presented to the predator is, therefore, essentially enigmatic or confusing'. Chance did not go on to test this hypothesis or to cite examples from more convincing predator–prey confrontations.

Such unpredictable behavioural diversity need not be polyethic in the genetic sense, as should be clear from our account of the escape reactions of the small tortoise-shell butterfly (p.37). Robinson (1969*b*), in his extensive studies of the defence behaviour of orthopteroid insects, notes that many species, particularly phasmids (stick insects and leaf insects) have an escape repertoire of several alternative responses with an irregular order of deployment. He comments 'it seems quite possible that there would be survival value in having escape responses that the predator cannot predict with accuracy. Prediction in these circumstances would imply anticipation, which in turn could reduce the efficacy of the escape behaviour'. To produce a protean effect, therefore, it is not necessary for alternative behavioural responses to be determined genetically. All that is required is that in a standard and repeated stimulus situation an individual does not always make the same response. It would also seem that the less frequently an escape response is shown, and consequently the less experience the predator has in dealing with it, the more likely it is to succeed.

In an analogous fashion to the development of interspecific diversification in appearance, it would seem that a polyphagic predator will induce in its various prey species a tendency to evolve a more diverse overall repertoire of escape behaviour. Any new behavioural variation will automatically be restricted to one species of prey and will therefore be rare and enjoy the advantage of deployment against a relatively unprepared predator. Escape behaviours shared by a number of species

— other factors being equal — will be more likely to be overcome due to the greater experience of the predator in dealing with them.

Since we wrote this passage we have seen the work of Schall and Pianka (1980) on escape behaviour diversity in whiptail lizards, in which they note that 'Sympatric whiptail species differ significantly in escape behaviour. Escape behaviours are more divergent than if each species evolved its escape tactics independently of others in a random fashion.' Also '. . . syntopic species frequently differ greatly in escape behaviors.'

Schall and Pianka (1980) suggest that 'predators which form search images or learn to predict prey escape behavior should be hindered by diverse prey morphologies and escape behaviors. Thus, among conspecific prey populations, escape tactic diversities should vary positively with predation pressure and escape tactics should diverge among similar sympatric species that share predators.' They tested their 'escape behavior diversity' hypothesis by quantifying the escape behaviors of six species of whiptail lizards, *Cnemidophorus* spp. Populations of *C. tigris* were studied at twelve different flatland desert sites in western North America, and five other sympatric species were studied at 'approximately 50' sites in southwestern Texas. Unfortunately, the results were inconclusive, probably because relative predation pressure was estimated by frequency of broken tails. The authors themselves note that several factors confound the use of this index — as we ourselves would have predicted.

This interesting experiment reveals how complex is the problem of escape behaviour variability and how difficult to study. It also marks up the importance of a broad general knowlege of the natural history of an animal in the planning of investigations.

Display diversification — a universal strategy?

In this chapter we have been considering certain systematic strategies that predators tend to follow in seeking prey, and certain anarchistic tactics which prey animals may utilize in combatting these strategies. Just as in direct confrontations anarchization in a sequence of escape behaviour can defeat the systematic strategies of anticipation and prediction in a predator attempting to make a capture, so can anarchization in dispersions and in colouration and display deployment within and between prey species defeat a predator's searching strategies. Viewed as protean phenomena, polymorphism, polyethism, and both intra- and interspecific diversification of appearance and escape behaviour repertoires would seem to have much in common. Irregular diversity is such a prominent and universal feature of the armament of populations, as well as individuals and small groups, that we are forced to recognize a universal strategy of *display diversification*. It would seem to be a significant demonstration of the protean principle that anarchization increases armament.

Display diversification in avian vocal mimicry

As with other categories of protean phenomena, the processes of display diversification seem so logical once one has perceived the nature of the protean principle. And it becomes clear that intra- and interspecific variations may have survival significance in more ways than have been thought possible. In polymorphism and polyethism we see elaborations of Darwin's fundamental dictum that no two individuals are alike. Phenotypic differences may involve any part of an animal's constitution, from enzyme patterns to skin colour. Biochemical variations make possible, for example, ontogenetic feeding adaptations of insect larvae to new food plants (Maynard Smith 1958). Behavioural variations are seen — or heard — in the song repertoires of oscine birds such as great tits (Krebs 1976), and it has even been suggested (Krebs 1977) that one singing individual producing different songs may give the impression of several individuals, thus preventing potential rivals from settling in his territory. Such a situation is ripe for anarchization.

Vocal mimicry in birds would seem to merit serious consideration in this context. The phenomenon is widespread among oscines (songbirds) and is found most notably — but not exclusively — in species which live on or near the ground in dense rain forests or reed beds. As a result of vocal mimicry birds develop extended repertoires which enhance the maintainance of contact and the recognition of individuals in habitats where visual contact is difficult.

We have already seen (p.106) how vocal mimicry in bower birds may act as a protean antipredator device, deterring would-be predators when the vocalizations unexpectedly include the calls of birds of prey. No two individual bower birds have the same song structure and, in a number of genera, for example *Amblyornis*, *Ptilonorhynchus*, and *Chlamydera*, several individuals call within the same general area and certainly all within earshot. We have already commented on the remarkable performances of *C. maculata*. Holyoak (1972) makes essentially similar comments on the vocalizations of *Amblyornis* spp. — 'ventriloquial and imitative of many sounds of the forest, as well as whistling, crackling and rattling noises'. In addition to any deterrent effect the occasional 'dangerous' call may have upon predators, the overall effect must be one of almost total auditory confusion.

Such display diversification is not limited to bower birds. A high degree of polyethism is seen in the song repertoires of starlings, *Sturnus vulgaris*, which may have hundreds of song types in a population (Järvi *et al.* 1977). One individual may incorporate accurate mimicry of more than 20 other bird species in its song, and the species mimicked need not be from the same area (Driver 1969). The incorporation of many specifically different bird vocalizations in the already diversified starling song must have a confusing effect, particularly when many starlings are singing together, as is commonly the case. Additionally, the 'caw' of the carrion crow and the magpie chatter which are frequently but irregularly included in starling territorial song could act as a deterrent device, as in bower birds.

Blood parasites and beech nuts

So far we have considered the application of the protean principle to a wide range of free-living animals and a few ectoparasites. But all free-living animals are parasitized, as are many parasites themselves. If the protean principle has the universal applicability which we claim, we may therefore expect it to operate for and against parasitic populations. Even endoparasites, hidden, sometimes, literally, in the bowels of their hosts, may utilize the principle of anarchization. They use it as a means of combatting the host's biochemical armaments. If we consider polymorphism in parasites and their hosts, we can see how this has come about.

As with so many insights into the complexities of biological systems, it was Haldane (1949) who first pointed to the advantages a host might derive from polymorphism in its fight against its endoparasites. Specifically, a host with a rare biochemical phenotype would have a selective advantage in its resistance to a parasite finely-attuned to the usual phenotype. A successful parasite would of course counter this biochemical polymorphism by developing variants itself. The host would then be obliged to retaliate if possible, and Clarke (1976) has shown how polymorphisms can be maintained in both host and parasite. These polymorphisms, though fluctuating, are balanced and are equivalent to regular and predictable armaments in free-living animals.

Sometimes, however, the principle of anarchization comes into play. For example, certain blood parasites may show unpredictable variations in their biochemical population structure as a counter to the immune responses of the host. Such counters, of course, should not be too effective, or the host may die, taking the parastie with it. For our non-biological readers we outline the ground rules of the game in the words of Bloom (1979):

If parasites totally eluded the immune response and were sufficiently virulent, they would kill their hosts on whom their survival depends and preclude their own survival. Conversely, if they were too easily destroyed by the immune response, their survival would be similarly jeopardised. Consequently, immunity and escape from surveillance are relative phenomena which exist in balance and tension.

In some blood parasites it appears that this balance is disturbed by their development of a protean weapon, thereby being replaced by population fluctuations as extreme — though not as regular — as those of the small mammals of the tundra and their predators (e.g. Elton 1942; Krebs *et al.* 1973).

The African trypanosomes are protozoan blood parasites which cause, among other things, sleeping sickness in man and nagana in cattle (Wilson 1979). Human sleeping sickness is caused by *Trypanosoma brucei gambiense* and *T.b. rhodesiense*, and untreated infections are usually fatal. Infection is followed by rapid multiplication of the parasites, followed by a decline in their numbers due to immune responses of the hosts. However, a resurgence follows and the situation fluctuates in this way until the parasites invade the nervous system and untreated hosts die. The resurgence of the parasite is due to its development of new antigens so that the host's antibodies, developed against the original antigens, are ineffective (Bloom 1979). As the host antibodies are antigen-specific each succeeding wave of parasites has to be countered anew. The number of varieties of antigens creates a protean problem for the host. The parasite may produce a long sequence of variant antigens — at least 22. There is no evidence of a variant antigen occurring twice in an infection and, moreover, the order of appearance of variant antigens may vary. This final protean factor makes the problem for the host very much more difficult.

An effectively similar situation is seen in malaria, where man is infected with various species of *Plasmodium*, another group of protozoan blood parasites. The most dangerous type of malaria is malignant tertian fever caused by infection with *P. falciparum* in which the attacks occur quite unpredictably at irregular intervals of up to two days, a circumstance which undoubtedly adds to the host's response problem. In malaria, the host's immune responses can often cope with the infection, but it is not until early adulthood that the host has encountered all the strains in his geographical area and is relatively immune from debilitating symptoms.

It would seem that the unpredictable appearance of new parasite antigens is a phenomenon very difficult to counter — at least in the mammalian hosts of blood parasites. The large number of host deaths

— including millions of humans annually — which result from infection with *Trypanosoma* and *Plasmodium* species could probably be more readily avoided and an equilibrium between host and parasite could be more nearly approached if the protean principle did not come into operation.

Since we wrote this section the AIDS virus has leaped to the fore as a formidable destroyer of human lives, equated by some with the 'black death' and the 'great plague' of earlier times. Much of the 'success' of the AIDS virus — or Human Immunodeficiency Virus — stems from its unpredictably changing protection mechanisms. As noted by Koch (1987) HIV has the ability to

> change the structure of its outer protein coat, the envelope. This ability to display so-called antigenic drift makes it difficult for the body's defence systems, the antibodies, to identify and attack these proteins, the antigens. It also makes it difficult to make a vaccine for the same reasons. HIV also sheds these proteins from the envelope, so making it difficult for antibodies to stick to the main structure of the virus. This antigen-shedding is analogous to a lizard losing its tail when attacked by a predator. These free antigens may also become attached to other cells, thereby directing the body's own antibodies to attack and destroy these cells.

It is this complex anarchization of structure and behaviour that makes HIV so formidable an opponent in the fight for the integrity of the human body. We never dreamed, when we began this work, that one of the most potentially dangerous threats to mankind would turn out to be one based on the protean principle.

From the irregular production of antigens in blood parasites to the irregular crops of fruit in beech trees, *Fagus sylvatica*, would seem a curious step. But they have in common the production of large numbers of individuals in such an unpredictable sequence from generation to generation that their enemies cannot effectively deal with those numbers. The blood parasites benefit through the unpredictable occurrence of their antigens. Beech trees benefit through the unpredictable occurrence of large crops of fruit — the nuts known as beech mast. These nuts are a favoured food of many rodents and some birds. But the crop is highly dependent on a preceding warm summer and equable spring. In the UK, at least, these factors are notoriously unpredictable, and so therefore is the beech mast crop. Animals such as mice, voles, squirrels, and nutcrackers which avidly devour, and store, beech nuts are therefore unable to obtain optimum benefit from a bumper crop because they can never be fully prepared for it. Even the most efficient food storer — the nutcracker, *Nucifraga caryocatactes* — cannot benefit beyond the next spring when it may feed young on stored fruits

(Swanberg 1951). Also, just as greater black-backed gulls, *Larus marinus*, can only take so many sea bird chicks per day (Darling 1938), so can a nut-eater handle only so many nuts. A superabundance therefore results in proportionally more seedlings next year. And next year is not likely to produce another bumper crop, so any increased nut-eating population will usually decrease again, or emigrate.

The general phenomenon of unpredictable production booms is well known, but, largely as a consequence of the unpredictable seed crop, not as an occurrence in itself. It is seen in the occasional mass movements of birds from northern forests (and a few from steppe regions) named 'irruptions' by naturalists of the countries receiving the birds. The irruption phenomenon in birds has been well documented (e.g. Lack 1954; Cornwallis 1964) and it is clear that these irruptions are irregular and unpredictable in occurrence, and vary considerably in magnitude of flocks and distances travelled. Many species of birds in both nearctic and palaearctic regions are irruptive, and the majority are food specialists. Most are specially adapted — the crossbills *Loxia* spp. uniquely so — for feeding upon hard fruits or seeds. This is particularly the case with birds of the northern coniferous forests and the more northerly parts of broad-leaved and mixed forests. Cornwallis (1964) notes that climate extremes, food specialization, and irregular fruiting

> combine to cause acute imbalance at times between the levels of the bird populations and the supply of their special foods. This is often reinforced by the fact that fine weather at flowering time favours a sucessful nesting season, and the following heavy fruit crop favours survival during that winter. In the following autumn, when the food supply is low through "tiredness" of the trees, bird populations are at a peak and the imbalance is at its most acute. In this situation the irregular migrants fly in search of more plentiful food supplies.

Considerable discussion has taken place as to the reasons for these irruptions. Wynne-Edwards' group selection explanations (1962) have been discounted — and indeed withdrawn — though much of his comment is still highly pertinent, particularly his summary of evidence for similar behaviour of springbok in recent historical times. It is generally agreed that the ultimate response of the birds is to food shortage, but Lack's suggestion (1966) that the proximate cause is high density of birds as well as food problems may be open to challenge. Svärdson (1957) suggests that irruptive species, like true migrants, begin to migrate annually, the movement only stopping when sufficient food is found. The proximate cause may be the same as for the regular migrants.

All the studies of irruptions and emigrations — we have quoted only a few — though valuable have, we believe, not sufficiently considered the fundamental point that the phenomena would not occur at all but for the protean nature of the animals' food source. The unpredictability of food supply does not allow the seed-eaters to adapt over a period to take full advantage of the situation. The benefit to the trees thus, though apparently fortuitous, is firmly based on the principle of anarchization.

A similar phenomenon occurs in animals. For example, there are many unexplained instances of variations in times of hatching in animals such as turtles, where the young are immediately and particularly vulnerable to predation. The unpredictable appearance of the hatchlings could well result in increased survival in such cases. This suggestion is strongly supported by the increased survival that almost certainly accrues from the effectively unpredictable appearance of 13 year and 17-year cicada adults, and other insects. In the periodical cicadas *Magicicada septendecim* and *M. cassini* of eastern North America, both 13-year and 17-year races are well established, sometimes in enormous numbers and in the same area. With relatively rare exceptions the final nymphal stage of each race emerges from the soil to quickly produce the adults with amazing precision. In some areas 13-year and 17-year races of both *M. septendecim* and *M. cassini* are exactly in step. The primary goal achieved by all this would seem to be

> emancipation from all specialist parasites and predators: no exploiting species is ever likely to succeed in locking its time-scale to theirs. Their emergence is invariably therefore a surprise to potential "enemies", and the vast horde that suddenly appears on the scene completely floods the capacity of all predators and parasites to take serious advantage of them, or endanger their reproductive success.
>
> (Wynne-Edwards 1962)

This explanation was first put forward by Marlatt in 1907. We warmly support it, as a well-established and continuing — if intermittent — illustration of our thesis.

General comments

It remains to be noted that the coverage of dispersion and display diversification in this chapter is far from exhaustive. Some aspects of the matter relating directly, sometimes awesomely, to humanity, are considered in the final chapter. But there must be many more related biological topics, known and unknown, where the principle of anarchization has played its part in the evolutionary success of the phenomena and the concealment of the process from the searching gaze of humans.

We therefore humbly ask our fellow-scientists to consider the processes of anarchization most seriously. Our suggestions in some instances may seem outrageous, but so were the suggestions of Darwin, and of Einstein in their time. We do not compare our work with theirs, but we have, like them, attempted a new, wide-ranging, and challenging synthesis of already-known phenomena — a procedure which is anathema to many scholars, and not just scientists. One can easily be carried away with one's own ideas of course, particularly once they have been shown to be fruitful, but we have tried hard to avoid overenthusiasm and sweeping generalizations such as the recent biological adventures into group selection, gene-dominion, and the super-natural. And it is very pleasing to us that out thesis, clearly an incredible one to some, has over the years been progressively supported by new studies in a variety of fields. We instance the research on bat/moth encounters, the chorus line hypothesis for wader flock manoeurability, the work on naturally-produced brain opiates, the tit-for-tat computer program, frequency-dependent selection, and much of the work on the behavioural ecology of African mammals.

And, after all, in the words of Kuhn (1970) 'Discovery commences with the awareness of anomaly, i.e., with the recognition that nature has somehow violated the paradigm-induced expectations that govern normal science'. We would emphasize therefore, with all due humility, the comment of Charles C. Colton (1825) that 'we owe almost all our knowledge not to those who have agreed, but to those who have differed'. This, we note, particularly in the study of natural history, may apply both to the observer and the observed.

Summary

We consider area-concentrated search and searching image behaviour strategies of predators as targets for prey anarchization tactics.

Scattered distribution is seen to be an effective protean device for reducing the hunting efficiency of terrestrial predators. In aquatic prey, clumped dispersions reduce predation by increasing the effective inter-prey distance. As soon as a clump is discovered, erratic displays will take all but a few prey individuals out of range.

Searching image behaviour is seen to involve an increase in the efficiency of prey detection. Responsiveness to key features of the prey may be raised. We consider some protean counters to searching image behaviour. Change of colour and form, and erratic dispersions are operative here. Cryptic fragility may also hinder the development of searching images.

We consider the anarchistic properties of certain polymorphisms in more detail. Intraspecific diversity of appearance in, for example, caterpillars, leaf-hoppers, spittle-bugs, and many molluscs, would seem to function, at least in part, as a protean countermeasure to searching image behaviour. Heterosematism may act in the same way. Apostatic polymorphism and reflexive selection are seen as further manifestations of the strategy, and the extreme polymorphism of adult male ruffs and murres' eggs may also be included.

Interspecific diversity of appearance may act upon some predators in the same way as intraspecific polymorphisms. Anomaly and aspect diversity can be seen as adaptations which force predators to develop more than one searching image. Polyethism is, in effect, the behavioural equivalent of structural polymorphism.

As irregular diversity is such a universal feature of populations, as well as individuals and small groups, we suggest that there is a universal strategy of *display diversification.* This strategy is reflected in unexpected places, including avian vocal mimicry.

Blood parasites sometimes use the protean principle in their unpredictable production of antigens for countering hosts' antibodies. Well-known examples are the protozoans which cause sleeping sickness and malaria. The HIV (or AIDS virus) has now come to the fore as an example of the great effectiveness of anarchization in this context.

Many organisms have developed a system for the unpredictable production of enormous numbers of individuals — potential or actual adult forms. For example, every few years, irregularly, beech trees produce great numbers of nuts.And every 17 years or so certain cicadas hatch to the adult form in myriads. Those who would devour such abundance can eat no more than their fill, and they cannot adapt their own numbers to this phenomenon because it is effectively unpredictable.

9
The Integration of Irregularity

> We often think that when we have completed our study of one we know all about two, because 'two' is 'one and one.' We forget that we still have to make a study of 'and.'
>
> *Eddington*

> The time for the closed mind in biology is finished.
>
> *Peter Stefan*

In successive chapters we have unfolded a new view of controlled irregularity and diversity of behaviour in conflict encounters between animals in the natural environment. We have taken the reader from the introductory scene–setting through a new analysis of the features of behavioural defence to an integration of our findings with established theory. This integration has itself generated further perspectives on the function and evolution of certain apparently unsystematic phenomena at individual, population, and ecological/community levels. We have drawn copiously from published field data, experimental evidence, and theoretical generalizations of a variety of disciplines, particularly zoology, natural history, ecology, ethology, psychology, and population genetics.

In exploring the significance of previously neglected or unrecognized features in the patterns, deployment, and phylogenesis of escape displays we have found it necessary to invent a number of new terms. We believe these terms, unlike much scientific jargon, will help to integrate rather than confuse biological thinking. Nevertheless, in the face of such an apparently protean display of supporting data, subordinate contextual theory, and novel terminology, it is possible that our systematic message has escaped complete understanding. In order to clarify our argument and findings we have therefore summarized their main features in tabular form (Table 9.1).

It can be seen from Table. 9.1 that in phylogenetic perspective three major categories of protean development are evident in the interactions between predators and their prey: anarchization of the individual display; intraspecific and interspecific display diversification; and the development of a covert dispersion. Within each category the protean effect may be achieved in more than one way, the different methods being in some cases alternative, in others complementary. Thus the

Table 9.1. Summary: protean phenomena in biological armament — tabulation, with apparent effects.

Phylogenetic process	Protean phenomenon		Employment	Target
Anarchization of the individual display	Disruptive colouration and shape		Features break up visual outlines; effective during predators' foraging (p.252)	Perception
	Cryptic fragility		Disintegration upon attack and handling by predator (p.217)	Perceptual learning
	Behavioural irregularity and complex rapid transformation		In display — usually initiated by prey's detection of predator — including single erratic, multiple erratic, united erratic, scatter, convulsoid and distraction displays (Chapter 4)	One or more of orientation, perception, motivation
Intraspecific and interspecific Display diversification	Display Diversity	Reaction Diversity	Serial diversity of reaction in or between individuals over extensive periods. Includes polyethism and deployment inconsistency (pp.237–40) Simultaneous diversity of reaction in multiple-lúre distraction, scatter and multiple erratic displays (pp.72–106)	Motivation and/ or orientation
		Aspect diversity	Differences between prey individuals' colour and/or pattern during sampling by common predator. Includes apostatic polymorphisms (p.223)	Searching image behaviour Area-restricted search
Development of a covert dispersion	Individual spacing out with high crypticity (mainly in terrestrial and bottom-living forms)		Effective against foraging predators only periodically in touch with prey (pp.208–12)	Area-restricted search. Searching image behaviour Perception
	Clumped dispersion, often with scatter display (mainly in open water forms)		Effective against foraging predators only periodically in touch with prey (pp.212–14)	Perception, orientation, area-restricted search

Effect	Result
Blocks recognition of body form and other give-away regular features	Fewer prey discovered per unit time. Longer intervals between captures hinder development of searching image behaviour
Reduces time available for learning prey appearance	Hinders development of searching image behaviour
Induction of conflict in orientation, perception and motivation system	Induces startle or avoidance reaction producing temporary or long-term blocking of attack on the prey individual or type. Lengthens reaction time and delays attack and interception manœuvres, increasing prey's chances of escape. Increases orientation errors.
Creates uncertainty and thus conflict	Lengthens initial reaction time after detection. Reduces probability of predator developing skill in countermeasures In group displays tends to interfere with attack completion
All but one morph at a time gains temporary relative immunity from individual predators	Percentage of prey which are searching image targets is reduced
Increases effective inter-prey distance	Hinders development of searching image behaviour. Fewer prey discovered per unit time
At moderate and low population densities, time for which predator is out of contact with prey is increased	Fewer captures per unit time
Reduces chances of a rapid sequence of captures	Hinders development of searching image behaviour
At moderate population density, increases time for which predator is out of contact with prey. Scatter display compensates against area-restricted search by limiting period of vulnerability of group of prey and by inducing orientation conflict	Captures occur in short bursts — on average there are fewer captures per unit time

effects of anarchization may be evidenced in the disruptive nature of coloration, in reduced structural strength, or in the irregular, rapid, and complex transformations of active display. Further, an individual animal may employ different methods on different occasions. A small tortoise-shell butterfly, for example, may exhibit complex irregularities in a single erratic display, or transformation with crypticity, or irregularity plus transformation in the convulsive skipping display.

Display diversification within prey populations may be evident in colour and pattern or arise from behavioural repertoire differences or from inconsistencies in an individual's deployment of particular defences. We see display diversification not only within a single species, appearing as polyethic variation and colour polymorphisms, but also within whole sets of cryptic prey species sharing the same predators. Different arrangements of colour patterns whereby various animal species achieve effectively the same camouflage are therefore no longer to be regarded as fortuitous unaddressed variations, but as arising via natural selection — or at least being maintained by it — for the differences as such.

Though an individual predator will normally experience reaction diversity as isolated events in a series of times and places, there are some instances where intraspecific irregularity of deployment of defensive behaviour forms the basis of group displays. Examples include multiple erratic displays, multiple-lure displays, and scatter.

Covert dispersions — that is, dispersions which tend to ensure that the antagonist, upon discovery of one individual, is provided with the minimum information as to the location of others — differ in pattern according to the visual properties of the medium and background, which determine the relative importance of crypticity and distance in affecting probability of visual detection by the predator. These factors are differently balanced in terrestrial as opposed to pelagic habitats, a difference reflected in the effectively random and over-dispersions of terrestrial cryptic prey, and the clumped dispersions of their aquatic counterparts. We include disruptive colouration in Table 9.1 as an indication of new possibilities of interpretation. Here the patterns of form and behaviour are of course formalized, yet anarchization is involved in the confusion of regular features and in individual variations.

We now illustrate how the study of protean phenomena may assist our understanding of categories of adaptation in which anarchization is not a major factor but does have some role to play. There are many such categories; indeed, the influence of anarchization may be so widespread that our understanding of biological systems in general may need reassessment. Here however we have limited space for such matters. We therefore select for comment four disparate categories of biological import: altruism, deflection, human affairs, and human evolution.

Proteus and altruism

The problem of altruism is a subject of some importance in sociobiology. Here we can treat the subject only superficially, but this should be enough to indicate the potential contribution of protean behaviour theory.

Within biological science altruism is generally regarded as 'the surrender of personal genetic fitness for the enhancement of genetic fitness in others' (Wilson 1975). This element of personal 'sacrifice' may indeed be present in the many examples of altruistic behaviour shown by social insects (Wilson 1971) but there we are dealing with a colony or 'super-organism,' parts of which are expendable as long as some reproductives survive.

In primates such as chacma baboons (Hall 1960), vervet monkeys (Struhsaker 1967), hamadryas baboons (Kummer 1968), and yellow baboons (Altmann and Altmann 1970), senior ranking males lead the battle against predators or competing troops of conspecifics. These dominant males, their genes strongly represented in their troops, are, however, the strongest and most alert and therefore least likely to be killed or injured. The danger to the participating animals is also reduced through the protean influences which accrue from attack in numbers — as in mobbing (see p.101).

Distraction displays, particularly in birds, are similar to the examples of male primates defending their troops in that the birds are protecting their nearest relatives — their own young. In birds however, the actors have approximately equal physical potential, and seem to put themselves in some danger while displaying. We have seen how this vulnerability is reduced, frequently to insignificance, by the protean nature of injury-feigning displays.

Protean factors also operate in the apparently altruistic actions of birds such as hornbills or passerines when mobbing predators, or other prey species such as bustards when diverting eagles (pp.100–3). If mobbing and diversionary displays were not protean in form they would indeed be 'sacrificial' with respect to many of the actors.

Warning calls in birds, particularly those given in response to flying raptors are well known to be evolved acoustically to make them difficult to locate (Marler 1957; Marler and Hamilton 1966). But they do, at first sight, seem to involve a degree of altruism. Various suggestions have been advanced to explain this situation (Wynne-Edwards

1962; Maynard Smith 1965; Williams 1966; Trivers 1971). We favour Trivers' suggestion that warning calls benefit the caller by denying the predator a meal and thereby encouraging him to move on. But we would point out that, in addition, as the calls are non-locatory they are likely to add a protean element to the brief encounter between predator and prey. The raptor, appearing suddenly near or amongst a group of birds, has to decide which one to attack. Often the choice is simple — one bird will be a much more suitable target than the others. But if, at the same time, a warning call is heard but cannot be placed in space with reference to the visible prey, the predator's reaction time is likely to be increased, with obvious benefit to the prey. Furthermore, as the warning calls are non-locatory and are produced first by those birds which happen to see the predator first, the pattern of warning call production will be unpredictable from one occasion to another — protean in time as well as in space, and as effective as primary protean displays.

The essence of these warning calls and mobbing displays is that they involve the genetic inheritance of appropriate behaviours, thus ensuring that all members of the group will, sooner or later, behave in a reciprocal manner. 'In social animals [natural selection] will adapt the structure of each individual for the benefit of the community, if each in consequence profits by the selected change' (Darwin 1859). Warning calls are classic examples of the inbuilt action pattern — that member of the hierarchy of innate mechanisms which involves reaction of the whole individual animal.

We cannot see how these behaviours differ from the original paradigm of reciprocal altruism offered by Trivers (1971), that of 'good samaritan' behaviour. If the risk involved in producing a warning call is low, the situation would seem to be essentially similar to that in which a good swimmer, with little risk to himself, rescues a drowning man.

The same principles may also apply to behaviours such as stotting in Thomson's gazelle. We have seen that stotting, far from being a disadvantage, benefits the stotter by alerting the herd, the movements of which may make it more difficult for the predator to follow its victim (Kruuk 1972). Stotting, pronking, and similar behaviours in other antelopes, are classic inbuilt action patterns, and all involve protean elements which reduce the dangers accruing from their performance.

It is important to remember that the signals of non-human animals are never deliberately descriptive or analytical; they are innate behaviours which evoke responses of survival value to the actor or its genotype. Non-human animals cannot deliberately describe things, let alone form concepts. As Haldane (1954) put it 'Les signalisations animals sont toujours évocatrices et jamais descriptives'. Thus, examples of animal altruism are not altruistic in the human sense of conscious sacrifice. This adds to their significance, for if an animal innately puts itself at some extra risk by 'altruistic' behaviour, then such behaviour must have been selected for. And such selection can only be explained by the advantages accruing to itself or its close kin through the incorporation into the behaviour of protean and other elements. And so, even in this cursory glance at the problem of altruism, we can see that the principle of anarchization has a relevance which has been unsuspected hitherto.

Clues to the actual evolution of reciprocal altruism may come from a study of mobbing displays. As Barash (1977) has noted, following Trivers (1971), reciprocity 'can select for altruism with no assumptions whatever concerning genetic relatedness'. The basic requirement is that 'the return to the altruist must be greater than the decrement in inclusive fitness imposed by the original altruistic act'. Barash also suggests conditions that would have to be met for reciprocal altruism to work. First, the altruistic act should involve low risk for the altruist and high benefit for the recipient, because the *origin* of reciprocity would rely largely upon chance and would therefore be unlikely. Second, there would have to be a high probability that the situation could be reversed so that any one altruist could benefit in turn. Third, the reciprocal altruists would have to recognize each other 'so as to reciprocate appropriately'. Some mobbing displays meet the first two suggested requirements and indicate that the third may be unnecessary.

In mobbing displays in which, for example, bustards or gulls defeat eagles, and hornbills or mongooses overcome snakes, the actors are protected by the erratic nature of their movement (p.101). The reactor is required to continually and unpredictably shift its attention from one mobbing animal to another. There is very little risk to the actors. Furthermore, on Barash's second point, reciprocity is built in; mobbing is unsuccessful if there are too few actors. A single pair of crows or gulls will often succeed in stealing eggs from under a sitting bird by their simple protean mobbing effect, but when the victim is a predator there are usually more actors.

All that is required for reciprocity is that individuals of mobbing species should have an innate drive or tendency to attack predators when it is relatively safe for them to do so, i.e. when they are in sufficient numbers. There will always be alternation of internal states in the actors as they switch from attack to escape with the switching of the predator's attention from one to another.

Any recognition requirement is unnecessary. Just as passerines of different species — who, presumably, do not recognize each other — will gather to mob owls in daylight, so will gulls gather to mob eagles. And although the members of small groups of bustards or hornbills doubtless recognize each other, all that matters is that they should attack the predator when feasible and retreat when necessary. The alternation continually provides the weapon, regardless of who wields it. Unpredictability is universally effective.

In this brief foray into the subject of altruism, we already see how protean theory may indeed help to clarify some of the theoretical problems involved. Now, finally in this connection, we are pleased to be able to note the concurrence between our analysis of mobbing behaviour and the Tit-for-Tat computer strategy noted by Axelrod (1985). This strategy was developed as a result of attempts to solve the game theory problem known as the Prisoner's Dilemma. This is one species of the class of 'nonzero-sum' games in which the interests of the players are partly coincident and partly opposed (see Luce and Raiffa 1957). We here outline the game in general terms only. Detailed studies can be found in Rapoport (1959), Rapoport and Chammah (1965), and Rapoport and Moshowitz (1966).

Broadly stated, each of the two players has to choose repeatedly in a formalized situation between two alternative strategies, one co-operative and one defecting, and is given a stated reward as a result. If both choose to defect (D), each has a small negative payoff. If both choose to co-operate (C), each has a very small positive payoff. If one chooses C and one D, then D has a large positive payoff while C has an equally large negative payoff. The trials are repeated any given number of times and changes in choice, and of course in score, are recorded. The aim of the game is to score optimally. It is interesting to note that this situation resembles some real-life problems faced by many kinds of animals, and all seem equally intractable.

Rapoport and Moshowitz (1966) note that as game theory concepts are insufficient or inadequate to solve such problems, 'psychological, sociological, and perhaps ethical notions have to be introduced and incorporated into any theory trying to account for the players' behavior.'

Axelrod (1984) invited computer strategies for tackling the Prisoner's Dilemma, and of the 63 submissions, Tit-forTat (Anatol Rapoport) was the simplest and the most successful. Basically, its strategy is to begin co-operatively, and then to copy its opponent. Its characteristics are that it is 'nice', i.e. it is never the first to defect; it is co-operative, not envious; and it is forgiving. As long as the series of trials is not a short one the Tit-for-Tat strategy will always succeed (in a short series, cheats will dominate).

As we have described them, the innate bases for mobbing, and the protean elements which accrue from its occurrence in an unpredictably varying context, result in mobbing behaviour showing the essential features of the Tit-for-Tat strategy. The mobbers share the danger from the predator more or less equally. No-one defects unless in real danger, and then all are equally likely to do so. They are all co-operative and not envious. They are forgiving, i.e. they do not retaliate by opting out. The behaviour is relatively simple. And it can all be repeated an indefinite number of times. Thus, mobbing behaviour, with its built-in protean protections, is self-fulfilling, like the Tit-for-Tat strategy.

If we are correct, protean theory again blends well with recent developments in other areas of biological investigation. And, most interestingly, we may now see that the biology of altruism has a different perspective. We no longer need to assume 'the surrender of personal genetic fitness', or indeed any kind of unusual personal disadvantage in an animal that, with others, is mobbing a predator, or is engaged in injury-feigning. The apparent risks taken by such animals are offset by built-in protean elements.

Dawkins (1986) likens the Tit-for-Tat strategy to reciprocal altruism and suggests the naturally occurring examples are very few, noting only the sharing of food by vampire bats, and the relationship between cleaner fish and their clients. It seems to us that reciprocal altruism in the sense of Trivers (1971) may apply much more widely. It is seen in Trivers' three examples — cleaning symbioses, warning cries in birds (which we have also suggested), and human cases — but also mobbing, in pronking and stotting, in multiple erratic displays, and probably in other examples where the mutual benefit is immediate. If we allow consideration of the trajectory of a genotype over several generations, we can also include injury-feigning, and more human examples, for each individual will both receive and provide benefit — apparently at some risk to itself, but protected in fact by some aspect of the protean principle.

Proteus and deflection

In our second example we apply our analysis to various aspects of distraction and deflection which hitherto have been seen as rather puzzling examples of adaptation involving a variety of biological principles but with no common factor. We refer particularly to the development of flight streamers in birds and insects, and to various anteroposterior confusion devices in several animal groups. To liken racquet-tailed motmots to leaf-footed bugs, and hair-streak butterflies to lobsters may seem arbitrary and capricious, but not so! Each pair is following the protean principle in its own way while acting upon the reactors' perception systems in the same general manner.

Birds of several families shows unusual developments of wing or tail streamers which are generally considered to have epigamic (or unknown) functions. The feathers in these instances are not simply elongated. They may be structured so that the elongations act as moving streamers in flight, extreme examples being seen in nightjars (Caprimulgidae), hummingbirds (Trochilidae), and birds of paradise (Paradisaeidae). Several other families provide less extreme examples. In a number of families the streamers have expanded tips, often termed 'racquets'. Examples are found in the families already named and in drongos (Dicruridae), motmots (Momotidae), and parrots (Psittacidae) (Fig. 9.1).

Several observers have noted that the racquets of, for example, the standard-winged nightjar, *Macrodipteryx longipennis*, (Fry, 1969), seem to follow the bird in flight like independent objects, especially when viewed in poor light. Such an appearance must be distracting and confusing to an observer and will therefore act as an antipredator device through protean principles already explained. The racquets will act in an analogous manner to the flares dropped by aircraft to distract and deflect heat-seeking missiles (Fig. 9.2).

Tail-streamers are also seen in insects, particularly Lepidoptera. In the saturniid moth *Eudaemonia brachyura* the streamers are twice as long as the body, pale in colour and twisted, so that in flight the moth appears to be followed by something else — as in the *Macrodipteryx* nightjars. Presumably the hindlimb extensions of leaf-footed bugs (Fig. 9.2) will, in certain circumstances, have the same effect. Even in aposematic species such as the Peruvian *Anisoscelis foliaceae*, the extensions will be distracting to inexperienced predators and confusing to all in poor light.

To adduce these adaptations as instances of the protean principle does not mean that we consider them to be primarily protean in nature; in that case they would appear in Chapter 4. But they certainly seem to have some protean application and this may well have influenced their development. For example, the well-developed tail-streamers in certain tyrant-flycatchers (Tyrannidae) and the sugarbirds (Meliphagidae), while clearly of epigamic function, could, by their rather extreme development, also have a confusing effect when the birds are in flight. Potential predators could thereby be so influenced that the tail-streamer effect could be enhanced through natural selection. Thus, protean factors could add to epigamic considerations in producing extreme adaptations such as those of the racquet-tailed species.

Fig. 9.1. Some racquet-tailed birds. 1, turquoise-browed motmot *Eumomota superciliosa*; 2, king bird of paradise *Cicinnurus regius*; 3, standard-winged nightjar *Macrodipteryx longipennis*; 4, racket-tailed drongo *Dicrurus paradiseus*; 5, marvellous spatuletail *Loddigesia mirabilis*. In all cases the tail streamers have some protean function.

Fig. 9.2. Protean deflection devices in flying targets. 1, Standard-winged night-jar (enlarged second primaries); 2, leaf-footed bug (expanded tibiae); 3, RAF Phantom (flares). All these deflectors move in protean fashion in relation to the true target, both distracting and confusing.

Anteroposterior confusion devices in animals are well covered in the literature (for example, Wickler 1968; Edmunds 1974), but an understanding of the protean principle enables us to see them in a clearer light and to draw larger conclusions.

Many species of butterflies and moths have extensions of the hind wings which give them a 'tailed' appearance. This is seen particularly in hairstreak butterflies (Lycaenidae), swallow-tail butterflies (Papilionidae), and moon moths (Saturniidae). In many cases these extensions are 'eyed', brightly-coloured, or both. It is well established (see Edmunds 1974) that insectivorous birds are attracted to such markings and direct their attacks at them with the result that the insect proper is missed.

This development is taken further in certain lycaenids. In the tropical genus *Thecla* the tail extensions are so fashioned as to give the appearance of a head — complete with eyes and antennae — larger than the animal's true head and facing backwards.

Lycaenids are already unusual in that many of them rest on vertical surfaces facing downwards, rather than upwards as in most butterflies. The tail extensions of course add to the confusing effect. Experienced predators attacking them, aiming slightly 'ahead', will therefore miss — the animal takes flight in the opposite direction. In the South

American *Thecla togarna* this effect is enhanced by the animal — alighting head up but immediately turning through 180° — briefly waggling its false antennae.

In this complex adaptation we have yet another example of an animal behaving in a quite unpredictable manner. The behaviour is protean with respect to any predator's previous experience with other butterflies — even other lycaenids. And it is essentially similar to the examples of 'back to front' animals of many groups so ably surveyed by Cott (1940) and Edmunds (1974). They all have behavioural and structural adaptations which seem to reverse their anteroposterior axis and deflect an attacker's attention from vulnerable parts.

With the protean element of this reversal in mind we now turn our attention to Decapoda (Crustacea). Prawns and lobsters are elongated anteroposteriorly and normally walk or swim slowly forwards. When attacked — or just threatened — however, they flex the well-developed abdomen forward beneath them and shoot backwards at speed, often in a zigzag fashion and into cover. The general direction of escape, as well as the detailed flight path, is effectively protean.

Crabs have used this principle of unpredictable change of direction in a different way — they move sideways with equal facility in either direction. Some can even move in any direction in the horizontal plane, thus greatly increasing the protean nature of their movements when attacked. The answer to the question 'why do crabs walk sideways?' therefore involves the protean principle yet again.

Proteus and human activities

We now indicate some of the ways in which the protean thesis may have relevance to certain aspects of the human state and contribute to the development of new and fruitful lines of research.

A rapidly developing human science which might develop a mutualistic association with protean theory is ergonomics — the study of the operational relationships between man and machine. The practical and theoretical significance of this work is considerable in this age of computer-based technology at the work-station. According to a report in *Newsweek*, 10 Nov, 1986, up to 30 per cent of the health problems of modern workers may be due to conditions at the machine/user interface.

Humans in many situations where accurate decision making is of vital importance are often bombarded with complex and somewhat irregular patterns and sequences of stimuli. One example is the airline pilot with his cargo of human lives. Another is the air traffic controller with his multiplicity of signals which have to be continually monitored on a life or death basis. Yet another is the automobile driver in modern city traffic, faced not only with the ever-changing complex of stimuli from traffic signs and other vehicles, but also with the persistent intrusion of attention-drawing advertisements and illuminations plus auditory signals and 'noise' of great variety. And it seems that the so-nearly tragic Three Mile Island nuclear breakdown in 1979 presented monitoring personnel with confusing information as well as straightforward information overload, compounding the difficulties of reacting quickly and appropriately. At the time of writing (1986) the pathological consequences of this disaster are still being felt. In the far more serious Chernobyl disaster (1986) the protean factors apparently extended to unpredictable variations in management standards.

It would seem that there is considerable common ground between contemporary research into the ergonomics of information presentation and the problems posed in the effect of protean displays upon predators. Whereas the ergonomist is concerned with designing information display systems which will be optimally suited to the procedures employed by the reactor's perceptual and decision-making mechanisms, natural selection — and occasionally learning — have created an ergonomist's nightmare: displays totally unsuited for efficient operation of the reactor's information-processing systems.

In defining certain conditions which hinder information processing, ergonomics gives pointers to the functional value of some characteristic aspects of protean displays. Reciprocally, we may speculate that closer examination of the protean principle may help to draw attention to human situations in which complex environmental stimuli have a hazardously confusing effect.

Many protean displays are characterized by the rapidity of sequential change in the orientation stimuli presented to the predator. Ergonomic investigations by Bertelson (1961) and Hale (1967) revealed that human subjects, required to react as quickly as possible with the appropriate response to each of two stimuli presented in irregular sequence, showed longer reaction times the shorter the interval between the previous response and the onset of the next stimulus. An incentive for faster reactions will shorten reaction time but only at the expense of a steep increase in errors. Extrapolating these findings to protean displays we conclude that rapid transition between many brief stimuli is likely to delay reactions to each, or to produce more errors. Either, or both consequences would help the prey to survive.

Likewise, the complexity of protean displays requires that the predator particularly, in orientation of its attack, should select one from many possible responses. Again, ergonomic research (Bertelson 1961; Leonard, *et al.* 1966) indicates that the larger the number of alternatives, the greater the relative increase in reaction time when dissimilar stimuli follow each other compared with similar stimuli following each other. In the protean display field it would seem that it not only pays to be inconsistent but that inconsistency becomes more effective the more complex the display in which it is embedded. And there is evidence (Bertelson and Renkin 1966; Hale 1967) that a similar enhancement of delay due to inconsistency occurs when stimuli more quickly follow previous responses. Thus, ergonomic research implies that both rapid transformation from one response to another, and inconsistency (irregularity) in sequence are likely to interfere with information-processing in predators — assuming that this processing is generally similar to that in man. We are not aware of any facts that negate that assumption.

Why is it that predators, and man, should be so vulnerable to irregularity and rapid transformations in stimulus input? In Chapter 6 we argued that such characteristics induce conflict in the reactor. But why should these characteristics act in this manner? Again, recent ergonomic studies appear to be pertinent, for, in man, vulnerability to irregularity and to narrow separation between successive stimuli appears to arise from the nature of the mechanisms of skill development. Many animal predators depend to a large extent upon the ontogenetic development of skills in their feeding techniques (Driver 1974), some to quite a remarkable degree (Norton-Griffiths 1967, 1968).

Human decision-making seems to be a single-channel process (Welford 1968). This generally restricts reaction time to a little less than 0.2 sec when a single predetermined response is given as soon as possible after a single known stimulus, with reaction time rising logarithmically (Hick 1952) to 0.5 or 0.6 sec for eight or more alternative stimuli, the arrival of any of which must be matched with an appropriate one of eight alternative responses. The channel capacity is thus limited, a fact seen additionally in the 'line busy' delayed reaction to a stimulus which arrives too soon — under 0.5 sec — after a preceding stimulus. The basic decision-making system would thus appear to be highly susceptible to information overload.

In skilled performances these deficiences are partly circumvented, for in tasks involving more than eight alternatives, and time intervals of much less than 0.5 sec, reaction time may apparently average 0.15 sec or less. Skill may thus be regarded, in part, as a strategy whereby the operator overcomes his own neurophysiological limitations. This strategy depends particularly on input 'buffer stores' (Whitfield 1966) in which, while the central decision-making mechanism is busy with earlier information, incoming information can be stored, or coded into a group for processing as one unit; and also output 'buffer stores' in which instructions for set sequences of responses can be stored without need for the sequential position of each component to be decided every time by the central mechanism. These learned groupings of responses and stimulus information are built up by long experience of systematic features in the sequential relationships between events. They enable, for example, a typist to strike consecutive keys at intervals much less than the reaction time which would otherwise apply; but if the order of the letters in the material to be copied is reversed, reaction time and error probability increase significantly as the grouping skill strategies fail.

It would seem that the biological design of protean displays is excellently adapted to defeat any such skill strategies developed by predators, for skill depends on the learning of systematic features of stimulus patterns, and it is these systematic features which make possible error-free, time-saving short cuts in decision making.

We have selected ergonomics as a category of human affairs in which enough research has already been carried out for us to be able to demonstrate clearly the relevance of the protean thesis to problems of human behaviour. We are sure there are many other aspects of human activities where such a comparison would be profitable. For the present we will point to just a few potential areas for study.

First, we would connect the generally confusing effect of protean displays and the induction of 'nervous breakdowns' in rats by protean manipulations of operant training procedures (Maier 1948). After learning that a particular performance is consistently rewarded, rats subjected to a protean regime of punishment and reward have their performance in the operant situation virtually destroyed and display symptoms similar to those seen in humans who can no longer cope with environmental pressures.

One is reminded here of Toffler's perceptive thesis on *Future Shock* (1970) which describes the stresses involved in contemporary man's forced adaptation to rapid unpredictable change. Among many examples he notes the emotional exhaustion suffered by combat soldiers and disaster victims, and 'culture shock, the profound disorientation suffered by the traveller who has plunged without adequate preparation into an alien culture'. The culture-shocked person, like the soldier and disaster victim, 'is forced to grapple with unfamiliar and unpredictable events, relationships and objects. Signs, sounds and other psychological cues rush past him before he can grasp their meaning. The entire experience takes on a surrealistic air. Every word, every action is shot through with uncertainty'. Toffler notes that combat soldiers, disaster victims, and culturally dislocated people all face rapid change, high novelty, or both, and all are required to adapt rapidly and repeatedly to unpredictable stimuli. And their responses are strikingly similar.

'First, we find the same evidences of confusion, disorientation, or distortion of the reality. Second, there are the same signs of fatigue, anxiety, tenseness, or extreme irritability. Third, in all cases there appears to be a point of no return — a point at which apathy and emotional withdrawal set in'. Clearly, there is similarity with some of the processes involved in 'information overload' (see for example Miller 1963, 1964 *b*). The problem in all these cases is exacerbated by protean factors.

We suggest that any human subject, who, like the experimental rats, is rewarded or punished in an unpredictable manner, will be unable to develop or maintain a stable relationship with the social environment and is therefore likely to be an unstable or rebellious member of the community. This will apply to infants at home and at school and may well be a factor in the causation of delinquency.

An infant's environment may become effectively protean in a number of ways. The unpredictable incidence of punishment and reward is probably much more common today at a time when those becoming parents or teachers are themselves less and less likely to have experienced the benefits of logical routine and reasoned discipline. This certainly seems to apply to Anglo-Saxon countries and is probably true elsewhere. All of the mature teachers we have spoken with — and we have many years of teaching experience ourselves — agree that standards of discipline in general, and in the home and in school in particular, have noticeably declined in the last 20–30 years — with exceptions of course. Also, as the stresses upon parents become greater they are more likely to show unpredictable outbursts of re-directed aggression, sometimes with tragic results, as reported all too frequently in the media.

In the adult human situation protean relationships between crime and punishment or work and reward can also result in considerable individual and social stress and discord and, we believe, contribute greatly to the general deterioration in social stability.

There are many protean factors operating in the general area of crime and punishment. Disregarding arguments as to the definition of criminal behaviour, the likelihood of a criminal being caught is relatively unpredictable. The detection of all kinds of unsocial activity, from minor traffic offences through theft and vandalism to the most serious crimes, depends to a degree on the skill of the offender, but also on variations in policing activities, and on the chances of being seen and reported. When caught, the offender experiences the protean principle in his handling by the police and his sentencing by the court. The police may handle him gently or roughly, according not only to his own behaviour but also to individual or local variations in attitude, personality, or instructions. Similarly, the allocation of penalties will vary according to the psyches of judge and prosecutors, the status of the defendant in society and his attitude in court, the sociological biases of the time, and so on. Add to all this the increasing incidence of proven corruption in the police forces of many countries, and amongst high officials, and it is no wonder that 'the law' is increasingly regarded with contempt. To the man in the street any connections between justice, law, and order are effectively protean.

In the area of work and reward there are two general kinds of problem where there is a protean influence. First, there is the problem of protean behaviour in others. Both workers and management behave in a protean manner from time to time according to the shifting influences upon them. But of much greater import is the considerable and neglected problem of reward for effort. To many workers there is no systematic relationship between effort and reward. However conscientious, intelligent, and hard-working a person may be, there is still a large element of chance in success: being in the right place at the right time or having a face that fits. When the thinking man sees others earning ten or twenty times as much money for no more overall effort he may be annoyed. And when he hears of the inherited or chance-won fortunes of others he may be jealous. All of these discrepancies in income involve protean factors which make their alleviation more difficult.

Before moving on we must note that there is one aspect of the work scene where a protean influence has been effectively dealt with. This concerns the unpredictably varying background noise in offices. Obtrusive auditory disturbance can be dealt with by the production of 'pink' noise — a continual hiss which smothers the usual noise intrusions. As a result, background noises are much easier to habituate to.

Now we must refer to the successful employment of protean strategies in human warfare. Sir Winston Churchill in his history of *The Second World War* (1948) shows that both he and the German Admiralty strove to reduce the predictability of their military moves, not only by secrecy (which we might regard as a parallel of crypsis) but also by the deliberate creation of uncertainty. He notes, for example, the use of decoy fires to prevent the previous night's bombing conflagrations being used as reliable markers for further attacks; the far greater effectiveness of blocking the Kiel Canal not with a single large explosion but with many smaller bombs of scattered placement and fuse times varying from a day to a month or more; the orders of the German Admiralty to the pocket battleships *Deutschland* and *Graf Spee* that frequent changes of position in the operational areas would create uncertainty and more effectively restrict enemy merchant shipping; the attack by the *Scheer* on allied convoy H.X. 84, in which a 'scatter' tactic was largely instrumental in saving all but five ships; and the way in which the effectivenss of a *Blitzkrieg* arises from the speed with which events develop and the smashing of communications so that the defender is unable to discover what is happening and thereby predict and meet the attacker's next move.

These are but a few of many examples. In more 'advanced' warfare the protean principle continues to be used in, for example, the programming of patterns of missile dispersal on the ground and their dispatch into space. There is now a considerable literature on the subject — for example Cruickshank (1979) — and, doubtless, much more on the secret list.

'Irregular' warfare such as terrorism, armed civil disturbance, guerrilla attack, and other forms of social disruption also depend to a considerable degree upon protean elements in their make-up — probably much more so than in 'regular' battles. As Napoleon recognized (1804–15) 'Anarchy is the stepping stone to absolute power.' But this is not the place to survey the role of the protean principle in irregular, or regular, warfare. We hope to give the matter more extensive treatment in a more specialized work.

On a happier note we see that protean principles enter into many sports and pastimes where two individuals or groups compete against each other. Both instinctive and learned protean tactics may be seen, the latter often reinforcing the former. In rugby football for example, it would seem partly instinctive, at least, to 'jink' at the appropriate moment before an impending tackle, though the effectiveness of this can be enhanced considerably by learning. On the other hand, the 'plays' in American football are systematically planned to be unpredictable. Other examples of protean ploys are the feint in boxing or martial arts, concealed spin in tennis or squash, concealed spin also by the bowler in cricket, and similar disguises by the pitcher in baseball, dummy runs to the penalty kick in association football, and dummy passes or kicks in rugger.

Chess and poker of course provide numerous opportunities for the employment of protean techniques, both in strategy and tactics. And the widespread fascination for gambling owes much to the protean principle. Football pools and bingo, betting shops and lotteries, casinos and fruit machines, all reflect a deeply-rooted fascination for the unpredictable. This is even seen in TV quiz games, some of which have become so protean in structure that no one can possibly predict their outcome. And TV comedy shows have begun to incorporate unpredictability as a normal part of their entertainment structure.

Unpredictability in broadcast entertainment goes back, in fact, to

'The Goons' who began their radio programmes in 1951. Almost anything could — and did — happen in *The Goon Shows*, largely through the seemingly anarchic scripts of Spike Milligan. Without the restrictions of television, the listener, in the mind's eye, could experience an advanced form of situational anarchy. Such experiences, as we shall see later, are significantly related to the optimal functioning of human intelligence, and it is satisfying to note that *The Goon Show* has played 'a highly significant role in the history of British comedy' (Wilmut and Grafton 1977).

A protean problem which is current (1986) in the UK, and doubtless will still be so as this is being read, is that of motorway matrix signals to drivers. These signals are seldom complied with fully, apparently through experience of occasions when the signals have been left on after they are needed, or have malfunctioned, or have been switched on unnecessarily. Experience shows drivers that there is no predictable relationship between the signalled instructions and the actual conditions. For example, speed reductions are signalled when there is no hazard, or diversions are signalled when there is no reason to turn off. The situation is classically anarchistic. The urgency of most motorway journeys thus comes to override the advisability of complying with the signals just in case they are meaningful. Such anarchization in motorway signalling has undoubtedly contributed to some accidents.

It would seem that unpredictability affects most areas of human activity. The majority of dentists are aware of problems arising from unpredictable variations in the course of the trigeminal nerve, which may result in missing the spot with the hypodermic and the tooth remaining unanaesthetized. Medical practitioners are familiar with the effectively protean occurrence of sensitivity to drugs and other chemicals. Ecologists have learned that biological control organisms may behave in a quite unpredictable manner when introduced into a new habitat. Even the horrific problems associated with drug-taking are exacerbated by the protean effects of drugs upon users.

In close interpersonal relationships, protean factors are most potent. One example must suffice. Marilyn Monroe was not just a Hollywood Star with a beautiful face and an exquisite figure. The director Billy Wilder has said 'She was an absolute genius as a comic actress, with an extraordinary sense for comic dialogue. . . Nobody else is in that orbit; everyone else is earthbound by comparison'. And yet she was impossible to work with, and actors and directors alike were made ill by the experience. As Wilder put it 'When you are a director you try to be a psychologist to some extent. You must establish communication with your people. Normally I learn quickly how to size them up. Marilyn was so difficult because she was totally unpredictable' (Summers 1985).

Proteus is everywhere. From planning (or avoiding) a missile strike, to playing cricket on a West Indian pitch; from teaching a class, to greeting a spouse from work; from buying a new car, to selling oneself at interview, protean factors play roles of ever-increasing importance. The more complex life becomes, the greater the incidence of the unpredictable.

Proteus and human evolution

We have shown that protean phenomena have a significant, often crucial part to play in the conflicts and competitions of many — perhaps most — animal groups, including man. Now we must finally consider the possible significance of the protean principle in human evolution. Clearly, this is a subject that merits extensive treatment, but for the present we must be content to outline our conclusions to date.

It is an understatement to say that a clear understanding of the factors that governed the emergence of modern *Homo sapiens* from the hominid stock still eludes us. Notwithstanding the knowledge built up from the foundations laid so soundly by Charles Darwin and T.H. Huxley, man is still 'a little-known and rather aberrant primate' (Barash 1977). Much is known about when and where and in what anatomical form contemporary man became established (see, for example, Leakey 1981). It is clear that around one hundred thousand years ago man was already a cultural animal with traditions and rituals and a necessarily stable — if simple — social structure (Calder 1984). He used fire and made tools. He had a complex gestural and (probably) vocal language. And he must have been able to think — at least in a simple way. All that remained for him to become fully human — modern *Homo sapiens sapiens* — was the acquisition of spiritual, technological, syntactical, and social riches through the passage of generations and the compound interest of cultural inheritance.

This was far from being a simple problem. The transformation of a relatively weak and generalized biped into a 'hyperspecies' so vastly superior in scope and power from all the other animals that he would seem to be quite different in kind, would require an unique and transcending innovation, or at least some additional feature that — when added to the existing inventory of adaptations — would result in the whole being much more than the sum of the parts.

What was this crucial factor? What was it that guided this particular hominid through hundreds of generations of physical vulnerability to full humanity and unique success? This question may be a vitally important one, for we are now faced with problems as serious as those with which our ancestors must have been confronted at the most critical phase of human development.

Despite early man's advanced biological characteristics he was very much at risk from his environment. How many people today could go into the African bush with only what they could carry, and survive — let alone flourish? Man at his beginning, that most important period, had some special faculty, some crucial quality, which transcended all problems, overcame all difficulties, and counteracted all dangers.

It was not any — or even all — of the recognized advances that proved so crucial. Bipedalism, stereoscopic vision, facial expression, manual dexterity, vocalization, and parental solicitude all had their part to play in the evolution of *H. sapiens*. But all these features have been developed, variously, by other animals; not to such a high degree, to be sure, and not all in one organism, but even operating together all these advanced features could hardly make man superior to the raw threats of a fickee environment. Or could they? And if not, what hidden advantages might accrue from the incorporation of a syntactical language? We must look at these questions in just a little more detail.

A significant factor in human evolution has been man's development as what Konrad Lorenz (1950, 1971) has called 'the specialist in non-specialisation'. In illustration he notes that another such 'specialist' — the cosmopolitan and very successful brown rat, *Rattus norvegicus* — lacks the swimming abilities of the beaver, the burrowing abilities of the vole, the climbing abilities of the squirrel, and the jumping abilities of the jerboa. However, in each case the rat is superior to that species of fellow-rodent in the three activities which are not that species' speciality. And Lorenz draws a parallel with man, suggesting that 'any average citizen' could pass a threefold test of physical ability involving a twenty-mile march, a climb up a freely suspended twenty-foot rope, and a dive sixty feet long at twelve feet depth to retrieve a sunken object. Perhaps Lorenz could pass that test when he wrote the passage, and maybe the present authors could at the same age, but we fancy it would be too much for 'the average citizen' nowadays! However, that is perhaps not the point, which is, rather, that man, over the crucial period of his early evolution — and as a whole, not just as a small specialized group — must have been just such a paragon of physical virtues — not so much a noble savage as an early Tarzan. In fact his enormous versatility, which includes all the physical abilities we have mentioned, makes man by far the most advanced animal in morphological terms alone.

Even so, early man's lack of some advanced and necessarily restricting specialized defensive feature must have meant that he was vulnerable to attack from stronger, fiercer, faster, better armoured, and more numerous competitors — of which there were many. His versatility was not enough. And his relatively advanced intelligence was of little use without the accumulation of knowledge and technology which would take hundreds of generations.

Yet he triumphed! How he did so remains one of the most serious gaps in the accumulated knowledge of mankind. Certainly, man's success must have been helped by the development of a syntactical language which makes complex organization possible, and facilitates the invention, manufacture, and utilization of new tools, weapons, and apparatus. But this in itself cannot be the answer. Man has made enormous technological advances since then, yet he repeatedly displays a penchant for cataclysmic strife, with a manic disregard for his overall well-being. Contemporary man is so destructive, despite his intelligence, that it is difficult to see how he could have evolved at all. The seminal influence which brought the enormous human potential to fruition has still not been identified. We believe we can remedy this. The answer, we suggest, involves proteus yet again.

Let us outline our thinking on this matter. The strongest and most recent evidence indicates that *Homo sapiens sapiens* appeared rather suddenly about forty-five thousand years ago. It could be that contemporary man 'came in with a bang' through a genetic mutation 'almost certainly associated with the perfection of the machinery of the brain required for language' (Calder 1984). We must note however that what is effective biologically is not so much a mutation *per se* as the progressive dominance of a new combination of alleles, including any possible mutant allele, through natural selection over many generations. And this involves the whole organism. As de Beer (1972) put it in a lucid exposition of evolution: 'it is all the other genes forming the gene-complex of an organism which are responsible for turning any given gene into a dominant or a recessive,... and it is natural selection which selects in favour of those gene-complexes which turn disadvantageous genes into recessives'. Or into dominants, of course. But what did this transformation actually *do*?

We believe that it was not the development of a verbal language *per se* which was crucial, but the strategy of thinking which is necessary for the optimal use of that language as seen in recent populations. Or rather in sections of recent populations. For the exceptions prove our rule. Large numbers, probably the majority of people today, certainly do not use their brains to anything like their full potential, yet they could do so given an appropriate early environment. They have been deprived, throughout childhood and adolescence, of the stimulation and opportunity to explore and express themselves in a controlled and disciplined way within a naturally varied and *unpredictably varying* environment. The exquisite machinery of the mind, so enormously in advance of anything man-made or anything evolved hitherto, is allowed to lie unformed, untried, and unappreciated. The vast capacity for self-development is doomed to stagnation.

Nowhere is this seen more clearly than in the use of language. The average English-speaker has a vocabulary maybe one-third that of a good communicator — a novelist, say, or one of the better 'popularizers' of science — and a syntactic ability to match. This is not, as some would maintain, irrelevant, for as Aldous Huxley put it (1937), 'Words form the thread on which we string our experiences'. Clearly, if this thread is wanting, our experiences will be relatively meaningless. The average person is thus unable to cope either with the varied and varying complexities of language, or with the varying and various complexities of everyday life.

These suggestions are supported by a considerable body of experimental work. It is now well known that mammals deprived of necessary experiential variety when young are limited in their behaviour when adult. Catalysed by Hebb's observations (1949) that rats reared as pets later performed much better in 'intelligence' tests than rats reared under restricted laboratory conditions, a number of workers have studied the influence of early environment on later learning abilities. Both rats and dogs reared unrestrained in a rich environment later perform significantly better in learning tests than their fellows reared under restricted circumstances (see, for example, Bingham and Griffiths 1952; Forgays and Forgays 1952; Thompson and Heron 1954; Woods 1959). In humans, maternal deprivation is an increasingly serious environmental factor in the retardation or blocking of development (Bowlby 1952, 1953, 1960, 1973). The fundamental significance of this is that maternal love and care provide the infant human with the proper stable environment for the acquisition of experiential information and the development of a balanced, enquiring mind. Skeels (1966), in a classic study, revealed quite remarkable improvements in the performance of orphan children when moved to richer (i.e. more complex, stimulating, and loving) environments than the overcrowded and understaffed orphanage in which they were originally confined.

Clearly, while man is the supreme manifestation of generalized physical ability, an early environment which is both 'loving' (caring, disciplined, reactive) and rich in variety and innovation is necessary for the development of that ability. Put another way, we (i.e. humans) demonstrate a highly evolved physical adaptability which necessitates a complex, controlled, early development for each individual.

But all this is still not enough! In order to survive 'the slings and arrows of outrageous fortune' we also require a psychological ability which will enable each individual to function behaviourally across a varied and varying range of circumstances. Each individual must, in his own lifetime, operate effectively across a range of behavioural adaptability even more complex and more varied than the range of physical adaptability which took the species thousands of generations to perfect. And (s)he must be able to respond appropriately however novel or unexpected the stimulus situation may be.

(S)he must not be deterred by the unpredictable. Here lies the key to it all — the unique element in man which makes positive, adaptive reaction to protean experience possible, even enjoyable. This must have been a major, maybe crucial, contributor to man's success, and it will be deeply rooted in human nature. Its proper deployment does of course involve a complex mix of genetic and environmental factors, some of which we have noted already. It should not seem surprising, therefore, that an individual who is deprived of appropriate experiences when young cannot cope with protean situations when adult. *Yet much of life is protean*!

So the crucial factor — giant step — *was the cerebral development which made possible an attunement to the unpredictable.* We believe that this was an essential precondition for the perfection of language, and that its progressive development was reciprocally driven by natural selection for the gradual emergence of freely-integrated, semantically open linguistic communication. The process probably began over two million years ago in *Homo habilis*, as evidenced by a suddenly accelerated increase in brain size, and culminated in *H. sapiens sapiens*. The ability to freely integrate sets of information or sets of linguistic symbols, to make disconnections and new connections, is central to human intelligence. It makes it possible to derive maximum benefit from a coherent train of experience, and it permits the psychosocial accumulation of shared language and knowledge. Thus *the final necessary stage in human development was reached when natural selection resulted in a protean facility becoming an integral part of man's intelligence.*

Now we must consider the meaning of 'intelligence', which is, at best, not very precise (Hebb 1958). 'IQ' tests are the best known and most widely used of all measures of intelligence, and certainly have some value, but there has long been considerable disagreement as to their biological respectability (Hebb 1949; Smith 1984). By their very nature, IQ tests suggest what *psychologists* think the paradigms of intelligence should be. But, as Storr (1974) put it 'Psychologists who specialise in the field of intelligence seldom have an adequate background of biology. Yet it is impossible to understand human intelligence unless one thinks biologically'.

Fortunately for us all, Donald Hebb and Ward Halstead were notable exceptions to this critical generalization. Hebb always attempted to maintain a biological perspective — to splendid effect — and Halstead's approach to the measurement of intelligence is a landmark in the study of man's higher faculties.

'Man's strategic position in the animal kingdom' wrote Halstead (1951) 'derives in part at least from the unusual range and quality of the perceptions of which he is capable, from his capacity for controlled adaptability, and from the degree of freedom through which orientation can be achieved and maintained'. Such qualities are indeed essential for the effective functioning of the human intellectual capacities in all their richness of multiple skills and adaptive variability, and combine to produce the faculty which Halstead very reasonably called *biological intelligence*.

The especial significance of biological intelligence was revealed as a result of Halstead's sophisticated psychological test procedures designed to bring out the mode of operation of the brain as 'an evolutionary machine'. Russell and Russell (1961), who survey much of Halstead's work, note that

> 'An evolutionary machine is made up of a very large number of component units, richly interconnected, the relations between which are *not* predetermined. As various inputs impinge on the machine, the small units become linked together into larger transient combinations, which shift and merge in kaleidoscopic fashion, like political parties in the parliamentary assemblies of some modern countries. The question can therefore arise, which of these transient functional groupings will *survive* as larger single components, and this may depend on a process of *selection*. The process is thus comparable with the natural selection of successful organisms and species in the grand procession of organic evolution, and we can apply to events in the individual brain all the principles observed on that more spacious stage.'

Here we have the objective clarification of what we *mean* by intelligence — the highest manifestation of a universal evolutionary process, the parameters of which are defined. It is entirely logical to assume that there is a basic evolutionary process common to all levels of operation (Russell 1961; Driver 1968). The essence of this process involves an objective measurement of all pertinent information, and an adaptive response. Halstead's programme to test for such a process was much more reasonable than operations based on the widespread assumption that intelligence is what is revealed by intelligence tests!

Halstead's tests were well-tried and proven procedures, and were given to hundreds of subjects, of many kinds, including brain-operates. Sophisticated analysis of the many thousands of responses suggested that four distinct principles, or intelligence factors, were involved in their production, and these were explored in more detail by further testing. For the sake of general clarity we will call these factors Exploration, Extraction, Integration, and Expression, according to their function.

'Expression' applies to the specific expression of an individual's abilities, i.e. the way in which his faculties are especially manifested. Many popular terms refer to this factor, including 'green fingers', 'greasy thumb', 'gift of the gab'. The overt expression of intelligence necessarily involves the use of some particular skill or skills, and this of course varies considerably between individuals. Some people can perform many tasks well; others may perform poorly, except for one outstanding skill. A well-developed Expression factor will thus mediate a special flair or talent.

Before an individual can express himself, or even register information, he has to have an input on which to work. This is governed by the 'Exploration' factor, by means of which a continual flow of new information is provided, through internal or external exploration or both. Exploration is thus the means of introducing variation, as well as monitoring the environment. Clearly, the great explorers — of matters both external and internal to man — have possessed a well-developed Exploration factor. But further, to explore is to appreciate and tolerate uncertainty and to realize that something better may turn up. As the Russells pointed out (Russell and Russell 1961) the gift that makes the great explorer, scientist, or artist is not just the capacity to frame an hypothesis or mode of expression, but also the ability to scrap them and start again. Internal exploration, of course, involves the recall and rearrangement of information previously received, and is connected with imaginative insight — a faculty almost entirely restricted to man.

'Extraction', or abstraction, is the factor controlling the recognition and isolation of the specific from the general, finding the odd one out, seeing the trees as well as the wood. By its means one abstracts the individuals in a complex, the principles which operate in a mass. It is to this Extraction ability that the detective story appeals, necessitating the separation of red herrings from the true path. Without it one could never see things as they really are, and so the accuracy of one's view of the world depends very largely upon this factor.

Exploration and Extraction bring in information. 'Integration' assesses that information and enables an adaptive response to be made. It allows new data to be compared with old, and makes it possible to qualify a generalization. It involves a capacity to learn *and* unlearn, with neurological flexibility achieved by free communication within the brain. Any fact is thus freely available for recall and comparison with others. A highly developed Integration factor is thus closely connected with balanced open-mindedness and psychological adaptability. This constant ability to re-arrange and re-organize is a fundamental of a progressive evolutionary machine.

Integration is closely connected with wit and a sense of humour which, as is often said, is an appreciation of the incongruous. It requires an ability to put two ideas or thoughts — including one's own thoughts — in juxtaposition, and see how crazily they clash. We have mentioned the Goons in this context. They stood splendidly if anarchically alone until Douglas Adams (1985) was able to integrate space-age science fiction with a Goon-like sense of humour. On this matter of wit and humour we will refer to just one other source. J.B.S. Haldane, in his delightful collection of writings *The inequality of man.* (1932), reviewing Bertrand Russell's book *The scientific outlook*, wrote

> Like all Russell's books, including *Principia Mathematica*, this is exceedingly witty. Wit consists in the unexpected but appropriate juxtaposition of ideas, and it was just the capacity for such a juxtaposition which made him a great mathematician. Thus we read of physicists: 'Only mathematics and mathematical logic can say as little as the physicist means to say', and of psychoanalysts: 'I suppose that for practical purposes "phantasy" is what the patient believes, and "reality" what the analyst believes.'

True wit and humour act as a mirror upon society, a kind of feedback report on the state of integration of our experience. With Exploration continually bringing in information, and Extraction selecting and incorporating the data, Integration is constantly updating its intellectual capacity. When functioning efficiently (many psychopathologies impair this) it is quite objective, and automatically involves the capacity to unlearn. The key properties of human intelligence, therefore, as seen in this bio-evolutionary light, are freedom of choice, free information flow, and adaptability — which are also key factors in natural selection. Much work of high quality has been done on the statistical aspects of 'IQ', but none of it is as significant as this biological approach to intelligence.

It will be clear to the perceptive reader that biological intelligence — and particularly Integration — is closely connected with the ability to adapt to the unpredictable. The tolerance of uncertainty, the positive appreciation of change, and the ability to unlearn are all necessary for a facility with protean phenomena. And further, the finding that the operation of biological intelligence is closely connected with the proper functioning of the cerebral neocortex is entirely to be expected, from the evolutionary point of view. The neocortex is the most recent evolutionary brain development, and its extreme expansion is by far the most striking difference between man and his nearest relatives. Furthermore, the 'highest' intelligence factor, Integration, would seem to be intimately connected with the frontal lobes, the most recently developed structures of all. Indeed, Nielsen (1951) remarked 'Dr. Halstead is the only psychologist I have ever heard of who can tell by his psychological tests that the frontal lobes have been taken off.'

Halstead's conclusions, drawn from a vast amount of quantitative data, blend quite well with the recent theory of multiple intelligences (Gardner 1984). This holistic approach — developmental, cross-cultural, evolutionary and historical — suggests that there are seven relatively autonomous intellectual competencies or intelligences: logical–mathematical, musical, spatial, linguistic, bodily-kinaesthetic, interpersonal, and intrapersonal. These intellectual competencies could well represent the principal modes of specific Expression, and would presumably be governed in their development by Exploration, Extraction, and Integration.

It seems that all that was lacking for 'pre-sapiens' man to fulfil his potential was the ability to integrate the unpredictable. The crucial point was not simply the ability to incorporate information; apes and computers can be taught that, albeit primitively. *The facility to rapidly unlearn as well as learn, the readiness to adapt to changing circumstances, a mind attuned to the likelihood of variation, and a positive interest in differences*, combined to make possible a super-organism that can cope with the unpredictable. Here is the essence of the big biological bang of forty-five thousand years ago, the prime mover of the great leap forward. The deciding factor was the ability to face up to the protean features of an anarchically diverse, often unpredictable environment.

This is such a reasonable, even obvious point when seen clearly. As Desmond Morris has noted (pers. comm., 1981) early man had to be attuned to the unpredictable in order to survive. He had to have a positive interest in a highly varied and continually varying potential supply of animals and plants. As a hunter–gatherer he had to be competent to deal with a great variety of species occurring in changing circumstances. In this respect he was probably similar to the Kung of the Kalahari who, almost to this day, have foraged for one hundred or more species of plants and hunted fifty or more species of animals (Leakey 1981). Or he may have resembled the North American Indians who, as a whole, are reliably estimated to have consumed over 2500 species of animals and plants. 'This suggests tremendous knowledge of the habits of the animals and an equally vast acquaintance with the nutritional qualities of the plants' (Driver 1969). And this does not take into account the many organisms which were utilized in other ways: for medicine, clothing, utensils, tools, fuel, transport, housing, pleasure, and religion.

Early man also had a varying complex of enemies and competitors to cope with, each separate entity requiring its own special treatment. He was necessarily in tune with the unexpected — he could not have survived otherwise. The adaptive strength and evolutionary resilience of the early hunter–gatherers is represented to this day in a few remote groups who, for one reason or another, have retained the old successful way of life involving little technology but a great variety of skills. Such groups include some of the Australian Aborigines, and the Waorani of Ecuador (Ortiz de Villalba 1980). Anthropological studies have indicated what a small step it is from such a generalized utilization of natural produce to a concentration upon either animal or plant-based nutrition. Animals form the bulk of the diet (still!) of some Inuit (Eskimaux) hunters, and of the Dinka cattle-herders (Ryle *et al.* 1982). The Akha of the forests of Eastern Asia, on the other hand, while retaining self-sufficiency and independence, rely heavily on their rice fields in hill forest clearings (Grunfeld *et al.* 1982). All retain a variety of skills which put the average Western city-dweller to shame, however.

With the development of agriculture and a more complex social organization, child rearing and other practices became restrictive, producing the conformist adult upon whom the continuity of such societies depends. Industrialization too in many ways encourages conformity, for it requires strict formalization and predictable behaviour in its complex productive and administrative organizations. Thus, necessarily, acquiescence gradually took precedence over adventure; rules and regulations came to be followed without question. Now, all too often, regimentation is preferred to experimentation. But by itself this leads inexorably to stultification. As J.F. Kennedy (or his speech-writer!) put it 'Conformity is the jailer of freedom and the enemy of growth'. When rigidly adhered to it is counter-adaptive, anti-evolutionary, and ultimately fatal. Stultification of an interest in the unpredictable leads to dislike of the unusual and fear of the unknown, and thence to an inability to change.

Fortunately, in industrial society the need to generate new markets and to satisfy new demands has increasingly encouraged non-conformity, leading to innovations in thought and hence in organization, products, and consumer behaviour. Further, with the recent development of significant leisure time the 'nine-to-five' conformist can express his natural fascination with the anarchic, and commercial interests encourage this and profit from it. Thus an interest in the unpredictable is seen universally in sporting activities and in gambling, and is reflected in the enormous contemporary surge of interest in the unknown: in things mystical, supernatural, occult, paranormal, astrological, and extra-terrestrial, and in the corresponding boom in sales of associated literature and software. It lies beneath much eccentricity, and the general interest in eccentrics (see, for example, Caulfield 1982). As we noted earlier, unpredictability appears more and more in media games, puzzles, and quizzes. And over the last fifty years or so it has played an increasing role in radio, TV, and big-screen humour. Inside every conformer the anarchist is trying to escape.

Finally, we have to note that while an interest in the unpredictable reflects the highest, uniquely human intelligence factor, it is a two-edged sword. Recognition of the protean principle can provide an offensive unit — individual, communal, or national — with a weapon which confers immediate superiority in the arms race; but fortunately it is just as useful in defence. Protean factors appear repeatedly in human battles, but it is to be hoped that an understanding of the protean principle will make war less likely rather than more efficient. It might lead to a better understanding of biological systems, and hence to an effective appreciation of the stupidity of immutability and unnecessary conflict.

We may summarize our analysis of human evolution as follows. Over a period, from more than two million years ago to around 50000 years ago, man developed many advanced features, yet physically he was relatively unspecialized and potentially able to adapt to almost any ecological niche that confronted him. But because of his still limited intelligence, his unspecialized form left him vulnerable to many environmental problems. He was, as it were, a star waiting to be born.

And then, potentiated by the increasingly free integration of information, signals, and concepts demanded by the evolution of language, came the most important transformation of all time. It involved the mechanism of intelligence and made it possible to instantaneously unlearn. Man was already capable of insight learning and the formation of simple concepts, as well as being a skilled tool-user. He probably lived in small groups, with a sexual division of labour — the standard hunter–gatherer way of life. He had taken the principal modes of behavioural evolution — instinct, learning, and intelligence — to their highest logical state of development. Now there came a mutant group suddenly capable of *unlearning*, of adaptively modifying any aspect of their behavioural repertoire *immediately*, not by trial and error through the passage of generations. They were not surprised by the unexpected, or thrown by the 'odd one out'. This group was in tune with the unpredictable.

Such is the power of this attunement that it must have rapidly made its possessors superior to most enemies, competitors, and natural hazards. Natural selection of this fitter genotype, reinforced by psychosocial evolution, would have achieved this in a few generations.

Man came thus to apply his extensive but unspecialized abilities across a vastly increased spectrum of interaction with environment. Whatever the problem, no matter what change, he was able to cope. He could continually rearrange and modify (re-integrate) his strategies and tactics. True intelligence, after all, means seeing things as they really are.

Conclusion

We believe Charles Darwin and Alfred Russell Wallace would have enjoyed this theory of anarchization with its broad spectrum of controlled observation, explanations of otherwise confusing examples of variability, and the synthesis of a wide range of organic phenomena in one relatively simple idea. Indeed, through our work in this area we believe we may echo Darwin's immortal understatement (1859) that 'Light will be thrown on the origin of man and his history'. The human developments Darwin uncovered — culminating, we suggest, in the incorporation of a protean facility — established man as the all-time dominant organism; so dominant that, at the species level, he has recently been able to effectively ignore the lessons of his past. From time to time this neglect has brought about the downfall of this nation or that empire, this class or that movement. People of course have sought the reasons for these catastrophes, but always in the wrong places. People as a whole, *scientists* as a whole, even — we have to say — *biologists* as a whole do not understand organic evolution, let alone accept it as a fundamental in human existence. Yet it seems to us, as it did to Darwin (1871)

> that man with all his noble qualities, with sympathy which feels for the most debased, with benevolence which extends not only to other men but to the humblest living creature, with his god-like intellect which has penetrated into the movements and constitution of the solar system — with all these exalted powers — Man still bears in his bodily frame the indelible stamp of his lowly origin.

Clearly, if the principle of anarchization has been as important in human evolution as we suggest, then it may be a subject of more than academic interest.

References

Adams, D. (1985). *The hitch hiker's guide to the galaxy*. Heinemann, London.

Alder, J. and Hancock, A. (1864). Notice of a collection of nudibranchiate Mollusca made in India by Walter Elliot, Esq., with descriptions of several new genera and species. *Trans. Zool. Soc. Lond.* **5**, 113–47

Allee, W.C., Emerson, A.E., Park, O., Park, T., and Schmidt, K.P. (1949). *Principles of animal ecology*. Saunders, Philadelphia.

Allen, J.A. (1972*a*). Apostatic selection: the responses of wild passerines to artificial polymorphic prey. Ph.D. thesis, University of Nottingham.

Allen, J.A. (1972*b*). Evidence for stabilising and apostatic selection by wild blackbirds. *Nature*, London **237**, 348–9.

Allen, J.A. (1974). Further evidence for apostatic selection by wild passerine birds: training experiments. *Heredity, Lond.* **33**, 361–72.

Allen, J.A. (1975). Further evidence for apostatic selection by wild passerine birds: 9:1 experiments. *Heredity, Lond.* **36**, 173–80.

Allen, J.A. and Clarke, B. (1968). Evidence for apostatic selection by wild passerines. *Nature, London* **220**, 501–2.

Allen, W.E. (1920*a*). Behavior of loon and sardines. *Ecology* **1**, 309–10.

Allen, W.E. (1920*b*). Behavior of feeding mackerel. *Ecology* **1**, 310.

Altmann, S.A. (1956). Avian mobbing behaviour and predator recognition. *Condor* **58**, 241–3.

Altmann, S.A. and Altmann, J. (1970). *Baboon ecology: African field research*. University of Chicago Press, Chicago.

Andersson, M. and Wiklund, C.G. (1978). Clumping versus spacing out: experiments on nest predation in fieldfares (*Turdus pilaris*). *Anim. Behav.* **26**, 1207–12.

Andrew, R.J. (1956). Some remarks on behaviour in conflict situations with special reference to *Emberiza* spp. *Br. J. Anim. Behav.* **4**, 41–5.

Andrew, R.J. (1961*a*). The motivational organisation controlling the mobbing calls of the blackbird (*Turdus merula*). L. Effects of flight on mobbing calls. *Behaviour* **17**, 224–46.

Andrew, R.J. (1961*b*). The motivational organization controlling the mobbing calls of the blackbird (*Turdus merula*). II. The quantitative analysis of changes in the motivation of calling. *Behaviour* **17**, 228–321.

Andrew, R.J. (1961*c*). The motivational organisation controlling the mobbing calls of the blackbird (*Turdus merula*). III. Changes in the intensity of mobbing due to changes in the effect of the owl or to the progressive waning of mobbing. *Behaviour* **18**, 25–43.

Andrew R.J. (1961*d*). The motivational organisation controlling the mobbing calls of the blackbird (*Turdus merula*). IV. A general discussion of the calls of the blackbird and certain other passerines. *Behaviour* **18**, 161–76.

Annandale, N. (1900). Observations on the habits and natural surroundings of insects made during the Skeat Expedition to the Malay Peninsula, 1899–1900. *Proc. Zool. Soc. London* **70**, 837–68.

Ansell, A.D. (1969*a*). Leaping movements in the Bivalvia. *Proc. Malac. Soc. London* **38**, 387–99.

Ansell, A.D. (1969*b*). Defensive adaptations to predation in the Mollusca. *Proceedings of Symposium on Mollusca*, Part 2, 487–512. Marine Biological Association of India.

Anthony, H.E. (1928). *Field book of North American mammals.* Putnam, New York.

Armstrong, E.A. (1947). *Bird display and behaviour.* Lindsay Drummond, London.

Armstrong, E.A. (1949*a*). Diversionary display, Part I. Connotation and terminology. *Ibis* **91**, 88–97.

Armstrong, E.A. (1949*b*). Diversionary display, Part II. The nature and origin of distraction display. *Ibis* **91**, 179–88.

Armstrong, E.A. (1954). The ecology of distraction display. *Br. J. Anim. Behav.* **2**, 121–35.

Armstrong, E.A. (1956). Distraction display and the human predator. *Ibis* **98**, 641–54.

Arnold, R. (1969). The effects of selection by climate on the land-snail *Cepaea nemoralis* (L.). *Evolution* **23**, 370–8, Lancaster, Pa.

Attneave, F. (1959). *Applications of information theory to psychology.* Holt, New York.

Atwal, A.S. and Seth, S.L. (1963). Predation by *Coccinella septempunctata* L. on the cabbage aphid *Lipaphis erysimi* (Kalt.) in India. *J. Anim. Ecol.* **32**, 481–8.

Audubon, J.J. (1827–1838). *Birds of America.* Audubon, London.

Avery, M.I. (1984). Lekking in birds: choice, competition and reproductive constraints. *Ibis* **126**, 177–88.

Axelrod, R. (1984). *The evolution of co-operation.* Harper and Row, London.

Baerends, G.P. (1950). Specialisations in organs and movements with a releasing function. *Symp. Soc. Exp. Biol.* **4**, 337–60.

Baerends, G.P. and Baerends-van Boon, J.M. (1950). An introduction to the ethology of cichlid fishes. *Behaviour Suppl.* **1**, 1–242.

Baier, L.F. (1930). Contribution to the physiology of the stridulation and hearing of insects. *Zool. Jb.* **47**, 151–248.

Baird, W. (1850). *Natural history of the British Entomostraca.* Ray Society, London.

Baker, R.B. and Bibby, C.J. (1987). Merlin *Falco columbarius* predation and theories of the evolution of bird coloration. *Ibis* **129**, 259–63.

Baker, R.R. (1985). Bird coloration: in defence of unprofitable prey. *Anim. Behav.* **33**, 1387–8.

Baker, R.R. and Parker, G.A. (1979). The evolution of bird coloration. *Phil. Trans. Roy. Soc. B.* **287**, 63–130.

Banks, C.J. (1957). The behaviour of individual coccinellid larvae on plants. *Br. J. Anim. Behav.* **5**, 12–24.

Barash, D.P. (1976). Some evolutionary aspects of parental behaviour in animals and man. *Am. J. Psychol.* **89**, 195–217.

Barash, D.P. (1977). *Sociobiology and behavior*. Elsevier, New York.

Barber, B. (1961). Resistance by scientists to scientific discovery. *Science* **134**, 596–602.

Bartholomew, G.A. (1942). The fishing activities of double-crested cormorants on San Francisco Bay. *Condor* **44**, 13–21.

Bastock, M., Morris, D., and Moynihan, M. (1953). Some comments on conflict and thwarting in animals. *Behaviour* **6**, 66–84.

Batten, L. (1969). Dalmatian pelican. In *Birds of the world* Vol. 1, pp. 80–1. IPC Magazines, London.

Beebe, W. (1935). *Half Mile Down.* Harcourt, Brace, New York.

Bell, T.R.D. and Scott, F.B. (1936). *The Fauna of British India including Ceylon and Burma.* Vol. 5: *Moths.* Taylor and Francis, London.

Bergman, G. 1941. Om simfåglars och vadares hackning i måsfågel-kolonier. *Dansk Orn. Foren. Tidsskr.* **35**, 120–3.

Berlyne, D.E. (1975*a*). Uncertainty and conflict: a point of contact between information-theory and behaviour-theory concepts. *Psychol. Rev.* **64**, 329–39.

Berlyne, D.E. (1957*b*). Conflict and choice time. *Br. J. Psychol.* **48**, 106–18.

Berlyne, D.E. (1958). The influence of complexity and novelty in visual figures on orienting responses. *J. exp. Psychol.* **55**, 289–96.

Berlyne, D.E. (1960). *Conflict, arousal and curiosity.* McGraw-Hill, London.

Bernard, C. (1865). *An introduction to the study of experimental medicine.* 1957 Dover Edition, New York.

Bertelson, P. (1961). Sequential redundancy and speed in a serial two-choice responding task. *Q. J. Exp. Psychol.* **13**, 90–102.

Bertelson, P. and Renkin, A. (1966). Reaction times to new versus repeated signals in a serial task as a function of response — signal time interval *Acta Psychol.* **25**, 132–6.

Bertram, B.C.R. (1978). Living in groups: predators and prey. *In: Behavioural ecology: an evolutionary approach.* Blackwell, Oxford.

Bertram, B.C.R. (1979). Serengeti predators and their social systems. In *Serengeti: the dynamics of the ecosystem* (ed. M. Norton-Griffiths and A.R.E. Sinclair). Chicago University Press, Chicago.

Bertsch, H. (1968). Effects of feeding by *Armina californica* on the bioluminescence of *Renilla Koellikeri. Veliger* **10**, 440–1.

Beukema, J.J. (1968). Predation by the three-spined stickleback (*Gasterosteus aculeatus L.*): the influence of hunger and experience. *Behaviour* **31**, 1–126.

Bevan, W. (1955). Sound-precipitated convulsions: 1947–1954. *Psychol. Bull.* **52**, 473–504.

Beven, G. (1946). "Injury-feigning" of a blackbird. *British Birds.* **39**, 248.

Bingham, W.E. and Griffiths, W.J. (1952). The effect of different environments during infancy on adult behavior in the rat. *J. comp. physiol. Psychol.* **45**, 307–12.

Blanford, W.T. (1888–91). *The fauna of British India: Mammalia.* Taylor and Francis, London.

Blaxter, J.H.S. and Parrish, B.B. (1958). The effect of artificial lights on fish and other marine organisms at sea. *Marine Research, Scotland, Dept. of Agriculture and Fisheries,* No. 2.

Blest, A.D. (1957*a*). The function of eyespot patterns in the Lepidoptera. *Behaviour* **11**, 209–55.

Blest, A.D. (1957*b*). The evolution of protective displays in the Saturnoidea and Sphingidae (Lepidoptera). *Behaviour* **11**, 257–309.

Blest, A.D. (1961). The concept of ritualization. In *Current problems in animal behaviour* (ed. W.H. Thorpe and O.L. Zangwill). Cambridge University Press, Cambridge.

Blest, A.D. (1963). Relations between moths and predators. *Nature, Lond.* **197**, 1046–7.

Blest, A.D. (1964). Protective display and sound production in some New World arctiid and ctenuchid moths. *Zoologica, N.Y.* **49**, 161–81.

Blest, A.D., Collett, T.S., and Pye, J.D. (1963). The generation of ultrasonic signals by a New World arctiid moth. *Proc. R. Soc. Lond. (B)* **158**, 196–207.

Bloom, B.R. (1979). Games parasites play; how parasites evade immune surveillance. *Nature Lond.* **279**, 21–6.

Bowen, E.S. (1931). The role of the sense organs in aggregations of *Ameiurus melas. Ecol. Monogr.* **1**, 1–35.

Bowlby, J. (1952). Critical phases in the development of social responses in man and other animals. In *Prospects in psychiatric research* (ed. J.M. Tanner). Oxford University Press, Oxford.

Bowlby, J. (1953). *Child care and the growth of love.* Penguin Books, Harmondsworth.

Bowlby, J. (1960). Separation anxiety. *Int. J. Psycho-anal.* **41**, 89–113.

Bowlby, J. (1973). *Separation: anxiety and loss.* Basic Books, New York.

Boycott, B.B. (1958). The cuttlefish — *Sepia. New Biology* **25**, 98–118.

Boycott, B.B. (1965). A comparison of living *Sepioteuthis sepioidea and Doryteuthis plei* with other squids and with *Sepia officinalis. Proc. zool. Soc. Lond.* **147**, 344–51.

Breder, C.M. (1959). Studies on social groupings in fishes. *Bull. Am. Mus. nat. Hist.* **117**, 393–481.

Breder, C.M. (1967). On the survival value of fish schools. *Zoologica, NY* **52**, 25–40.

Breder, C.M. and Halpern, F. (1946). Innate and acquired behavior affecting the aggregation of fishes. *Physiol. Zool.* **19**, 154–92.

Briggs, G.E. and Swansea, J.M. (1971). Encoding, decoding, and central functions in human information processing. *J. exp. Psychol.* **86**, 296–308.

Bristowe, W.S. (1958). *The world of spiders.* Collins, London.

Bristowe, W.S. (1972). Spiders. In *Encyclopedia of the animal world.* Elsevier International, London.

British Trust for Ornithology. Bird ringing comm. (not dated). *Rules and instructions.*

Broadbent, D.E. (1958). *Perception and communication.* Pergamon Press, London.

Brock, V.E. and Riffenburgh, R.H. (1960). Fish schooling: a possible factor in reducing predation. *J. Cons. perm. into Explor. Mer.* **25**, 307–17.

Broughton, W.B. (1965). Song learning among grasshoppers. *New Scientist*, Aug. 5, 1965, 338–41.

Brower, L.P. (1958). Bird predation and food-plant specificity in closely related procryptic insects. *Am. Nat.* **92**, 183–7.

Brower, L.P. (1979). Ecological chemistry. *Scient. Am.* **220**, 22–9.

Brower, L.P. (1971). Prey coloration and predator behaviour. In *Topics in the study of life. The BIO source book*, section 6, Animal behavior. Harper and Row, New York.

Brown, R.G.B. (1962). The aggressive and distraction behaviour of the western sandpiper *Ereunetes mauri. Ibis.* **104**, 1–12.

Brown, R.H. (1939). Notes on the lapwing and curlew breeding populations on a Cumberland farm. *British Birds* **33**, 12–15.

Brown, J.S. and Farber, I.E. (1951). Emotions conceptualised as intervening variables with suggestions towards a theory of frustration. *Psychol. Bull.* **38**, 465–95.

Bruner, J.S. and Postman, L.J. (1949). On the perception of incongruity: a paradigm. *J. Personality* **18**, 206–23.

Buck, G.B. (1954). Bioluminescence in *Renilla* in relation to nerve net physiology. Unpublished report (see Bullock 1958).

Buck, J.B. (1978). Functions and evolutions of bioluminescence. In *Bioluminescence in action* (ed. P.J. Herring). Academic Press, London.

Bullis, M.R. (1961). Observations on the feeding behavior of whitetip sharks on schooling fishes. *Ecology* **42**, 194–5.

Bullock, T.H. (1953). Predator recognition and escape responses of some intertidal gastropods in presence of starfish. *Behaviour* **5**, 130–41.

Bullock, T.H. (1958). Evolution of neurophysiological mechanisms. In *Behaviour and evolution* (ed. A. Roe and G.G. Simpson). Yale University Press, New Haven.

Burgers, J.M. (1963). On the emergence of patterns of order. *Bull. Am. math. Soc.* **69**, 1–25.

Burke, V.E.M. and Brown, L.H. (1970). Observations on the breeding of the pink-backed pelican *Pelecanus rufescens. Ibis* **112**, 499–512.

Burton, P.J.K. and Thurston, M.H. (1959). Observations on arctic terns in Spitsbergen. *British Birds* **52**, 149–61.

Burtt, E.T. and Catton, W.T. (1966). Image formation and sensory transmission in the compound eye. *Adv. Ins. Physiol.* **3**, 2–46.

Butterfield, H. (1949). *The origins of modern science, 1300–1800.* G. Bellow and Sons, London.

Butynski, T.M. (1979). Reproductive ecology of the Springhaas, *Pedetes capensis* in Botswana. *J. Zool., London* **189**, 221–32.

Cain, A.J. (1963). *Animal species and their evolution.* Hutchinson, London.

Cain, A.J. (1977). The uniqueness of the polymorphism of *Cepaea* (Pulmonata: Helicidae) in Western Europe. *J. Conchology* **29**, 129–36.

Cain, A.J. (1979). Introduction to general discussion. In: The evolution of adaptation by natural selection. *Proc. R. Soc. London (B)* 205, 599–604.

Cain, A.J. (1982). Heterosematism in snails. *Malacologia* **25**, 161–72.
Cain, A.J. and Sheppard, P.M. (1954). Natural selection in *Cepaea. Genetics Princeton* **39**, 89–116.
Calder, N. (1984). *Timescale.* Chatto and Windus, London.
Campbell, A.J. (1901). *Nests and eggs of Australian birds.* Sheffield.
Campbell, B. (1969). Frog's screams. In *The countryman wildlife book* (ed. B. Campbell). David & Charles, Newton Abbot.
Caraco, T. and Wolf, L.L. (1975). Ecological determinants of group sizes of foraging lions. *Am. Nat.* **109**, 343–52.
Carpenter, G.D.H. (1921). Experiments on the relative edibility of insects, with special reference to their coloration. *Trans. ent. Soc. London* **69**, 1–105.
Carpenter, G.D.H. (1933). Resemblance to a snake. In *Mimicry* (ed. G.D.H. Carpenter and E.B. Ford), Methuen, London.
Carrington, H. and Meader, J.R. (1911). *Death: Its causes and phenomena.* Rider, London.
Caryl, P.G., (1979). Communication by agonistic displays: what can game theory contribute to ethology? *Behaviour* **68**, 136–69.
Caulfield, C. (1982). *The emperor of the United States of America and other magnificent British eccentrics.* Corgi Books, London.
Chance, M.R.A. (1954). The suppression of audiogenic hyperexcitement by learning in *Peromyscus maniculatus. Br. J. Anim. Behav.* **2**, 31–5.
Chance, M.R.A. (1956). Social structure of a colony of *Macaca mulatta. Brit. J. Anim. Behav.* **4**, 1–13.
Chance, M.R.A. (1957). The role of convulsions in behaviour. *Behav. Sci.* **2**, 30–45.
Chance, M.R.A. (1956). Polyethism — cryptic behaviour. Proc. 15th. Int. Congress Zoology. Section II (1).
Chance, M.R.A. and Russell, W.M.S. (1959). Protean displays: a form of allaesthetic behaviour. *Proc. zool. Soc. London* **132**, 65–70.
Chance, M.R.A. and Yaxley, D.C. (1949). New aspects of the behaviour of *Peromyscus* under audiogenic hyperexcitement. *Behaviour* **2**, 96–105.
Chandler, A.E.F. (1969). Locomotory behaviour of first instar larvae of aphidophagous Syrphidae (Diptera) after contact with aphids. *Anim. Behav.* **17**, 673–8.
Chapman, A. (1897). *Wild Norway.* Arnold, London.
Cherry, C. (1957). *On human communication.* Wiley, New York.
Churchill, W.S. (1948). *The Second World War.* I. The gathering storm. 2. Their finest hour. Cassell, London.
Claridge, M.F. and Edington, J.M. (1965). Bilateral asymmetry: a larval polymorphism in some argid sawflies (Hymenoptera, Symphyta). *Nature, London* **208**, 301–2.
Clarke, A.H. (1965). The scallop superspecies *Aequipecten irradians* (Lamarck). *Malacologia* **2** (2), 161–88.
Clarke, A.H. (1978). Polymorphism in marine mollusks and biome development. *Smithsonian Contributions to Zoology*, No. 274. Smithsonian Institution, Washington.

Clarke, B.C. (1962*a*). Balanced polymorphism and the diversity of sympatric species. In *Taxonomy and geography* (ed. D. Nichols). Systematics Association, Oxford.

Clarke, B.C. (1962*b*). Natural selection in mixed populations of two polymorphic snails. *Heredity* **17**, 319–45.

Clarke, B.C. (1964). Frequency-dependent selection for the dominance of rare polymorphic genes. *Evolution* **18**, 364–9.

Clarke, B.C. (1969). The evidence for apostatic selection. *Heredity London* **24**, 347–52.

Clarke, B.C. (1976). The ecological genetics of host-parasite relationships. In *Genetic aspects of host-parasite relationships* (ed. A.E.R. Taylor and R. Muller). Blackwell, Oxford.

Clarke, B.C. (1979). The evolution of genetic diversity. *Proc. R. Soc. London (B)* **208**, 453–74.

Clarke, C.A. and Sheppard, P.M. (1971). Further studies on the mimetic butterfly *Papilio memnon* L. *Phil. Trans. R. Soc. London* **B263**, 35–70.

Colyer, C.N. and Hammond, C.O. (1951). *Flies of the British Isles*. Warne, London.

Conway, N.J. (1972*a*). Cheetah. In *Encyclopedia of the animal world*. Elsevier International, London.

Conway, N.J. (1972b). Hyaenas. In *Encyclopedia of the animal world*. Elsevier International, London.

Cook, L.M. and Miller, P. (1977). Density-dependent selection on polymorphic prey — some data. *Am. Nat.* **111**, 594–8.

Cooke, A.H. (1895). Molluscs. In *The Cambridge natural history* (Vol. 3). (ed. S.F. Harmer and A.E. Shipley). Macmillan, London.

Coppinger, R.P. (1970). The effect of experience and novelty on avian feeding behavior with reference to the evolution of warning coloration in butterflies. II. Reactions of naive birds to novel insects. *Am. Nat.* **104**, 323–35.

Cornwallis, R.K. (1964). Irruption. In *A new dictionary of birds* (ed. Sir A. Landsborough Thomson) Nelson, London.

Cott, H.B. (1940). *Adaptive coloration in animals*. Methuen, London.

Coues, E., (1875). Habits of the prairie hare. *Bull. Essex Inst.*, 7, 73–85.

Coward, T.A., (1950). *The birds of the British Isles and their eggs*. Warne, London.

Cramp S. (1969). White pelican. *Birds of the World* Vol. pp. 3, 74–8. IPC Magazines, London.

Crook, J.H. (1965). The adaptive significance of avian social organisations. *Symp. zool. Soc. London* **14**, 181–218.

Crook, J.H. and Gartlan, J.S., (1966). Evolution of primate societies. *Nature London* **210**, 1200–3.

Croze, H.J. (1970). Searching image in carrion crows. *Z. Tierpsychol.* **5**, 1–85.

Cruickshank, C. (1979). *Deception in World War II*. Oxford University Press, Oxford.

Curio, E. (1959). Verhaltensstudien am Trauerschnapper. *Z. Tierpsychol.* Beiheft **3**, 1–118.

Curio, E. (1965). Die Schutzanpassungen dreier Raupen eines Schwärmes (Lepidopt., Sphingidae) auf Galapagos. *Zool. Jb. Syst.* **92**, 487–522.
Curio, E. (1970*a*). Die Messung des Selektionswertes einer Verhaltensweise. *Verh. dt. zool. Ges.* **64**, 348–52.
Curio, E. (1970*b*). Die Selektion dreier Raupen former eines Schwärmers (Lepidopt., Sphingidae) durch einen *Anolis* (Rept., Iguanidae). *Z. Tierpsychol.* **27**, 899–914.
Curio, E. (1970*c*). Validity of the selective coefficient of a behaviour trait in hawkmoth larvae. *Nature, London* **228**, 382.
Curio, E. (1976). *The ethology of predation.* Springer-Verlag, Berlin.
Cushing, D.H. and Harden-Jones, F.R. (1968). Why do fish school? *Nature, London* **218**, 714–920.
Debelsteen, T. (1978). An analysis of the song-flight of the lapwing (*Vanellus vanellus* L.) with respect to causation, evolution and adaptations to signal function. *Behaviour* **66**, 136–77.
Darling, F. Fraser (1937). *A herd of red deer.* Oxford University Press, Oxford.
Darling, F. Fraser (1938). *Bird flocks and the breeding cycle.* Cambridge University, Press.
Darwin, C. (1859). *The origin of species.* Murray, London.
Darwin, C., (1871). *The descent of man.* Murray, London.
Davies, N.B. (1978). Territorial defence in the speckled wood butterfly *Pararge aegeria*, the resident always wins. *Anim. Behav.* **26**, 138–47.
Davies, N.B. and Krebs, J.R. (1978). Ecology, natural selection and social behaviour. In *Behavioural ecology — an Evolutionary Approach* (ed. J.R. Krebs and N.B. Davies). Blackwell, Oxford.
Davis, J.M. (1980). The co-ordinated aerobatics of dunlin flocks. *Anim. Behav.* **28**, 668–73.
Dawkins, M. (1971*a*). Perceptual changes in chicks: another look at the 'search image' concept. *Anim. Behav.* **19**, 566–74.
Dawkins, M. (1971*b*). Shifts of 'attention' in chicks during feeding. *Anim. Behav.* **19**, 575–82.
Dawkins, R. (1976*a*). *The selfish gene.* Oxford University Press. Oxford.
Dawkins, R. (1976*b*). Hierarchical organisation: a candidate principle for ethology. In *Growing points in ethology* (ed. P.P.G. Bateson and R.A. Hinde). Cambridge University Press.
Dawkins, R. (1986). The nice way to survive. *The Listener* 17 April 1986.
Dawkins, R. and Krebs, J.R. (1978). Animal signals: information or manipulation? In *Behavioural ecology — an evolutionary approach* (ed. J.R. Krebs and N.B. Davies). Blackwell, Oxford.
de Beer, G.R. (1972). Evolution. In *Encyclopedia of the animal world.* Elsevier International, London.
Denton, E.J. and Nicol, J.A.C. (1965*a*). Studies on reflexion of light from silvery surfaces of fishes, with special reference to the bleak *Alburnus alburnus. J. mar. Biol. Ass. UK* **45**, 683–703.
Denton, E.J. and Nicol, J.A.C. (1965*b*). Reflexion of light by external surfaces of the herring, *Clupea harengus. J. mar. Biol. Ass. UK* **45**, 711–38.

Denton, E.J. and Nicol, J.A.C. (1966). A survey of reflectivity in silvery teleosts. *J. mar. Biol. Ass. UK* **46**, 685–722.

Desai, M.M. (1939) *Surprise. Br. J. Psychol.* Monogr. Suppl. No. 22.

Dewar, D., (1913). *Glimpses of Indian birds.* John Lane, London.

Dice, L.R., Barto, E., and Clark, P.J. (1963). Modifications of behaviour associated with inherited convulsions or whirling in three strains of *Peromyscus. Anim. Behav.* **11**, 40–50.

Ditmars, R.L. (1954). *Snakes of the world.* MacMillan, New York.

Diver, C. (1929). *Nature, London,* **124**, 183.

Dixon, A.F.G. (1959). An experimental study of the searching behaviour of the predatory coccinellid beetle, *Adalia decempunctata* (L.). *J. anim. Ecol.* **28**, 258–81.

Dobzhansky, T. (1950). Evolution in the tropics. *Am. Scientist* **38**, 209–21.

Dobzhansky, T. (1956). What is an adaptive trait? *Am. Nat.* No. 855, 337–47.

Drever, J. (1952). *A dictionary of psychology.* Penguin Books, Harmondsworth.

Driver, H.E. (1968). *Indians of North America* (2nd edn). University of Chicago Press, Chicago.

Driver, P.M. (1958). Biological studies in Ungava during 1958. (Note.) *Arctic* **11**, 191–3.

Driver, P.M. (1960). Behaviour studies in sea-ducklings. PhD Thesis. McGill University, Montreal.

Driver, P.M. (1968). An ethological approach to the problem of mind. In *The mind: biological approaches to its function* (ed. W.C. Corning and M. Balaban). Wiley, New York.

Driver, P.M. (1969). Starling virtuoso. In *The countryman wildlife book* (ed. B. Campbell) David and Charles, Newton Abbot.

Driver, P.M. (1974). *In search of the eider.* Saturn Press, London.

Driver, P.M. and Corning, W.C. (1968). The nature of psychology. *J. biol. Psychol.* **10** 47–73.

Driver, P.M. and Humphries, D.A. (1966). Protean behaviour; systematic unpredictability in interspecific encounters. Preprint 197, Mental Health Research Institute, Ann Arbor.

Driver, P.M. and Humphries, D.A. (1969). The significance of the high-intensity alarm call in captured passerines. *Ibis* **111**, 243–4.

Driver, P.M. and Humphries, D.A. (1970). Protean displays as conflict inducers. *Nature, Lond.* **226**, 968–9.

Driver, P.M. and Midwinter, H.J. (N.D.). Voices of Birdland. *E.P. disc No. 101.* Midriver Recording Co., Andoversford, Glos.

Duffey, E., Creasey, N., and Williamson, K. (1950). The rodent-run distraction behaviour of certain waders. *Ibis* **92**, 27–33.

Dumortier, B. (1964). Ethological and physiological study of sound emissions in Arthropoda. In *Acoustic behaviour of animals* (ed. R.G. Busnel). Elsevier, Amsterdam.

Dunning, D.C. and Roeder, K.D. (1965). Moth sounds and the insect-catching behavior of bats. *Science, NY* **147**, 173–4.

Earle, S.A. (1979). The gentle whales. *National Geographic* **155**, 2–25.

Edmunds, M. (1966*a*). Protective mechanisms in the Eolidacea (Mollusca Nudibranchia). *J. Linn. Soc. (Zool)*. **46**, 27–71.

Edmunds, M. (1966*b*). Defensive adaptations of *Stiliger vanellus* Marcus, with a discussion on the evolution of 'nudibranch' molluscs. *Proc. malac. Soc. Lond.* **37**, 73–81.

Edmunds, M. (1971). Opisthobranchiate Mollusca from Tanzania (suborder Doridacea). *Zool. J. Linn. Soc.* **50**, 337–96.

Edmunds, M. (1972). Defensive behaviour in Ghanaian praying mantids. *Zool. J. Linn. Soc.* **51**, 1–32.

Edmunds, M. (1974). *Defence in Animals*. Longman England.

Eibl-Eibesfeldt, I. (1950*a*). Beitrage zur Biologie des Haus- und der Ahrenmaus nebst einiger Beobachtungen an anderen Nagern. *Z. Tierpsychol.* **7**, 558–87.

Eibl-Eibesfeldt, I. (1950*b*). Über der Jugendentwicklung des Verhaltens eines männlichen Daches (*Meles meles*) unter besonderer Berücksichtigung des Spieles. *Z. Tierpsychol.* **7**, 327–55.

Eibl-Eibesfeldt, I. (1961). The interactions of unlearned behaviour patterns and learning in mammals. In *Brain mechanisms and learning* (ed. J.F. Delafresnaye). Blackwell, Oxford.

Eibl-Eibesfeldt, I. (1962). Freiwasserbeobachtung zur Dentung des Schwärmverhaltens verschiedener Fische. *Z. Tierpsychol.* **19**, 165–82.

Eisenberg, J.F., Muckenhirn, N.A., and Rudran, R. (1972). The relation between ecology and social structure in primates. *Science, NY* **176**, 863–74.

Eisher, T., Alsop, R., and Ettershank, G. (1964). Adhesiveness of spider silk. *Science, NY* **146**, 1058–61.

Elton, C.S. (1927). *Animal ecology*. Sidgwick and Jackson, London.

Elton, C.S. (1942). *Voles, mice and lemmings*. Clarendon Press, Oxford.

Elton, R.A. and Greenwood, J.J.D. (1970). Exploring apostatic selection. *Heredity, London* **25**, 629–33.

Evans, G. (1962). The jump of the click beetle (Coleoptera, Elateridae) — a preliminary study. *J. Zool. London* **167**, 319–36.

Evans, G. (1975). *The life of beetles*. Allen and Unwin, London.

Evans-Pritchard, E. (1940). *The Nuer*. Oxford University Press, Oxford.

Ewer, R.F. (1972). The devices in the web of the West African spider *Argiope flavipalpis*. *J. nat. Hist.* **6**, 159–67.

Feder, H.M. (1972). Escape responses in marine invertebrates. *Scient. Am.* **227**, 93–100.

Fentress, J.C. (1965). Aspects of arousal and control in the behaviour of voles. PhD thesis. Cambridge University.

Ficken, M.S. and Dilger, W.C. (1960). Comments on redirection with examples of avian copulations with substitute objects. *Anim. Behav.* **8**, 219–22.

Fisher, R.A. (1930). *The genetical theory of natural selection*. Clarendon Press, Oxford.

Fisher, R.A. (1958). *The genetical theory of natural selection* (2nd edn). Dover, New York.

Fitzpatrick, J.P. (1907). *Jock of the Bushveld*. Longmans and Green, London.

Fitzsimmons, F.W. (1920). *The natural history of South Africa: mammals.* Longmans and Green, London.

Flugel, J.C. (1948). L'appétit vient en mangeant: some reflexions on the self-sustaining tendencies. *Br. J. Psychol.* **38**, 171–90.

Ford, E.B. (1964). *Ecological genetics.* Methuen, London

Ford, E.B. (1965). *Genetic polymorphism.* Faber and Faber, London.

Forgays, D.G. and Forgays, J.W. (1952). The nature of the effect of free-environmental experience in the rat. *J. comp. Physiol. Psychol.* **45**, 322–8.

Francq, E. (1968). Feigned death in the opossum. *Dissertation Abstracts* 28**B**, 2665.

Fretter, V. and Graham, A. (1962). *British prosobranch molluscs. Their functional anatomy and ecology.* Ray Society, London.

Fricke, H. (1971). Fische als Feinde tropischer Seeigel. *Marine Biol.* **9**, 328–38.

Friedmann, H. (1934). The instinctive emotional life of birds. *Psychoanal. Rev.* **21**, Nos. 3 and 4.

Fry, C.H. (1969). Structural and functional adaptation of display in the standard-winged nightjar *Macrodipteryx longipennis. J. Zool., London* **157**, 19–24.

Gadgil, M. (1972). Male dimorphism as a consequence of sexual selection. *Am. Nat.* **106**, 574–80.

Gans, C. (1961). Mimicry in procryptically colored snakes of the genus *Dasypeltis. Evolution*, Lancaster, Pa., **15**, 72–91.

Gardner, H. (1984). *Frames of mind.* Heinemann, London.

Gibb, J.A. (1962). L. Tinbergen's hypothesis of the role of specific search images. *Ibis* **104**, 106–11.

Gibb, J.A. (1966). Tit predation and the abundance of *Ernarmonia conicolana* (Heyl) on Weeting Heath, Norfolk, 1962–1963. *J. Anim. Ecol.* **35**, 43–53.

Giddings, A. (1984). An incredible feasting of whales. *National Geographic* **165**, 88–93.

Gilbert, L.E. and Raven, P.E. (Eds) (1975) *Coevolution of animals and plants.* University of Texas Press, Austin.

Gilliard, E.T. (1962). On the breeding behavior of the cock-of-the rock (Aves *Rupicola rupicola*). *Bull. Am. Mus. nat. Hist.* **124**, 31–68.

Goodwin, D. (1953). The reactions of some nesting passerines towards live and stuffed jays. *British Birds* **46**, 193–200.

Gordon, B.L. (1977). *The secret lives of fishes.* Grosset and Dunlap, New York.

Goss-Custard, J.D. (1977). Optimal foraging and the size selection of worms by redshank *Tringa totanus. Anim. Behav.* **25**, 10–29.

Gould, S.J. and Lewontin, R.C. (1979). The spandrels of San Marco and the Panglossian paradigm: a critique of the adaptationist programme. *Proc. R. Soc. London* (B) **205**, 581–98.

Gramza, A.F. (1967). Responses of brooding nighthawks to a disturbance stimulus. *Auk* **84**, 72–86.

Grant, E.C. and Mackintosh J.H. (1963). A comparison of the social postures of some common laboratory rodents. *Behaviour* **21**, 246–59.

Graves, R. (1965). *Collected poems.* Cassell, London.

Greenwalt, C.H. (1955). The flight of the black-capped chickadee and the white-breasted nuthatch. *Auk* 72, 1–5.

Greenwood, P.J. and Harvey, P.H. (1978). Foraging and territory utilisation of blackbirds (*Turdus merula*) and song thrushes (*Turdus philomelos*) *Anim. Behav.* **26**, 1222–36.

Griffin, D.E. (1958). *Listening in the dark.* Yale University Press, New Haven, Conn.

Griffiths, W.J. (1942). The production of convulsions in the white rat. *Comp. Psychol. Monogr.* **17**, 1–29.

Groves, C.P. (1972). Impala. In *Encyclopedia of the animal world.* Elsevier International, London.

Grunfeld, F.V. and the editors of Time-Life Books (1982). *Wayfarers of the Thai forest.* Time-Life Books, Amsterdam.

Gurney, J.H. (1861). On some additional species of birds received in collections from Natal. *Ibis* **12**, 132.

Guthrie, R.D. (1971). A new theory of mammalian rump patch evolution. *Behaviour* **38**. 132–45.

Hailey, A. (1968). *Airport.* Michael Joseph, London.

Haldane, J.B.S. (1932). *The causes of evolution.* Longmans, Green and Co, London.

Haldane, J.B.S. (1949). Disease and evolution. *Ricerca scient.,* suppl. 19, 68–76.

Haldane, J.B.S. (1953). Animal populations and their regulation. *Modern Biology* **15**, 9–24.

Haldane, J.B.S. (1954). La signalisation animale. *Ann. Biol.* **30**, 89–98.

Haldane, J.B.S. (1955). On the biochemistry of heterosis and the stabilization of polymorphism. *Proc. R. Soc. London (B)* **144**, 217–20.

Hale, D.J. (1967). Sequential effects in a two-choice serial reaction task. *Q.J. Exp. Psychol.* **19**. 133–41.

Halkka, O., Heinover, L., Raatikainen, M., and Vasarainen, A. (1966). Crossing experiments with *Philaenus spumarius* (Homoptera). *Hereditas* **56**, 306–12.

Hall, B.P., Moreau, R.E., and Galbraith, I.C.J. (1966). Polymorphism and parallelism in the African bush-shrikes of the genus *Malaconotus* (including Chlorophoneus). *Ibis* **108**, 161–81.

Hall, D.F.N. (1956). Ink ejection by cephalopods. *Nature, London* **177**, 663.

Hall, K.R.L. (1960). Social vigilance behaviour in the chacma baboon *Papio ursinus. Behaviour* **16**, 261–94.

Hall, K.R.L. (1966). The wild patas monkey, *Erythrocebus patas*, in Uganda. *J. Zool.* **148**, 15–87.

Halstead, W.C. (1947). *Brain and intelligence.* University of Chicago Press, Chicago.

Halstead, W.C. (1951). Brain and intelligence. In *Cerebral mechanisms in behaviour* (ed. L.A. Jeffress). Wiley, New York.

Hamilton, W.D. (1971). Geometry for the selfish herd. *J. theor. Biol.* **31**, 295–311.

Hanby, J. (1983). *Lions share*. Collins, London.

Hardy, A.C. (1956). *The open sea: the world of plankton*. Collins, London.

Harrison, C.J.O. (1976). Feathering and flight evolution in *Archaeopteryx*. *Nature, London* No. 5580, 762–3.

Harrison Matthews, L. (1952). *British mammals*. Collins, London.

Harvey, E.N. (1952). *Bioluminescence*. New York.

Harvey, P.H. and Greenwood, P.J. (1978). Anti-predator defence strategies: some evolutionary problems. In *Behavioural ecology — an evolutionary approach* (ed. J.R. Krebs and N.B. Davies) Blackwell, Oxford.

Hassell, M.P. (1971). Mutual interference between searching parasites. *J. Anim. Ecol* **40**, 473–86.

Haviland, M.D. (1917). Notes on the breeding habits of the dotterel on the Yenesei. *British Birds* **11**, 6–11.

Hebb, D.O. (1946). On the nature of fear. *Psychol. Rev.* **53**, 259–76.

Hebb, D.O. (1949). *Organization of behavior. A neuropsychological theory.* Wiley, New York.

Hebb, D.O. (1958). *A textbook of psychology*. W.B. Saunders, London.

Heiligenberg, W. (1965). The suppression of behavioural activities by frightening stimuli. *Z. vergl. Physiol.* **50**, 660–72.

Heller, R. and Milinksi, M. (1979). Optimal foraging of sticklebacks on swarming prey. *Anim. Behav.* **27**, 1127–41.

Hemmings, C.C. (1965). Factors influencing the visibility of objects underwater. In *Symposium on light as an ecological factor* (ed. R. Bainbridge, G.C. Evans, and D. Rackham). Blackwell, Oxford.

Herrebout, W.M., Kuyten, P.J., and de Ruiter, L. (1963). Observations on colour patterns and behaviour of caterpillars feeding on Scots pine. *Archs. néerl. Zool.* **15**, 315–57.

Herring, P.J. (Ed.) (1979). *Bioluminescence in action*. Academic Press, London.

Hess, E.H. (1966). Origins of behaviour. *Science, NY* **154**, 1636.

Hick, W.E. (1952). On the rate of gain of information. *Q.J. exp. Psychol.* **4**, 11–26.

Hinde, R.A. (1954*a*). Factors governing the changes in strength of a partially inborn response, as shown by the mobbing behaviour of the chaffinch (*Fringilla coelebs*): I. The nature of the response and an examination of its course. *Proc. R. Soc. London (B)* **142**, 306–31.

Hinde, R.A. (1954*b*). Factors governing the changes in strength of a partially inborn response, as shown by the mobbing behaviour of the chaffinch (*Fringilla coelebs*): II. The waning of the response. *Proc. R. Soc. Lond. (B)* **142**, 331–58.

Hinde, R.A. (1960). Energy models of motivation. In *Models and analogues in biology. Sym. Soc. exp. Biol.* **XIV**, 199–213.

Hinde, R.A. (1970). *Animal behaviour: a synthesis of ethology and comparative psychology* (2nd edn). McGraw-Hill, New York.

Hingston, R.W.G. (1932). *A naturalist in the Guiana forest*. Longmans, New York.

Hingston, R.W.G. (1933). *The meaning of animal colour and adornment.* Edward Arnold, London.

Hoagland, H. (1928). On the mechanism of tonic immobility in vertebrates. Jl. Gen. Physiol. 715.

Hobson, E.S. (1965). Diurnal–nocturnal activity of some in-shore fishes in the Gulf of California. *Copeia* 3, 291–302.

Hobson, E.S. (1968). *Predatory behaviour of some shore fishes in the Gulf of California.* US Dept. of the Interior, Fish and Wildlife Service, Bureau of Sport, Fisheries and Wildlife, Research Report 73.

Hochbaum, H.A. (1944). *The canvasback on a prairie marsh.* American Wildlife Institute Washington, DC.

Hogan, J.A. (1965). An experimental study of conflict and fear; an analysis of behaviour of young chicks toward a mealworm. Part I. The behaviour of chicks which do not eat the mealworm. *Behaviour* **25**, 45–97.

Hohn, E.O. (1958). Observations on the behaviour of certain arctic birds. *Arctic* **11**, 93–101.

Holling, C.S. (1965). The functional response of predators to prey density and its role in mimicry and population regulation. *Mem. eng. Soc. Can.* **45**, 5–60.

Holling, C.S. (1966). The functional response of invertebrate predators to prey density. *Mem. ent. Soc. Can.* **48**, 1–86.

Holmes, W. (1940). The colour changes and colour patterns of *Sepia officinalis L. Proc. Zool. Soc. London (A)* **110**, 17–35.

Holyoak, D.T. (1972). Bower birds. In *Encyclopedia of the animal world.* Elsevier International, London.

Hoogland, J.L. and Sherman, P.W. (1976). Advantages and disadvantages of bank swallow (*Riparia riparia*) coloniality. *Ecol. Monogr.* **45**, 33–58.

Horsley, D.T. (1978). The role of avian predators in maintaining colour polymorphism. Ph.D. thesis, University of Nottingham.

Howard, L.O. (1901), *The insect book.* Doubleday and Page, New York.

Hudson, W.H. (1892). *The naturalist in La Plata.* Dent, London.

Hume, A.O. (1890). *Nests and eggs of Indian birds.* R.H. Porter, London.

Humphries, D.A. (1970). Erratic movement and a cataleptic posture in the escape behaviour of fleas. *Entomologist's mon. Mag.* **107**, 200–2.

Humphries, D.A., (1963). The behaviour of certain fleas in relation to their behaviour and ecology. PhD Thesis, University of Durham.

Humphries, D.A. and Driver, P.M. (1967). Erratic display as a device against predators. *Science, NY* **156**, 1767–8.

Humphries, D.A. and Driver, P.M. (1970). Protean defence by prey animals *Oecologia* **5**, 285–302.

Hutchinson, G.E. (1963). A note on the polymorphism of *Philaenus spumarius* (L.) (Homopt., Cercopidae) in Britain. *Entomologist's mon. Mag.* **99**, 175–8.

Huxley, J.S. (1925). The absence of 'courtship' in the avocet. *British Birds* **19**, 88–94.

Huxley, J.S. (1955). Morphism in birds. *Acta XI Congressus Internationalis Ornithologici* 309–28.

Huxley, T.H. (1860). Letter to Charles Kingsley, Sept 23, 1860. In *Life and letters of Thomas Henry Huxley* (ed. L. Huxley), 2 vols. Macmillan, London.

Huxley, T.H. (1884). *The crayfish*. Kegan Paul, Trench, London.

Hyman, L.H. (1940). *The invertebrata*. I. *Protozoa through Ctenophora*. McGraw-Hill, New York.

Iersel, J.J.A., van and Bol, A.C. Angela (1958). Preening of two tern species. A study on displacement activities. *Behaviour* **13**, 1–88.

Inglis, I.R., Fletcher, M.R., Feare, C.J., Greig-Smith, P.W., and Land, Sue (1982). The incidence of distress calling among british birds. *Ibis* **124**, 351–5.

Ivlev, V.S. (1961). *Experimental ecology of the feeding of fishes*. Yale University Press, New Haven, Conn.

Jarman, P.J. (1974). The social organisation of antelope in relation to their ecology. *Behaviour* **48**, 215–67.

Järvi, T., Radesäter, T., and Jakobsson, S. (1977). Song-repertoire: a way to transmit information. *Biophon.* **V** (3), 6–7.

Jefferies, R. (1879). *The amateur poacher*. Oxford University Press, Oxford.

Johnsgard, P.A. (1965). *Handbook of waterfowl behaviour*. New York.

Johnston, R.F. and Hardy, J.W. (1962). Behaviour of the purple martin. *Wilson Bull.* 74, 243–62.

Jones, P.H. (1961). Observations sur la migration a Beauduc, Camargue, aux printemps 1959 et 1960. *Alauda*. **29**, 118–33.

Jones, M.D.R. (1966). The acoustic behaviour of the bush cricket *Pholidoptera griseoaptera*. 2. Interaction with artificial sound signals. *J. exp. Biol.* **45**, 31–44.

Jones, J.S. Leight, B.H., and Rawlings, P. (1977). Polymorphism in *Cepaea*: a problem with too many solutions? *Ann. Rev. Ecol. Systematics*, **8**, 109–43.

Jourdain, F.C.R. (1936–7). The so-called "injury-feigning" in birds. *Ool. Rec.*, **16**, 25–37, **17**, 14–16, 71–2.

Kastendiek, J. (1976). Behaviour of the sea pansy *Renilla koellikeri* Pfeffer (Coelenterata, Pennatulacea) and its influence on the distribution and biological interactions of the species. *Biol. Bull. mar. Biol. Lab., Woods Hole* **151**, 518–37.

Kennedy, J.S. and Booth, C.O. (1963). Free flight of aphids in the laboratory. *J. exp. Biol.* **40**, 67–85.

Kinnear, P.K. (1978). *British Birds* **71**, 126–7.

Kipling, R. (1894). *The jungle book*. Macmillan, London.

Koch, M. (1987). The anatomy of the virus. *New Scientist* **1553**, 46–51.

Kohn, A.J. and Waters, V. (1966). Escape responses of three herbivorous gastropods to the predatory gastropod *Conus textile*. *Anim. Behav.* **14**, 340–5.

Kortlandt, A. and Kooij, M. (1963). Protohominid behaviour in primates. *Symp. zool. Soc. London* **10**, 61–88.

Kortright, F.H. (1943). *The ducks, geese and swans of North America*. Amer. Wildl. Inst., Washington, D.C.

Kratzig, H. (1940). Untersuchungen zur Lebensweise des Moorschneehuhns,

Lagopus lagopus, während der jugendentwicklung. *J. für Ornithol.* **88**, 139–66.

Krebs, C.J., Gaines, M.S., Keller, B.L., Myers, J.H., and Tamarin, R.H. (1973). Population cycles in small rodents. *Science, NY* **179**, 35–41.

Krebs, J.R. (1974). Colonial nesting and social feeding as strategies for exploiting food resources in the great blue heron (*Ardea herodias*), *Behaviour* **51**, 99–134.

Krebs, J.R. (1976). Habituation and song repertoires in the great tit. *Behav. Ecol. Sociobiol.* **I**, 215–27.

Krebs, J.R. (1977). The significance of song repertoires: The Beau Geste hypothesis. *Anim. Behav.* **25**, 475–8.

Krebs, J.R. (1978). Optimal foraging: decision rules for predators. In *Behavioural ecology: an evolutionary approach* (ed. J.R. Krebs and N.B. Davies).

Kruijt, J.P. (1964). Ontogeny of social behaviour in Burmese Red Jungle fowl (*Gallus gallus spadiceus*) Bonnaterre. *Behaviour Suppl.* No.12.

Krushinskii, L.V. (1962). *Animal behaviour: its normal and abnormal development.* Plenum, New York.

Kruuk, H. (1964). Predators and anti-predator behaviour of the black-headed gull (Larus ridibundus L.) *Behaviour Suppl.* **II**, 1–129.

Kruuk, H. (1972). *The spotted hyena.* University of Chicago Press, Chicago.

Kruuk, H. (1975). *Hyaena.* Oxford University Press, Oxford.

Krunk, H. (1976). The biological function of gulls' attraction towards predators. *Anim. Behav.* **24**, 146–53.

Kuhlmann, D.H.H. and Karst, H. (1967). Freiwasserbeobachtungen zum Verhalten von Tobiasfisch schwärmen (Ammodytidae) in der westlichen Ostsee. *Z. Tierpsychol.* **24**, 282–97.

Kuhn, T.S. (1970). *The structure of scientific revolutions.* University of Chicago Press, Chicago.

Kummer, H. (1968). *Social organization of hamadryas baboons: a field study.* Chicago University Press, Chicago.

Lack, D. (1941). Some aspects of instinctive behaviour and display in birds. *Ibis* **14**, 407–41.

Lack, D. (1943). *The life of the robin.* Witherby, London.

Lack, D. (1954). *The natural regulation of animal numbers.* Clarendon Press, Oxford.

Lack, D. (1966). *Population studies of birds.* Clarendon Press, Oxford.

Lack, D. (1968). *Ecological adaptations for breeding in birds.* Methuen, London.

Lagerweij, E., Neils, P.C., Wiegant, V.M., and van Ree, J.M. (1984). The twitch in horses: a variant of acupuncture. *Science* **225**, 1172.

Laven, H. (1940). Beitrage zur Biologie des Sandregenpfeifers (*Charadrius hiaticula* L.). *J. für Ornithol.* **88**, 183–288.

Leakey, L.S.B. (1969). *Animals of East Africa.* National Geographic Society, Washington, DC.

Leakey, R.E. (1981). *The making of mankind.* Michael Joseph, London.

Lees, H.T. (1964). *British Birds* 57, 202.

Leonard, J.A., Newman, R.C., and Carpenter, A. (1966). On the handling of heavy bias in a self-paced task. *Q.J. exp. Psychol.* **18**, 130–41.

Lettvin, J.Y., Maturana, H.R., McCulloch, W.S., and Pitts, W.H. (1959). What the frog's eye tells the frog's brain. *Proc. Inst. Radio Engr.* **47**, 1940–51.

Leyhausen, P. (1956). Verhaltensstudien an Katzen. *Z. Tierpsychol.* Suppl. 2.

Leyhausen, P. (1965). The communal organisation of solitary mammals. *Symp. zool. soc. London* **14**, 249–63.

Li, C.C. (1962). On 'reflexive selection'. *Science, NY* **136**, 1055–6.

Lind, H. (1961). Studies on the behaviour of the black-tailed godwit (*Limosa limosa* L.). *Meddelelse fra Naturfrednigsrådets reservatudvalg* Nr. 66.

Linné, C., von (1748–52). *Föreläsningar over djurrileet.* (ed, Einar Lonnberg).

Lindahl, T.P. (1979). Evolution of handling time: the functional response of a predator to the density of sympatric and allopatric strains of prey. *Evolution* **33**, 765–8.

Livingstone, D. (1865). *Missionary travels and researches in Southern Africa.* Murray, London.

Longley, W.H. (1915). Coloration of tropical reef fishes. *Yb. 14. Carnegie Instn. Wash.* 208–9.

Longley, W.H. (1916). The significance of the colours of tropical reef fishes. *Yb. 15, Carnegie Instn. Wash.* 209–12.

Lorenz, K. (1935). Der Kumpan in der Umwelt des Vogels. *J. für Ornith.* **83**, 137–213, 289–413.

Lorenz, K. (1941). Vergleichende Bewegungsstudien an Anatinen. *J. für Ornithol.* **89**, 194–294. Sonderheft (translated as Lorenz 1966*a*).

Lorenz, K. (1950). Ganzheit und Teil in der tierischen und menschlichen Gemeinschaft. *Studium Generale* 3/9, 1950.

Lorenz, K. (1965). Introduction in Darwin, C., *The expression of the emotions in man and animals.* University of Chicago Press, Chicago.

Lorenz, K. (1966*a*). Comparative studies on the behaviour of the Anatinae. London, Avicult. Soc. (translation of Lorenz 1941).

Lorenz, K. (1966*b*). *On aggression.* Methuen, London.

Lorenz, K. (1971). *Studies in animal and human behaviour.* 2 vols. Translated by Robert Martin. Methuen, London.

Louw, G. (1979). Letter on 'Biological Strategies'. *Science NY* No. 4384, 955.

Lovenskiold, H.L. (1954). Studies on the avifauna of Spitsbergen. *Norsk Polarinstitutt.* Skrifter No. 103.

Loveridge, A. (1933). Reports on the Scientific Results of an Expedition to the South Western Highlands of Tanganyika Territory. VII. Herpetology. *Bull. Mus. comp. Zool. Harvard* **LXXIV**, 197–416.

Luce, R.D. and Raiffa, H. (1957). *Games and decisions.* Wiley, New York.

Lynes, H. (1910). Manoeuvres of lapwing in defence of young. *British Birds* **4**, 157.

MacGinitie, G.E. and MacGinitie, N. (1949). *The natural history of marine animals.* McGraw-Hill, New York.

Mackie, G.O. (1976) *Coelenterate ecology and behaviour.* Plenum, New York.

Mackinnon, J. (1970). Indications of territoriality in mantids. Z. *Tierpsychol.* 27, 150–5.

Magnuson, J.J. and Prescott, J.H. (1966). Courtship, locomotion, feeding and miscellaneous behaviour of Pacific bonito (*Sarda chiliensis*). *Anim. Behav.* **14**, 54–67.

Maier, N.R.F. (1939). *Studies of abnormal behavior in the rat.* I. The neurotic pattern and an analysis of the situation which produces it. Harper, New York.

Maier, N.R.F. (1948). Experimentally induced abnormal behavior. *Scientific Monthly* **117**, 210–16.

Major, P.F. and Dill, L.M. (1978). The 3-dimensional structure of airborne bird flocks. *Behavl. Ecol. Sociobiol,* **4**, 111–22.

Manly, B.F.J., Miller, P., and Cook, L.M. (1972). Analysis of a selective predation experiment. *Am Nat.* **106**, 719–36.

Mantiefel, B.P. and Radakov, D.V. (1961). The adaptive significance of schooling behaviour in fishes. *Uspekhy sovremennoi biologii* **50**, 338–45

Marlatt, C.L. (1907). The periodical cicada. *US Dept. Agri. Div, Ent. Bull.* **71**, 181pp.

Marler, P. (1956). The voice of the chaffinch and its function as a language. *Ibis* **78**, 231–61.

Marler, P. (1957). Specific distinctiveness in the communication signals of birds. *Behaviour* **11**, 13–39.

Marler, P. (1959). Developments in the study of animal communication. In *Darwin's biological work: some aspects reconsidered* (ed. P.R. Bell). Wiley, New York.

Marler, P. and Hamilton, W.J. (1966). *Mechanisms of animals behaviour.* Wiley, New York.

Marples, B.J. (1969). Observations on decorated webs. *Bull. Br. Arachnol. Soc.* **1**, 13–18.

Marshall, A.J. (1954). *Bower birds.* Oxford University Press, Oxford.

Marshall, S.M. and Orr, A.P. (1955). *The biology of a marine copepod, Calanus finmarchicus* (Gunnerus). Oliver and Boyd, Edinburgh.

Marson, J.E. (1947). Some observations on the ecological variation and development of the cruciate zigzag camouflage device of *Argiope pulchella* (Thor). *Proc. zool. Soc. London* **117**, 219–27.

Martens, E. von (1867). *Die preussische Expedition nach Ost-Asien.* Zoologicher Theil. Zweites Band. 447pp. Decker, Berlin.

Martinez, D.R. and Klinghammer, E. (1970). The behavior of the whale *Orcinus orca*: a review of the literature. *Z. Tierpsychol.* **27**, 828–39.

Maskell, M., Parkin, D.T., and Verspoor, E. (1977). Apostatic selection by sticklebacks upon a dimorphic prey. *Heredity, Lond.* **39**, 83–9.

Mason, C.W. and Maxwell-Lefroy, H. (1912). The food of birds in India. *Mem. Dept. Agric. India. Ent. Ser.* III, 1–371.

Maynard Smith, J. (1958). *The theory of evolution.* Pelican Books, London.

Maynard Smith, J. (1965). The evolution of alarm calls. *Am. Nat.* **99**, 59–63.

Maynard Smith, J. (1974). The theory of games and the evolution of animal conflicts. *J. Theor. Biol.* **47**, 209–21.

Maynard Smith, J. (1976). Group selection. *Q. Rev. Biol.* **51**, 277–83.

Maynard Smith, J. (1979). Game theory and the evolution of behaviour. *Proc. R. Soc. London (B)* **205**, 475–88.

Mayr, E. (1950). Taxonomic categories in fossil hominids. *Cold Spring Harbour Symp. Quant. Biol.* **15**, 109–18.

Mech, L.D. (1970). *The Wolf: The ecology and behavior of an endangered species.* Natural History Press, New York.

Medawar, P.B. (1973). In *New York Review of Books*, 8th March 1973, p. 21.

Meesters, A. (1940). Über die organisation des Gesichtsfeldes der Fische. *Z. Tierpsychol.* **4**, 84–149.

Meinertzhagen, R. (1959). *Pirates and predators.* Oliver and Boyd, Edinburgh.

Melzack, R. (1952). Irrational fear in the dog. *Can. J. Psychol.* **6**, 141–47.

Milinski, M. (1977). Do all members of a swarm suffer the same predation? *Z. Tierpsychol.* **45**, 373–8.

Milinski, M. (1979) Can an experienced predator overcome the confusion of swarming prey more easily? *Anim. Behav.* **27**, 1122–6.

Milinski, M. and Curio, E. (1975). Untersuchungen zur Selektion durch Räuber gegen Vereinzelung der Beute. *Z. Tierpsychol.* **35**, 400–2.

Milinski, M. and Heller, R. (1978). Influence of a predator on the optimal foraging behaviour of a stickleback (*Gasterosteus aculeatus* L.) *Nature, London* **275**, 642–4.

Millais, J.G. (1895). *A breath from the veldt.* Sotheran, London.

Millais, J.G. (1913). *British diving ducks.* London.

Miller, G.A. (1956). The magical number seven, plus or minus two: some hints on our capacity for processing information. *Psychol. Rev.* **63**, 81–97.

Miller, J.G. (1963). Coping with administrators' information overload. In *Proc. First Institute on Medical School Administration.* Association of American Medical Colleges.

Miller, J.G. (1964*a*). *A theoretical review of individual and group psychological reactions to stress.* In *The threat of impending disaster* (ed. G.M. Grosser, H. Wechsler, and M. Greenblatt). MIT Press, Cambridge, Mass.

Miller, J.G. (1964*b*). Psychological aspects of communication overloads. In *International psychiatry clinic: communications in clinical practice* (ed. R.W. Waggoner and D.J. Carek). Little, Brown and Co., Boston.

Miller, J.G. (1965). Living systems: basic concepts. *Behav. Sci.* **10**(3), 193–411.

Miller, R.C. (1922). The significance of the gregarious habit. *Ecology* **3**, 122–6.

Moment, G.B. (1962). Reflexive selection: a possible answer to an old puzzle. *Science, NY* **136**, 262–3.

Monod, J. (1972). *Chance and necessity.* Collins, London.

Mook, J.H., Mook, L.J., and Heikens, H.S. (1960). Further evidence for the role of 'searching images' in the hunting behaviour of titmice. *Arch. néerl. Zool.* **13**, 448–65.

Morgan, C.T. and Waldman, H. (1941). Conflict and audiogenic seizures. *J. comp. Psychol.* **31**, 1–11.

Morgan, N.C. (1956). The biology of *Leptocerus aterrimus* Steph. with reference to its availability as a food for trout. *J. Anim. Ecol.* **25**, 349–65.

Morin, J.G. (1976). Probable function of bioluminescence in the Pennatulacea (*Cnidaria, Anthozoa*) In *Coelenterate ecology and behaviour* (ed. G.D. Mackie). Plenum, New York.

Morris, D. (1954). The snail-eating behaviour of thrushes and blackbirds. *British Birds* **47**, 33–49.

Morris, D. (1956). The feather postures of birds and the problem of the origin of social signals. *Behaviour* **9**, 75–113.

Mortensen, T. (1917). Observations on protective adaptations and habits, mainly in marine animals. *Saertryk af Vidensk, Medd. fra Dansk naturhist. Foren.* **69**, 57–96.

Mosebach-Pukowski, E. (1937). Über die Raupengesellschaften von *Vanessa io* und *Vanessa urticae. Z. Morch. Otiol-Tiere.* **33**, 358–80.

Moss, C. (1976). *Portraits in the wild.* Hamish Hamilton, London.

Mowat, F. (1963). *Never cry wolf.* Dell, New York.

Moynihan, M. (1958). Notes on the behaviour of some North American gulls: I. Aerial hostile behaviour. *Behaviour* **10**, 126–78.

Mueller, H.C. (1971). Oddity and specific searching image more important than conspicuousness in prey selection. *Nature, London* **233**, 345–6.

Müller, G. (1978). Üntersuchungen zur Selektion gegen Grössensonderlinge in Benterschwärm durch Räuber. *Staats examens arbeit.*, Abt. Biologie, Ruhr-Universität Bochum.

Munro, J. and Bédard, J. (1977). Gull predation and crêching behaviour in the common eider. *J. Anim. Ecol.* **46**, 799–810.

Murtagh, P. (1979). Strategy (Letter). *Science, NY* No. 4395, 795.

Neill, S.R. St. J. and Cullen, J.M. (1974). Experiments on whether schooling by their prey affects the hunting behaviour of cephalopods and fish predators. *J. Zool. London* **172**, 549–69.

Nelson, E.W. (1909). The rabbits of North America. *N. Am. Fauna,* No.29, US Biol. Surv., Wash.

Nice, M.M. (1943). Studies in the life history of the song sparrow. II. The behaviour of the song sparrow and other passerines. *Trans. Linn. Soc. NY* **6**, 1–328.

Nice, M.M. (1962). Development of behaviour in precocial birds. *Trans. Linn. Soc. NY* **8**, 1–211.

Nice, M.M., Nice, C., and Ewers, D. (1956). Comparison of behaviour development in snowshoe hares and red squirrels. *J. Mammal.* 37, 64–74.

Nicholson, E.M. (1939). Cited in Witherby et al. (1939).

Nicol, J.A.C. (1955). Observations on luminescence in *Renilla* (Pennatulacea) *J. exp. Biol.* **32**, 299–320.

Nicol, J.A.C. (1960). *The biology of marine animals.* Pitman, London.

Nicol, J.A.C. (1971). Physiological investigations of ocean's animals. In *Deep oceans* (ed. P.J. Herring and M.R. Clarke). Barker, London.

Nielsen, J.M. (1951). Comment in general discussion in *Cerebral mechanisms in behavior* (ed.) L.A. Jeffress. Wiley, New York.

Norman, J.R. (1931). *A history of fishes.* Benn, London.

Norris, R.A. and Stamm, D.D. (1965). Relative incidence of distress calls or 'squeals' in mist-netting birds. *Bird Banding* **36**, 83–8.

Norton, A.C., Beran, A.U., and Miszahy, G.A. (1964). Electroencephalography during feigned sleep in the opossum. *Nature London* **204**, 162.

Norton-Griffiths, M. (1967). Some ecological aspects of the feeding behaviour of the oystercatcher, *Haematopus ostralegus*, on the edible mussel, *Mytilus edulis*. *Ibis* **109**, 412–24.

Norton-Griffiths, M. (1968). The feeding behaviour of the oystercatcher, *Haematopus ostralegus*. D. Phil Thesis, Oxford University.

Novick, A. (1969). *The world of bats*. Edita, Lausanne.

O'Connor, R.J. and Brown, R.A. (1977). Prey depletion and foraging strategy in the oystercatcher *Haematopus ostralegus*. *Oecologia* **27**, 75–92.

O'Donald, P. (1968). Natural selection by glow-worms in a population of *Cepaea nemoralis*. *Nature, London* **217**, 194.

Ohguchi, O. (1978). Experiments on the selection against colour oddity of water fleas by three-spined sticklebacks. *Z. Tierpsychol.* **47**, 254–67.

Olson, F.C.W. (1964). The survival value of fish schooling. *J. Cons. perm. int. Explor. Mer.* **29**, 115–16.

Ortiz de Villalba, J.S. (1980). *Los ultimos Huaorani*. Centro de Investigaciones Culturales de la Amazonia Ecuatoriana, Quito, Peru.

Osgood, C.E. (1957). A behaviouristic analysis of perception and language as cognitive phenomena. In *Contemporary approaches to cognition* (ed. J.S. Bruner). Harvard University Press, Cambridge, Massachusetts.

Ostrom, J.H. (1972). *Proc. K. ned. Akad. Wet.* **B75**, 289–305.

Owen, D.F. (1963). Polymorphism and population density in the African land snail *Limicolaria martensiana*. *Science, NY* **140**, 666–7.

Owen, D.F. (1965). Density effects in polymorphic land snails. *Heredity*, **20**, 312–15.

Owen, D.F. (1967). The interpretation of polymorphism in the African bush-shrikes. *Ibis* **109**, 278–9.

Owen, D.F. (1969). Ecological aspects of polymorphism in an African land snail *Limicolaria martensiana*. *J. Zool. London* **159**, 79–96.

Owen, D.F. (1980). *Camouflage and mimicry*. Oxford University Press, Oxford.

Owen, D.F. and Wiegert, R.G. (1962). Balanced polymorphism in the meadow spittlebug, *Philaenus spumarius*. *Am. Nat.* **96**, 353–9.

Packard, A. (1960). The behaviour of *Octopus vulgaris*. *Bull. Inst. Oceanogr. Monaco*, 35–49.

Packard, A. and Sanders, G.D. (1971). Body patterns of *Octopus vulgaris* and maturation of the response to disturbance. *Anim. Behav.* **19**, 780–90.

Panceri, P. (1872). The luminous organs and light of the *Pennatulae*. *Q.J. Microsc. Soc.* **12**, 248–54.

Pantin, C.F.A. (1950). Behaviour patterns in lower invertebrates. *Symp. Soc. exp. Biol.* **4**, 175–95.

Parker, G.E. (1978). Searching for mates. In *Behavioural ecology — An Evolutionary Approach*. D.R. Krebs (ed. and N.B. Davies). Blackwell, Oxford.

Parker, H.N. (1963). *Snakes*. Hale, London.

Paulson, D.R. (1973). Predator polymorphism and apostatic selection. *Evolution* **27**, 269–77.

Payne, R.B. (1967). Interspecific communication signals in parasitic birds. *Am. Nat.* **101**, 363–75.

Payne, R.S., Roeder, K.D., and Wallman, J. (1966). Directional sensitivity of the ears of noctuid moths. *J. exp. Biol.* **44**, 17–31.

Payne, R.B. and Skinner, N.J. (1970). Temporal patterns of duetting in African barbets. *Ibis* **112**, 173–183.

Pernetta, J.S. (1973). Field and laboratory experiments to determine the feeding ecology and behaviour of *Sorex araneus* and *Sorex minutus*. D. Phil. Thesis, Oxford University.

Perrone, M. (1980). Factors affecting the incidence of distress calls in passerines. *Wilson Bull.* **92**, 404–8.

Peters, J.L. and Loveridge, A. (1936). Scientific results of an expedition to rain forest regions in Eastern Africa. IV. Birds. *Bull. Mus. comp. Zool. Harvard* **LXXIX**, 129–205.

Pickwell, G., and Smith, E. (1938). The Texas night-hawk in its summer home. *Condor*, **40**, 193–215.

Pierce, G.W. (1948). *Songs of Insects*. Harvard University Press, Cambridge, Massachusetts.

Pilcher, T.J. (1973). The 3-dimensional structure of schools in the minnow *Phoxinus phoxinus*. *Anim. Behav.* **21**, 673–86.

Pinhey, E. (1960). *Hawk moths of Central and Southern Africa*. Longman, Cape Town.

Pitt, F. (1931). *The intelligence of animals: studies in comparative psychology*. Allen and Unwin, London.

Pomeroy, H. and Heppner, F. (1977). Laboratory determinants of startle reaction time of the starling (*Sturnus vulgaris*). *J. Anim. Behav.* **25**, 720–6.

Popham, E.J. (1941). The variation of colour in certain species of *Arctocorixa* (= *Corixa*) and its significance. *Proc. zool. Soc., London (A)* **III**, 135–72.

Popper, K.R. (1968). *The logic of scientific discovery*. Hutchinson, London.

Portielje, A.F.J. (1925). Zur Ethologie bzw. Psychologie der *Rhea americana* L. *Ardea* **14**, 1–15.

Potts, G.W. (1970). The schooling ethology of *Lutianus monostigma* (Pisces) in the shallow reef environment of Aldabra. *J. Zool. London* **161**, 223–35.

Potts, Wayne K. (1984). The chorus-line hypothesis of manoeuvre coordination in avian flocks. *Nature, London.* **309**, 344–45.

Poulton, E.B. (1888). Notes in 1887 upon lepidopterous larvae, etc., including a complete account of the life-history of the larvae of *Sphinx convolvuli* and *Aglia tau*. *Trans ent. Soc. London* 515–606.

Poulton, E.B. (1890). *The colours of animals*. Kegan Paul, London.

Poulton, E.B. (1924). The terrifying appearance of *Laternaria* (Fulgoridae) founded on the most prominent features of the alligator. *Proc. ent. Soc. London.* **X/III–X/IX**.

Prosser, C.L. (1965). Levels of biological organization and their physiological significance. In *Ideas in modern biology* (ed. J.A. Moore). *Proc. XVI Internat. Congr. Zool.* Vol. 6.

Pumphrey, R.J. (1961). Sensory organs: vision. In *Biology and comparative physiology of birds*, Vol. II (ed. A.J. Marshall). Academic Press, London.

Pycraft, W.P. (1925). *Camouflage in nature.* Hutchinson, London.

Pyke, G.H., Pulliam, H.R., and Charnov, E.L. (1977). Optimal foraging: a selective review of theory and tests. *Q. Rev. Biol.* **52**, 137–54.

Räber, H. (1949). Das verhalten gefangener Waldrohreuen (*Asio otus otus*) und Waldhäuse, *Strix aluco aluco* zur Beute. *Behaviour* **2**, 1–95.

Radakov, D.V. (1973). *Schooling in the Ecology of Fish.* Wiley, New York.

Rand, A.S. (1967). Predator-prey interactions and the evolution of aspect diversity. *Atas do Simpósio sobrê a Biota Amazonica (Zoologia)*, **5**, 73–83.

Rapoport, Amnon, and Moshowitz, A. (1966). Stochastic models for the Prisoner's Dilemma. *Behav. Sci.* **11**, 444–58.

Rapoport, Anatol (1959). Critiques of game theory. *Behav. Sci.* **4**, 49–66.

Rapoport, Anatol (1962). The use and misuse of game theory. *Scient. Am.* **207**, 108–18.

Rapoport, Anatol, and Chammah, A.M. (1965). *Prisoner's dilemma: a study in conflict and cooperation.* University of Michigan Press, Ann Arbor.

Rasa, O.A.E. (1977). The ethology and sociology of the dwarf mongoose (*Helogale undulata rufula*). *Z. Tierpsychol.* **43**, 337–406.

Reed Hainsworth, F. (1981). *Animal physiology: adaptations in function.* Addison-Wesley, Reading, Massachusetts.

Regen, J. (1908). Das tympanale Sinnesorgan von *Thamnotrizon apterus* Fab. (männchen) als Gehörapparat experimentell nachgewiesen. *Sber. Akad. Wiss. Wien* **177**, 487–90.

Reichert, E.S. (1978). Games spiders play: behavioural variability in territorial disputes. *Behav. Ecol. Sociobiol.* **3**, No. 2.

Reighard, J. (1908). An experimental field-study of warning coloration in coral-reef fishes. *Publs Carnegie Instn.* **103**, 257–325.

Reist, J.D. (1983). Behavioral variation in pelvic phenotypes of brook stickleback *Culaea inconstans* in response to predation by northern pike *Esox lucius*. In *Predators and prey in fishes* (ed. D.L.G. Noakes). W. Junk, The Hague.

Ricklefs, R.E. and O'Rourke, K. (1975). Aspect diversity in moths: a temperate-tropical comparison. *Evolution* **29**, 313–24.

Robinson, M.H. (1969*a*). Predatory behaviour of *Argiope argentata*. *Am. Zool.* **9**, 161–73.

Robinson, M.H. (1969*b*). The defensive behaviour of some orthopteroid insects from Panama. *Trans. R. ent. Soc. London* **121**, 281–303.

Robinson, M.H. (1969*c*). Defences against visually hunting predators. In *Evolutionary biology* 3 (ed. T. Dobzhansky, M.K. Hecht, W.C. Steere). Meredith Corp., New York.

Robinson, M.H. (1973). Insect anti-predator adaptations and the behavior of predatory primates. *Acta Congr. Latinamericano Zool.* **4**(2), 811–36.

Robinson, M.H., Abele, L.G., and Robinson, B. (1970). Attack autotomy: a defense against predators. *Science, NY* **169**, 300–1.

Robinson, M.H. and Robinson, B. (1970). The stabilimentum of the orb-web spider *Argiope argentata*: an improbable defence against predators. *Can. Ent.* **102**, 641–55.

Roeder, K.D. (1962). The behaviour of free-flying moths in the presence of artificial ultrasonic pulses. *Anim. Behav.* **10**, 300–4.

Roeder, K.D. (1963). Echoes of ultrasonic pulses from flying moths. *Biol. Bull Mar. Biol. Lab., Woods Hole* **124**, 200–10.

Roeder, K.D. (1966*a*). Acoustic sensitivity of the noctuid tympanic organ and its range for the cries of bats. *J. Insect Physiol.* **12**, 843–59.

Roeder, K.D. (1966*b*). Auditory system of noctuid moths. *Science, NY* **154**, 1515–21.

Roeder, K.D. (1967*a*). Prey and predator. *Bull. ent. Soc. Am.* **13**, 6–9.

Roeder, K.D. (1967*b*). Turning tendency of moths exposed to ultrasound while in stationary flight. *J. insect Physiol.* **13**, 873–88.

Roeder, K.D. and Treat, A.E. (1957). Ultrasonic reception by the tympanic organ of noctuid moths. *J. exp. Zool.* **134**, 127–58.

Roeder, K.D. and Treat, A.E. (1960). The acoustic detection of bats by moths. *Proc. 11th Int. Congr. Ent.* **3**, 1–12.

Roeder, K.D. and Treat, A.E. (1961). The detection and evasion of bats by moths. *Am. Scient.* **49**, 135–48.

Rohwer, S., Fretwell, S.D., and Tuckfield, R.C. (1976). Distress screams as a measure of kinship in birds. *Am. Midl. Natur.*, **96**, 418–30.

Rosenthal, H. and Hempel, G. (1970). Experimental studies in feeding and food requirements of herring larvae (*Clupea harengus* L.). In *Marine food chains* (ed. J.H. Steele). Oliver and Boyd, Edinburgh.

Rothschild, M., and Haskell, P.T. (1966). Stridulation of the garden tiger moth *Arctia caja* L., audible to the human ear. *Proc. R. ent. Soc. London (A)* **41**, 167–70.

Rowell, C.H.F. (1961). Displacement grooming in the chaffinch. *Anim. Behav.* **9**, 38–63.

Rowley, I. (1962). 'Rodent-run' distraction display by a passerine, the superb blue wren, *Malurus cyaneus* (L.). *Behaviour* **19**, 170–6.

Royama, T. (1970). Factors governing the hunting behaviour and selection of food by the great tit (*Parus major* L.). *J. anim. Ecol.* **39**, 619–68.

Ruiter, L. de (1952). Some experiments on the camouflage of stick caterpillars. *Behaviour* **4**, 222–32.

Ruiter, L. de (1953). Camouflage in het dennenbos. 1. De dennepijlstaart. *Levende Nat.* **56**, 41–8.

Russell, Claire and Russell, W.M.S. (1961). *Human behaviour — a new approach*. André Deutsch, London.

Russell, W.M.S. (1961). Evolutionary concepts in behavioural science, III. The evolution of behaviour in the individual animal and the principle or combinatorial selection. *General Systems* **6**, 51–92.

Russell, W.M.S. (1954). Experimental studies of the reproductive behaviour of *Xenopus laevis*. *Behaviour* **7**, 113–88.

Ryle, J. and the Editors of Time-Life Books (1982). *Warriors of the White Nile*. Time-Life Books, Amsterdam.

Sargent, T.D. (1969). A suggestion regarding hindwing diversity among moths of the genus *Catocala* (Noctuidae). *J. lepid. Soc.* **23**, 261–4.

Sargent, T.D. (1973). Studies on the *Catocala* (Noctuidae) of southern New England. IV. A preliminary analysis of beak-damaged specimens, with discussion of anomaly as a potential anti-predator function of hindwing diversity. *L. lepid. Soc.* **27**, 175–92.

Sargent, T.D. (1976). *Legion of night.* Massachusetts University Press.

Sargent, T.D. (1978). On the maintenance of stability in hindwing diversity among moths of the genus *Catocala* (Lepidoptera: Noctuidae). *Evolution* **32**(2), 424–34.

Saunders, A.a. (1937). Injury-feigning by a wood duck. *Auk* **53**, 202.

Saxby, H.L. (1874). *The birds of Shetland.* Oliver and Boyd, Edinburgh and London.

Schall, Jos. J., and Pianka, E.R. (1980). Evolution of escape behaviour diversity. *Am. Nat.* **115**, 551–65.

Schaller, F. and Timm, C. (1950). Das Hörvermögen der Nachtschmetterlinge. *Z. vergl. Physiol.* **32**, 369–81.

Schaller, G.B. (1964). Breeding behaviour of the white pelican at Yellowstone Lake, Wyoming. *Condr* **63**, 3–23.

Schaller, G.B. (1967). *The deer and the tiger.* University of Chicago Press, Chicago.

Schaller, G.B. (1972). *The Serengeti lion.* University of Chicago Press, Chicago.

Schenkel, R. (1947). Ausdruckstudien an Wolfen. *Behaviour* **1**, 81–129.

Schenkel, R. (1966). On sociology and behaviour in impala (*Aepyceros melampus sucora* Matschie). *Z. Sangetierk.* **31**, 177–205.

Schneirla, T.C. (1966). Instinct and aggression. *Natural History* **75**, 10.

Selander, R.K. (1972). Sexual selection and dimorphism in birds. In *Sexual selection and the descent of man* (ed. B. Campbell). Ch. 8. Heinemann, London.

Selye, H. (1956). *The stress of life.* McGraw-Hill, New York.

Serventy, V.N. and White, S.R. (1944). Notes on the splendid blue wren. *Emu* **44**, 48–9.

Sevenster, P. (1961). A causal analysis of a displacement activity (fanning in *Gasterosteus aculeatus* L.). *Behav.* Suppl. No. 9, 1–170.

Shannon, C.E. and Weaver, W. (1949). *The mathematical theory of communication.* University of Illinois Press, Urbana.

Shaw, E. (1960). The development of schooling behavior in fishes. *Physiol. Zool.* **33**, 79–86.

Sheffield, F.D., Wulff, J.J., and Backer, R., (1951). Reward value of copulation without sex drive reduction. *J. comp. Physiol. Psychol.* **44**, 3–8.

Shelford, R. (1903). Bionomical notes on some Bornean Mantidae. *Zoologist* **7**, 293–304.

Sheppard, P.M. (1967). *Natural selection and heredity* (3rd edn.). Hutchinson, London.

Simmons, K.L. (1952). The nature of the predator-reactions of breeding birds. *Behaviour* **4**, 161–71.

Simmons, K.L. (1955). The nature of the predator-reactions of waders towards humans; with special reference to the role of the aggressive-, escape-, and brooding-drives. *Behaviour* **8**, 130–82.

Singer, M.C., Ehrlich, P.R., and Gilbert, L.E. (1971). Butterfly feeding on Lycopsid. *Science, NY* **172**, 1341–2.

Skeels, H.M. (1966). Adult status of children with contrasting early life experiences. *Monographs of the Society for Research in Child Development* **31**, 1–65.

Slobodkin, L.B. (1965). On the present incompleteness of mathematical ecology. *Am. Scient.* **53**, 347–57.

Slocum, J. (1900). *Sailing alone around the world.* Century Co, New York. (Republished by Dover 1956).

Smith, D.A.S. (1976). Disruptive selection and morph-ratio clines in the polymorphic snail *Littorina obtusata* (L.) (Gastropoda: Prosobranchia). *J. Molluscan Studies* **42**(1), 114–35.

Smith, M. (1951). *The British amphibians and reptiles.* Collins, London.

Smith, J.N.M. (1971). The searching behaviour and prey recognition of certain vertebrate predators. Oxford University D. phil. Thesis.

Smith, J.N.M. (1974*a*). The food searching behaviour of two European thrushes. I. Description and analysis of search paths. *Behaviour* **48**, 276–302.

Smith, J.N.M. (1974*b*). The food searching behaviour of two European thrushe. II. The adaptiveness of the search patterns. *Behaviour* **49**, 1–61.

Smith, J.N.M. and Dawkins, R. (1971). The hunting behaviour of individual great tits in relation to spatial variations in their food density. *Anim. Behav.* **19**, 695–706.

Smith, J.N.M. and Sweatman, H.P.A. (1974). Food searching behaviour of titimice in patchy environments. *Ecology* **55**, 1216–32.

Snow, C.P. (1964). *The two cultures and a second look.* Cambridge University Press, Cambridge.

Snow, D.W. (1963). The evolution of manikin displays. *Proc. XIII Internat. Orn. Congr.*, Ithaca NY, 1962, pp. 552–61.

Soane, I.D. and Clarke, B. (1973). Evidence for apostatic selection by predators using olfactory cues. *Nature, London* **241**, 62–4.

Sokolov, E.N. (1960). Neuronal models and the orienting reflex. In *The central nervous system and behaviour* (ed. M.A. Brazier). Joseph Macy, Jr Foundation, New York.

Sokolov, Y.N. (1963). *Perception and the conditioned reflex.* Pergamon Press, London.

Solomon, M.E. (1949). The natural control of animal populations. *J. Anim. Ecol.* **18**, 1–35.

Spencer, K.G. (1953). *The lapwing in Britain.* Brown, London.

Spinage, C.A. (1962). *Animals of East Africa.* Collins, London.

Spooner, G.M. (1931). Some observations on schooling in fish. *J. Mar. Biol. Ass. U.K.* **17**, 421–48.

Stamm, R.A. (1968). Zur Abwehr von Raubfeinden durch *Lobiger serradifalci*

(Calcara), 1840, und *Oxynoe olivacea* Rafinesque, 1819. (Gastropoda, Opisthobranchia). *Revue suisse Zool.* **75**: 661–5.

Starck, W.A. (1966). Marvels of a coral realm. *National Geographic Magazine* **130** 710–38.

Starck, W.A. and Davis, W.P. (1966). Night habits of fishes of Alligator Reef, Florida. *Ichthyologica* (the Aquarium Journal) **38**, 313–56.

Stark, A. and Sclater, W.L. (1900–1906). *The birds of South Africa.*

Stasek, C.R. (1967). Autotomy in the Mollusca. *Occ. Pap. Calif. Acad. Sci.* **61**. 1–44.

Stiefel, A. (1964). Vergleichende Betrachtungen zu Verhaltensähnlickeiten von Kiebitz (*Vanellus vanellus*) und Spornkiebitz (*Hoplopterus spinosus*). *Z. Ornith.* **105**, 468–74.

Stoltz, L.P. and Saayman, G.S. (1970). Ecology and behaviour of baboons in the Northern Transvaal. *Ann. Transv. Mus.*, **26**, 99–143.

Storr, A. (1974). The life of the mind (book review). *Sunday Times* 10 March 1974.

Street, P. (1952). *Between the tides.* University of London Press, London.

Stresemann, E. (1927–34) Aves. In *Kukenthals Handbuch der Zoologie.* Berlin and Leipzig.

Struhsaker, T. (1967). Social structure among vervet monkeys (*Ceropithecus aethiops*). *Behaviour* **29**, 83–121.

Summerhayes, V.S. and Elton, G.S. (1928). Further contributions to the ecology of Spitsbergen. *J. Ecol.* **16**, 193–268.

Summers, A. (1985). *Goddess.* Gollancz, London.

Summer, F.B. (1935). Evidence for the protective value of changeable colouration in fishes. *Am. Nat.* **69**, 245–66.

Svärdson, G. (1957). The 'invasion' type of bird migration. *British Birds*, **50**, 314–43.

Swanberg, P.O. (1951). Food storage, territory and song in the thick-billed nutcracker. *Proc. Xth Int. Ornithol. Congr.* Uppsala, 545–54.

Taverner, P.A. (1936). Injury-feigning by birds. *Auk* **53**, 366.

Taylor, K. (1978). Synchronous diving as feeding strategy. *British Birds* **71**, 598–9.

Thomas, G. (1974). The influence of encountering a food object on subsequent searching behaviour in *Gasterosteus aculeatus* L. *Anim. Behav.* **22**, 941–52.

Thompson, V. (1973). Spittlebug polymorphic for warning coloration. *Nature. London* **242**, 126–8.

Thompson, W.R., and Heron, W. (1954). The effects of restricting early experience on the problem-solving capacity of dogs. *Can. J. Psychol.* **8**, 17–31.

Thorpe, W.H. (1956). *Learning and instinct in animals.* Methuen, London.

Thurber, J. (1953). *The Thurber carnival.* Penguin Books, Harmondsworth, Middlessex.

Tinbergen, L. (1960). The natural control of insects in pinewoods. I Factors influencing the intensity of predation by songbirds. *Arch. néerl. Zool.* **13**, 265–343.

Tinbergen, N. (1942). An objectivistic study of the innate behaviour of animals. *Biblioth. biotheor.* **1**, 39–98.

Tinbergen, N. (1951). *The study of instinct.* Oxford University Press, Oxford.

Tinbergen, N. (1952). "Derived" activities, their causation, biological significance, origin, and emancipation during evolution. *Q. Rev. Biol.* **27**, 1–32.

Tinbergen, N. (1953). *The herring gull's world.* Collins, London.

Tinbergen, N. (1958). *Curious naturalists.* Country Life, London.

Tinbergen, N. (1959). Comparative studies of the behaviour of gulls (*Laridae*): a progress report. *Behaviour* **15**, 1–70.

Tinbergen, N. (1962). Bullfinch escaping from cat by "playing dead". *British Birds* **55**, 420.

Tinbergen, N. (1969). *The study of instinct.* Reprint with new introduction. Oxford University Press, Oxford.

Tinbergen, N. (1972). The Croonian Lecture 1972. *Proc. R. Soc. London (B).* **182**, 385–410.

Tinbergen, N., Impekoven, M., and Franck, D. (1967). An experiment on spacing-out as a defence against predation. *Behaviour* **28**, 307–21.

Toffler, A. (1970). *Future shock.* Bodley Head, London.

Townsend, C.H. (1929). Records of changes in colour among fishes. *Zoologica, NY* **9**, 321–42.

Treat, A.E. (1955). The response to sound in certain lepidoptera. *Ann. ent. Soc. Am.* **48**, 272–84.

Trivers, Robert L. (1971). The evolution of reciprocal altruism. *Q. Rev. Biol.* **46**, 35–57.

Tuck, L.M. (1960). *The Murres* The Queen's Printer, Ottawa.

Turner, E.L. (1920). Some notes on the ruff. *British Birds* **14**, 146–53.

Uspenski, S.M. (1956). *The Bird Bazaars of Novaya Zemlya.* USSR Acad. Sci. Moscow (*Canad. Wildl. Serv. Translations of Russian Game Reports* No. 4).

Van der Post, L. (1961). *Lost world of the Kalahari.* Hogarth Press, London.

van Rhijn, J.G. (1973). Behavioural dimorphism in male ruffs *Philomachus pugnax* (L.). *Behaviour* **47**, 153–229.

van Rhijn, J.G. (1983). On the maintenance and orgin of alternative strategies in the ruff *Philomachus pugnax. Ibis* **125**, (4), 482–98.

Vitz, P.C. (1966). Preference for different amounts of visual complexity. *Behavl. Sci.* **11**, 105–14.

Vowles, D.M. (1961). Neural mechanisms in insect behaviour. In *Current problems in animal behaviour.* (ed. W.H. Thorpe and O.L. Zangwill). Cambridge University Press, Cambridge.

Walker, T.J. (1964). Experimental demonstration of a cat locating orthopteran prey by the prey's calling song. *Fla. Ent.* **47**, 163–5.

Walter, W. Grey, (1953). *The living brain.* Duckworth, London.

Walther, F.R. (1969). Flight behaviour and avoidance of predators in Thomson's gazelle (*Gazella Thomsoni* Guenther 1884). *Behaviour* **34**, 184–221.

Warmke, G.L. and Almodovar, L.R. (1972). Observations on the life cycle and regeneration in *Oxynoe antillarum* Mörch, an ascoglossan opisthobranch from the Caribbean. *Bull. mar. Sci.,* **22**, 67–74.

Waterman, T.H. (1968). Systems theory and biology — view of a biologist. In *Systems theory and biology* (ed. M.O. Mesarović). Proc. IIIrd Systems Symposium at Case Institute of Technology. Springer-Verlag, New York.

Watson, Lyall (1973). *Supernature.* Doubleday, New York and Hodder and Stoughton, London.

Watson, Lyall (1974). *The Romeo error.* Hodder and Stoughton, London.

Watson, Lyall (1979). *Lifetide.* Hodder and Stoughton, London.

Webster, F.A. (1963). Bat-type signals and some implications. In *Human factors in technology* (ed. E. Barnett, J. Degan, and J. Spiegel). McGraw-Hill, New York.

Weihs, D. (1973*a*). Hydromechanics of fish schooling. *Nature, London* **241**, 290–1.

Weihs, D. (1973*b*). Optimal fish cruising speed. *Nature, London* **245**, 48–50.

Welford, A.T. (1968). *Fundamentals of skills.* Methuen, London.

Wells, M.J. (1962). *Brain and behaviour of cephalopods.* Heinemann, London.

Welty, J.C. (1934). Experiments in group behaviour of fishes. *Physiol. Zool.* 7, 85–128.

Welty, J.C. (1964). *The life of birds.* Saunders, Philadelphia.

White, G. (1788). *The natural history of Selbourne.* London.

White, T.H. (1939). *The once and future king.* Putnam, London.

Whitfield, D. (1966). High-speed human skills. *New Scientist* 5 May 1966, 300–2.

Whittaker, J.B. (1968). Polymorphism of *Philaenus spumarius* (L.) (Homoptera, Cercopidae) in England. *J. anim. Ecol.* **37**, 99–111.

Whittaker, R.H., Levin, S.A., and Root, R.B. (1973). Niche habitat and ecotope. *Am. Nat.* **107** (955), 321–38 (May–June 1973).

Wickler, W. (1968). *Mimicry in plants and animals.* McGraw-Hill, New York.

Wiley, R.H. (1974). Evolution of social organisation and life history patterns among grouse. *Q. Rev. Biol.* **49**, 201–27.

Williams, G.C. (1964). Measurements of consociation among fishes. *Biological Series, Michigan State University* 2(7), 351–83.

Williams, G.C. (1966). *Adaptation and natural selection.* Princeton University Press, Princeton, New Jersey.

Williamson, K. (1950). Interpretation of 'rodent run' display. *Ibis* **92**, 85–128.

Wilmut, R. and Grafton, J. (1977). *The Goon Show companion.* Sphere Books, London.

Wilson, E.O. (1971). *The insect societies.* Harvard University Press, Cambridge, Massachusetts.

Wilson, E.O. (1975). *Sociobiology — the new synthesis.* Harvard University Press, Cambridge, Masschusetts.

Wilson, R.A. (1979). *An introduction to parasitology.* Studies in Biology No. 4, 2nd edn. Edward Arnold, London.

Wingate, P. (1972). *The Penguin medical encyclopedia.* Penguin Books, Harmondsworth, Middlesex.

Witherby, H.F., Jourdain, F.C.R., Ticehurst, N.F., and Tucker, B.W. (1944). *The handbook of British birds* (revised edition). Witherby, London.

Wolda, H. (1963). Natural populations of the polymorphic land snail *Cepaea nemoralis* (L.). Factors affecting their size and their genetic constitution. *Archs. néerl. Zool.* **15**, 381–471.

Wolda, H. (1965). Some preliminary observations on the distribution of the various morphs within natural populations of the polymorphic landsnail *Cepaea nemoralis* (L.). *Archs. néerl. Zool.* **16**, 280–92.

Wolda, H. (1967). The effect of temperature on reproduction in some morphs of the landsnail *Cepaea nemoralis* (L.). *Evolution* Lancaster, Pa. **21**, 117–29.

Woods, P.J. (1959). The effects of free and restricted environmental experience on problem-solving behaviour in the rat. *J. comp. physiol. Psychol.* **52**, 399–402.

Woodward, M. (1944). *Adventures in woodcraft.* London.

Wynne-Edwards, V.C. (1962). *Animal dispersion in relation to social behaviour,* Oliver and Boyd, Edinburgh.

Wynne-Edwards, V.C. (1977). In *Population control by social behaviour* (ed. F.S. Ebling and D.M. Stoddart). Institute of Biology, London.

Wynne-Edwards, V.C. (1986). *Evolution through group selection.* Blackwell, Oxford.

Yonge, C.M. (1949). *The sea shore.* Collins, London.

Young, J.Z. (1957). *The life of mammals.* Oxford University Press, London.

Young, J.Z. (1950). *The life of vertebrates.* Oxford University Press, London.

Young, J.Z. (1959). Observation on *Argonauta* and especially its method of feeding. *Proc. Zool. Soc. Lond.* **133**, 471–9.

Young, J.Z. (1961). Learning and discrimination in the octopus. *Biol. Rev.* **36**, 32–96.

Young, J.Z. (1971). *An introduction to the study of man.* Oxford University Press, Oxford.

Zahavi, A. (1977). Reliability in communication systems and the evolution of altruism. In *Evolutionary ecology* (ed. B. Stonehouse and C.M. Perring). Macmillan, London.

Author index

Adams, D. (1985) 276
Alder, J. and Hancock, A. (1864) 94
Allee, W.E., Emerson, A.E., Park, O., Park, T., and Schmidt, K.P. (1949) 23, 25
Allen, J.A. (1972a) 215; (1972b) 215; (1974) 215; (1975) 215
Allen, J.A. and Clarke, B. (1968) 215
Allen, W.E. (1920a) 83, 153; (1920b) 83
Altmann, S.A. (1956) 100
Altmann, S.A. and Altmann, J. (1970) 253
Andersson, M. and Wiklund, C.G. (1978) 100
Andrew, R.J. (1956) 143, 165, 186; (1961a) 100; (1961b) 100, 171; (1961c) 100; (1961d) 100
Annandale, N. (1900) 128
Ansell, A.D. (1969a) 71; (1969b) 71
Anthony, H.E. (1928) 45, 182
Armstrong, E.A. (1947) 35, 92, 98, 111, 143, 228; (1949a) 92; (1949b) 92, 163; (1954) 99; (1956) 92
Arnold, R. (1969) 222
Attneave, F. (1959) 142
Atwal, A.S. and Seth, S.L. (1963) 156
Audubon, J.J. (1827–38) 86
Avery, M.I. (1984) 229
Axelrod, R. (1984) 174, 256

Baerends, G.P. (1950) 34
Baerends, G.P. and Baerends-van Roon, J.M. (1950) 83, 132, 133
Baier, L.F. (1930) 134
Baird, W. (1850) 155
Baker, R.R. (1985) 231
Baker, R.R. and Bibby, C.J. (1987) 231
Baker, R.R. and Parker, G.A. (1979) 231
Banks, C.J. (1957) 207
Barash, D.P. (1976) 92; (1977) 255, 269
Barber, B. (1961) 2
Bartholomew, G.A. (1942) 198
Bastock, M., Morris, D., and Moynihan, M. (1953) 143
Batten, L. (1969) 198
Beebe, W. (1935) 69
Bell, T.R.D. and Scott, F.B. (1937) 218
Bergman, G. (1941) 106
Berlyne, D.E. (1957a) 142, 144, 148; (1957b) 148, 184; (1958) 158; (1960) 131, 142, 144, 146, 147, 153, 156, 163, 172, 184
Bernard, C. (1865) 18
Bertelson, P. (1961) 262
Bertelson, P. and Renkin, A. (1966) 263
Bertram, B.C.R. (1979) 26, 27, 156
Bertsch, H. (1968) 112
Beukema, J.J. (1968) 215, 216, 234
Bevan, W. (1955) 183
Beven, G. (1946) 98
Bingham, W.E. and Griffiths, W.J. (1952) 272
Blanford, W.T. (1889–91) 181
Blaxter, J.H.S. and Parrish, B.B. (1958) 83
Blest, A.D. (1957a) 34, 35, 144, 152, 157; (1957b) 111; (1961) 172; (1963) 217; (1964) 128, 151
Blest, A.D., Collett, T.S., and Pye, J.D. (1963) 128
Bloom, B.R. (1979) 242, 243
Bowen, E.S. (1931) 88, 132
Bowlby, J. (1952) 272; (1953) 272; (1960) 272; (1973) 272
Boycott, B.B. (1958) 94; (1965) 67, 89
Breder, C.M. (1959) 84; (1967) 84
Breder, C.M. and Halpern, F. (1964) 83
Briggs, G.E. and Swansea, J.M. (1971) 143
Bristowe, W.S. (1958) 89; (1972) 235
Broadbent, D.E. (1958) 142
Brock, V.E. and Riffenburgh, R.H. (1960) 213
Broughton, W.B. (1965) 133, 134
Brower, L.P. (1958) 210, 233; (1969) 35; (1971) 26, 28, 29, 30, 35
Brown, R.G.B. (1962) 92
Brown, R.H. (1939) 98
Brown, J.S. and Farber, I.E. (1951) 148
Bruner, J.S. and Postman, L.J. (1949) 148
Buck, G.B. (1954) 112; (1978) 69, 112
Bullis, M.R. (1961) 84
Bullock, T.H. (1953) 71; (1958) 112, 128
Burgers, J.M. (1963) 13
Burke, V.E.M. and Brown, L.H. (1970) 113

Burton, P.J.K. and Thurston, M.H. (1959) 235
Burtt, E.T. and Catton, W.T. (1966) 137
Butynski, T.M. (1979) 47

Cain, A.J. (1963) 220; (1977) 221; (1979) 161; (1983) 220
Cain, A.J. and Sheppard, P.M. (1954) 220, 223
Calder, N. (1984) 269, 271
Campbell, A.J. (1901) 107
Campbell, B. (1969) 70
Caraco, T. and Wolf, L.L. (1975) 202
Carpenter, G.D.H. (1921) 125
Carrington, H. and Meader, J.R. (1911) 121, 192
Caryl, P.G. (1979) 15
Caulfield, C. (1982) 279
Chance, M.R.A. (1954) 36, 182; (1957) 36, 88, 113, 136, 182, 186; (1959) 37, 135, 136, 237
Chance, M.R.A. and Russell, W.M.S. (1959) 34, 35, 37, 38, 44, 112, 113, 136, 142
Chance, M.R.A. and Yaxley, D.C. (1949) 62
Chandler, A.E.F. (1969) 208
Chapman, A. (1897) 96
Cherry, C. (1957) 142
Churchill, W.S. (1948) 266
Claridge, M.F. and Edington, J.M. (1965) 218
Clarke, A.H. (1965) 222, 226; (1978) 193, 224
Clarke, B.C. (1962a) 31, 109, 223, 226, 234; (1962b) 223; (1964) 223; (1969) 28, 223; (1976) 242; (1979) 226, 227
Clarke, C.A. and Sheppard, P.M. (1971) 226
Colyer, C.N. and Hammond, C.O. (1951) 127
Conway, N.J. (1972a) 201; (1972b) 192, 201
Cook, L.M. and Miller, P. (1977) 215
Cooke, A.H. (1895) 71
Coppinger, R.P. (1970) 233
Cornwallis, R.K. (1964) 245
Cott, H.B. (1940) 20, 23, 30, 35, 37, 67, 69, 89, 93, 108, 125, 158, 227, 237, 261
Coues, E. (1875) 182
Coward, T.A. (1950) 230
Cramp, S. (1969) 198
Crook, J.H. (1965) 210
Crook, J.H. and Gartlan, J.S. (1966) 100
Croze, H.J. (1970) 31, 208, 210, 211, 215, 225, 230, 234
Cruickshank, C. (1979) 267
Curio, E. (1959) 100; (1965) 219; (1970a) 219; (1970b) 219; (1970c) 219; (1976) 23, 162, 198, 207
Cushing, D.H. and Harden-Jones, F.R. (1968) 175

Dabelsteen, T. (1978) 48, 76
Darling, F.F. (1937) 65, 66; (1938) 245
Darwin, C. (1859) 15, 41, 174, 188, 254, 280; (1871) 15, 35, 231, 280
Davies, N.B. (1978) 211
Davies, N.B. and Krebs, J.R. (1978) 163
Davis, J.M. (1980) 80
Dawkins, M. (1971a) 215, 216; (1971b) 216; (1976a) 22, 82; (1976b) 165; (1986) 257
Dawkins, R. and Krebs, J.R. (1978) 22, 228
de Beer, G.R. (1972) 273
Denton, E.J. and Nicol, J.A.C. (1965a) 129, 214; (1965b) 129, 214; (1966) 129
Desai, M.M. (1939) 146
Dewar, D. (1913) 96
Dice, L.R., Barto, E., and Clark, P.J. (1963) 135, 183, 188
Ditmars, R.L. (1954) 110
Diver, C. (1929) 220
Dixon, A.F.G. (1959) 207
Dobzhansky, T. (1950) 224; (1956) 161
Driver, H.E. (1969) 277
Driver, P.M. (1960) 53, 56, 87, 99, 179, 208; (1968) 14, 23, 274; (1969) 242; (1974) 34, 56, 263
Driver, P.M. and Corning, W.C. (1968) 16, 34, 183
Driver, P.M. and Humphries, D.A. (1966) 36, 146, 206; (1969) 36, 70, 108, 125, 136; (1970) 36, 142, 222, 233
Dumortier, B. (1964) 134
Dunning, D.C. and Roeder, K.D. (1965) 128, 151, 152

Earle, S.A. (1979) 197
Edmunds, M. (1971) 94; (1972) 191; (1974) 23, 24, 26, 30, 35, 46, 70, 93, 94, 114, 130, 158, 218, 235, 236, 260, 261
Eibl-Eibesfeldt, I. (1950a) 65; (1950b) 65; (1961) 34; (1962) 82, 83, 153
Eisenberg, J.F., Muckenhirn, N.A., and Rudran, R. (1972) 100

Eisher, T., Alsop, R. and Ettershank, G. (1964) 32
Elton, C.S. (1927) 23; (1942) 243
Elton, R.A. and Greenwood, J.J.D. (1970) 224
Evans, G. (1972) 126; (1975) 126
Evans-Pritchard, E. (1940) 200
Ewer, R.F. (1972) 236

Feder, H.M. (1972) 71
Fentress, J.C. (1965) 165
Ficken, M.S. and Dilger, W.C. (1960) 184
Fisher, R.A. (1930) 22, 224
Fitzpatrick, J.P. (1907) 4
Fitzsimmons, F.W. (1920) 45
Flugel, J.C. (1948) 164
Ford, E.B. (1964) 218, 219; (1965) 218, 219
Forgays, D.G. and Forgays, J.W. (1952) 272
Francq. E. (1968) 191
Fretter, V. and Graham, A. (1962) 223
Fricke, H. (1971) 32
Friedmann, H. (1934) 97
Fry, C.H. (1969) 258

Gadgil, M. (1972) 229
Gans, C. (1961) 114
Gardner, H. (1984) 277
Gibb. J.A. (1962) 215; (1966) 210
Giddings, A. (1984) 197
Gilbert, L.E. and Raven, P.E. (1975) 34, 162
Gilliard, E.T. (1962) 228
Goodwin, D. (1953) 170
Gordon, B.L. (1977) 69, 123, 200
Goss-Custard, J.D. (1977) 153
Gould, S.J. and Lewontin, R.C. (1979) 15, 161
Gramza, A.F. (1967) 92, 97
Grant, E.C. and Mackintosh, J.H. (1963) 178
Greenwalt, C.H. (1955) 80
Greenwood, P.J. and Harvey, P.H. (1978) 153
Griffin, D.R. (1958) 149
Griffiths, W.J. (1942) 183
Groves, C.P. (1972) 75
Grunfeld, F.V. *et al.* (1982) 278
Gurney, J.H. (1861) 101
Guthrie, R.D. (1971) 158

Haldane, J.B.S. (1932) 22; (1949) 242; (1953) 162; (1954) 255; (1955) 220, 223
Hale, D.J. (1967) 262, 263
Halkka, O., Heinonen, L., Raatikainen, M., and Vasarainen, A. (1966), 226
Hall, D.F.N. (1956) 94
Hall, K.R.L. (1960) 253; (1966) 95
Hall, B.P., Moreau, R.E., and Galbraith, I.C.J. (1966) 232
Halstead, W.C. (1951) 274
Hamilton, W.D. (1971) 82, 156
Hanby, J. (1983) 145, 178
Hardy, A.C. (1956) 214
Harrison, C.J.O. (1976) 195
Harrison Matthews, L. (1952) 62
Harvey, E.N. (1952) 69
Harvey, P.H. and Greenwood, P.J. (1978) 71, 100, 153, 162, 235
Hassell, M.P. (1971) 153
Hebb, D.O. (1946) 147; (1949) 164, 272, 273; (1958) 273
Heiligenberg, W. (1965) 147
Heller, R. and Milinski, M. (1979) 83, 133, 154, 181
Hemmings, C.C. (1966) 129
Herrebout, W.M., Kuyten, P.J., and de Ruiter, L. (1963) 34, 234
Herring, P.J. (ed.) (1979) 69
Hess, E.H. (1966) 14
Hick, W.E. (1952) 263
Hinde, R.A. (1954a) 100, 170; (1954b) 100; (1960) 165; (1970) 165, 184, 215
Hingston, R.W.G. (1932) 158; (1933) 231
Hoagland, H. (1928) 111
Hobson, E.S. (1965) 83; (1968) 82, 83, 84, 152, 154
Hochbaum, H.A. (1944) 98
Hogan, J.A. (1965) 145, 152
Hohn, E.O. (1958) 98
Holling, C.S. (1965) 35, 141, 215; (1966) 215
Holmes, W. (1940) 36, 67, 68, 94, 113, 128
Holyoak, D.T. (1972) 241
Hoogland, J.L. and Sherman, P.W. (1976) 100, 155
Horsley, D.T. (1978) 215, 216
Howard, L.O. (1901) 89
Hudson, W.H. (1892) 104, 110
Hume, A.O. (1890) 97
Humphries, D.A. (1963) 132; (1970) 36, 39, 60, 121, 132
Humphries, D.A. and Driver, P.M. (1967) 36, 44, 112, 237; (1970) 36, 92, 237

Hutchinson, G.E. (1963) 219
Huxley, J.S. (1925) 98
Huxley, T.H. (1860) 19; (1884) 94
Hyman, L.H. (1940) 112

Iersel, J.J.A. van and Bol, A.C.A: (1958) 165
Inglis, I.R., Fletcher, M.R., Feare, C.J., Greig-Smith, P.W., and Land, S. (1982) 125
Ivlev, V.S. (1961) 156

Jarman, P.J. (1974) 38, 155
Jarvi, T., Radesater, T., and Jacobssen, S. (1977) 242
Jefferies, R. (1879) 48
Johnsgard, P.A. (1966) 175
Johnston, R.F. and Hardy, J.W. (1962) 158
Jones, M.D.R. (1966) 134
Jones, P.H. (1961) 52
Jones, J.S., Leith, B.H., and Rawlings, P. (1977) 220
Jourdain, F.C.R. (1936–7) 98

Kastendiek, J. (1976) 112
Kennedy, J.S. and Booth, C.O. (1963) 56
Kinnear, P.K. (1978) 199
Kipling, R. (1894) 101, 114
Koch, M. (1987) 244
Kohn, A.J. and Waters, V. (1966) 71
Kortlandt, A. and Kooij, M. (1963) 100
Kratzig, H. (1940) 44
Krebs, C.J., Gaines, M.S., Keller, B.L., Myres, J.H., and Tamarin, R.H. (1974) 243
Krebs, J.R. (1974) 199; (1976) 241; (1977) 241; (1978) 153, 207, 231
Kruijt, J.P. (1964) 184
Krushinskii, L.V. (1962) 186
Kruuk, H. (1964) 48, 77, 100, 106, 136, 155, 168, 170; (1972) 26, 27, 46, 71, 72, 77, 85, 100, 145, 156, 181, 254; (1975) 189
Kuhlmann, D.H.H. and Karst, H. (1967) 82
Kuhn, T.S. (1970) 2, 247
Kummer, H. (1968) 253

Lack, D. (1941) 163; (1954) 245; (1966) 245; (1968) 211, 228
Lagerweij, E., Nelis, P.C., Wiegant, V.M., and van Ree, J.M. (1984) 190
Laven, H. (1840) 97
Leakey, L.S.B. (1969) 44, 45
Leakey, R.E. (1981) 277
Lees, H.T. (1964) 109
Leonard, J.A., Newman, R.C., and Carpenter, A. (1966) 262
Lettvin, J.Y., Maturana, H.R., McCulloch, W.S. and Pitts, W.H. (1959) 137
Leyhausen, P. (1956) 34, 145; (1965) 195
Lind, H. (1961) 49, 51, 52
Linné, C. von (1748–52) 212
Lindahl, T.P. (1979) 217
Livingstone, D. (1865) 109, 188
Lorenz, K. (1935) 165; (1941) 175; (1950) 270; (1966a) 175; (1966b) 22, 54, 88; (1971) 270
Louw, G. (1979) 21
Lovenskiold, H.L. (1954) 235
Loveridge, A. (1933) 94
Lynes, H. (1910) 98

MacGinitie, G.E. and MacGinitie, N. (1949) 67, 94
Mackinnon, J. (1970) 211
Magnuson, J.J. and Prescott, J.H. (1966) 175, 214
Maier, N.R.F. (1939) 183; (1948) 183, 264
Major, P.F. and Dill, L.M. (1978) 175
Manly, B.F.J., Miller, P., and Cook, L.M. (1972) 215
Mantiefel, B.P. and Radakov, D.V. (1961) 83
Marlatt, C.L. (1907) 246
Marler, P. (1956) 100, 170; (1957) 253; (1959) 100
Marler, P. and Hamilton, W.J. (1986) 253
Marples, B.J. (1969) 236
Marshall, A.J. (1954) 106
Marshall, S.M. and Orr, A.P. (1955) 61, 213
Marson, J.E. (1947) 235
Martens, E von (1867) 220
Martinez, D.R. and Klinghammer, E. (1970) 198
Maskell, M., Parkin, D.T., and Verspoor, E. (1977) 215
Mason, C.W. and Maxwell-Lefroy, H. (1912) 94
Maynard Smith, J. (1958) 241; (1965) 254; (1974) 15, 22, 229; (1976) 22
Mayr, E. (1950) 223

Mech, L.D. (1970) 156
Medawar, P.B. (1973) 11, 17
Meesters, A. (1940) 216
Meinertzhagen R. (1959) 88, 101, 104
Melzack, R. (1952) 147
Milinsky, M. (1977) 153; (1979) 83, 155
Milinski, M. and Curio, E. (1975) 154
Milinski, M. and Heller, R. (1978) 133, 155, 181
Millais, J.G. (1895) 45; (1913) 51
Miller, G.A. (1956) 142
Miller, J.G. (1963) 265; (1964) 265; (1965) 12, 13
Miller, R.C., (1922) 153
Moment, G.B. (1962) 225
Monod, J. (1972) 15, 16, 42, 194
Mook, J.H., Mook, L.J., and Heikens, H.S. (1960) 215
Morgan, C.T. and Waldman, H. (1941) 183
Morgan, N.C. (1956) 216
Morin, J.G. (1976) 112
Morris, D. (1954) 207; (1956) 186
Mortensen, T. (1917) 133
Mosebach-Pukowski, E. (1937) 154
Moss, C. (1976) 46, 75
Mowat, F. (1963) 63, 184
Moynihan, M. (1958) 143
Müller, G. (1978) 154, 155, 230
Munro, J. and Bédard, J. (1977) 88, 106
Murtagh, P. (1979) 21

Neill, S.R.St.J. and Cullen, J.M. (1974) 83, 133, 154
Nelson, E.W. (1909) 71
Nice, M.M. (1943) 65, 66; (1962) 64
Nice, M.M., Nice, C., and Ewers, D. (1956) 65
Nicholson, E.M. (1939) 98
Nicol, J.A.C. (1955) 112, 162; (1960) 67; (1971) 67
Nielsen, J.M. (1951) 277
Norman, J.R. (1931) 133
Norris, R.A. and Stamm, D.D. (1965) 125
Norton, A.C., Beran, A.U., and Miszahy, G.A. (1964) 191, 192
Norton-Griffiths, M. (1967) 263; (1968) 263
Novick, A. (1969) 70

O'Connor, R.J. and Brown, R.A. (1977) 153
O'Donald, P. (1968) 222
Ohguchi, O. (1978) 154, 155
Olson, F.C.W. (1964) 213
Ortiz de Villalba, J.S. (1980) 278
Osgood, C.E. (1957) 145
Ostrom, J.H. (1972) 195
Owen D.F. (1965) 215, 220; (1967) 232; (1969) 215, 220; (1980) 218
Owen, D.F. and Wiegert, R.G. (1962) 136, 219

Packard, A. (1960) 69
Packard, A. and Sanders, G.D. (1971) 69
Panceri, P. (1872) 112, 128
Pantin, C.F.A. (1950) 140
Parker, G.E. (1978) 229
Parker, H.N. (1963) 114
Paulson, D.R. (1973) 232
Payne, R.B. (1967) 232
Payne, R.S., Roeder, K.D., and Wallman, J. (1966) 179
Payne, R.B. and Skinner, N.J. (1970) 135
Perrone, M. (1980) 125
Peters, J.L. and Loveridge, A. (1936) 94
Pickwell, G. and Smith, E. (1938) 98
Pierce, G.W. (1948) 134
Pilcher, T.J. (1973) 175
Pinhey, E. (1960) 218
Pitt, F. (1931) 98
Pomeroy, H. and Heppner, F. (1977) 80
Popham, E.J. (1941) 223
Popper, K.R. (1968) 22
Portielje, A.F.J. (1925) 44
Potts, G.W. (1970) 153
Potts, W.K. (1984) 176
Poulton, E.B. (1888) 218; (1890) 223; (1924) 237
Prosser, C.L. (1965) 23
Pumphrey, R.J. (1961) 175
Pycraft, W.P. (1925) 129
Pyke, G.H. Pulliam, H.R., and Charnov, E.L. (1977) 153

Raber, H. (1949) 34
Radakov, D.V. (1973) 153
Rand, A.S. (1967) 234
Rapoport, A. and Moshowitz, A. (1966) 256
Rapoport, A. (1959) 256; (1962) 157
Rapoport, A. and Chammah, A.M. (1965) 256
Rasa, O.A.E. (1977) 101
Regen, J. (1908) 134
Reichert, E.S. (1978) 236

Reighard, J. (1908) 223
Ricklefs, R.E. and O'Rourke, K. (1975) 234
Robinson, M.H. (1969a) 32; (1969b) 35, 146, 239; (1969c) 191; (1973) 216
Robinson, M.H., Abele, L.G., and Robinson, B. (1970) 95
Robinson, M.H. and Robinson, B. (1970) 236
Roeder, K.D. (1962) 58, 59, 179; (1963) 151; (1966a) 150, 179; (1966b) 179; (1967a) 150, 151; (1967b) 151, 179
Roeder, K.D. and Treat, A.E. (1957) 149; (1960) 149; (1961) 149, 150, 151, 180
Rohwer, S., Fretwell, S.D., and Tuckfield, R.C. (1976) 125
Rosenthal, H. and Hempel, G. (1970) 208
Rothschild, M. and Haskell, P.T. (1966) 135
Rowell, C.H.F. (1961) 165
Rowley, I. (1962) 131, 163
Royama, T. (1970) 210
Ruiter, L. de (1952) 157, 216; (1953) 34, 35, 216
Russell, C. and Russell, W.M.S. (1961) 274, 275
Russell, W.M.S. (1961) 23, 274
Ryle, J. *et al.* (1982) 200, 278

Sargent, T.D. (1969) 233; (1973) 233; (1978) 125, 227, 233
Saunders, A.A. (1937) 96
Saxby, H.L. (1874) 97
Schall, J.J. and Pianka, E.R. (1980) 240
Schaller, F. and Timm, C. (1950) 149
Schaller, G.B. (1964) 113; (1967) 100; (1972) 26, 27, 46, 90, 145, 155, 156, 189, 201, 202
Schenkel, R. (1966) 46
Schneirla, T.C. (1966) 22
Selander, R.K. (1972) 228
Selye, H. (1956) 191
Serventy, V.N. and White, S.R. (1944) 98
Sevenster, P. (1961) 165
Shannon, C.E. and Weaver, W. (1949) 142
Shaw, E. (1960) 85
Sheffield, F.D., Wulff, J.J., and Backer, R. (1951) 164
Shelford, R. (1903) 127
Sheppard, P.M. (1967) 220
Simmons, K.L. (1952) 92, 163, 167, 170; (1955) 92, 96, 131
Singer, M.C., Ehrlich, P.R., and Gilbert, L.E. (1971) 210
Skeels, H.M. (1966) 272
Slobodkin, L.B. (1965) 17
Slocum, J. (1900) 86
Smith, A. (1984) 190, 273
Smith, D.A.S. (1976) 224
Smith, J.N.M. (1971) 31, 207, 208; (1974) 31
Smith, J.N.M. and Dawkins, R. (1971) 153
Smith, J.N.M. and Sweatman, H.P.A. (1974) 153
Snow, C.P. (1964) 18
Snow, D.W. (1963) 228
Soane, I.D. and Clarke, B. (1973) 219
Sokolov, E.N. (1960) 233
Sokolov, Y.N. (1963) 144, 147
Solomon, M.E. (1949) 156
Spencer, K.G. (1953) 49
Spinage, C.A. (1962) 44, 85
Spooner, G.M. (1931) 88, 133
Stamm, R.A. (1968) 94
Starck, W.A. (1966) 85
Starck, W.A. and Davis, W.P. (1966) 83
Stark, A. and Sclater, W.L. (1900–6) 94
Stiefel, A. (1964) 49
Stoltz, L.P. and Saayman, G.S. (1970) 100
Storr, A. (1974) 273
Street, P. (1952) 95
Stresemann, E. (1927–34) 198
Struhsaker, T. (1967) 253
Summerhayes, V.S. and Elton, C.S. (1928) 235
Summers, A. (1985) 268
Sumner, F.B. (1935) 34
Svardson, G. (1957) 245
Swanberg, P.O. (1951) 212, 255

Taverner, P.A. (1936) 98
Thomas, G. (1974) 208
Thompson, V. (1973) 219
Thompson, W.R. and Heron, W. (1954) 272
Thurber, J. (1953) 47
Tinbergen, L. (1960) 131, 207, 215, 217
Tinbergen, N. (1942) 177; (1951) 44, 163, 165, 177; (1952) 143, 163, 167, 172; (1953) 210; (1958) 34; (1959) 184; (1962) 109, 136; (1972) 18
Tinbergen, N., Impekoven, M., and Franck, D. (1967) 210

Toffler, A. (1970) 264
Treat, A.E. (1955) 149
Trivers, R.L. (1971) 229, 254, 255, 257
Tuck, L.M. (1960) 231, 232
Turner, E.L. (1920) 228

Uspenski, S. M. (1956) 231

Van der Post, L. (1961) 169, 181
Van Rhijn, J.G. (1973) 228; (1983) 229
Vitz, P.C. (1966) 131, 158, 183
Vowles, D.M. (1961) 178

Walker, T.J. (1964) 134
Walther, F.R. (1969) 71
Warmke, G.L. and Almodovar, L.R. (1972) 94
Watson, L. (1973) 110; (1974) 113, 194; (1979) 69
Webster, F.A. (1963) 150
Weihs, D. (1973) 83
Welford, A.T. (1968) 263
Wells, M.J. (1962) 69
Welty, J.C. (1934) 154; (1964) 175
White, T.H. (1939) 51
Whitfield, D. (1966) 263
Whittaker, J.B. (1968) 219
Wickler, W. (1968) 26, 30, 89, 91, 218, 226, 260
Wiley, R.H. (1974) 228
Williams, G.C. (1964) 82, 84; (1966) 15, 22, 82, 161, 254
Williamson, K. (1950) 131
Wilmut, R. and Grafton, J. (1977) 268
Wilson, E.O. (1971) 253; (1975) 82, 93, 169, 253
Wingate, P. (1972) 191
Witherby, H.F., Jourdain, F.C.R., Ticehurst, N.F., and Tucker, B.W. (1947) 48, 51, 52, 77, 230
Wolda, H. (1963) 222; (1965) 222; (1967) 222
Woods, P.J. (1959) 272
Woodward, M. (1944) 98
Wynne-Edwards, V.C. (1962) 22, 55, 158, 229, 245, 246, 254; (1977) 22; (1986) 174

Yonge, C.M. (1949) 94
Young, J.Z. (1957) 16; (1959) 113; (1961) 137; (1971) 16

Zahavi, A. (1977) 181

Species index (English)

adder 94
Afghan hound 202
African egg-eating snake 114
albacore 123
ant 194
antelope 41, 45, 116, 122, 123, 182, 193, 254
  dikdik 45, 46
  gazelle 27, 43, 45, 46, 71, 79, 91, 201, 202
    Thomson's 181, 189, 254
  gerenuk, 46
  impala 4, 5, 38, 40, 46, 72, 133, 155, 159
  Indian 181
  oribi 45
  springbok 45, 65, 71, 72, 133, 158, 181, 245
  wildebeest 27, 72, 77, 85, 133, 178, 189
arctic fox 63, 98
arctic tern 235
aphid 208
armadillo 32

baboons 253
badger 65
barbet 134, 135
barn owl 232
bar-tailed godwit 48, 81
bats 27, 39, 58, 123, 128, 129, 149, 150, 151, 152, 160, 179, 194, 196, 204, 235
  vampire 70, 257
beaver 270
bee 34
beetle 28, 29
  coccinellid 208
  glow-worm 222
  ladybird 208
  skipjack 125, 126
bird of paradise 258, 259
blackbird 109, 111, 153, 208
black-tailed godwit 49, 51, 52, 81
black-throated diver (loon) 96
blue tit 210
bobwhite quail 86
bonito 123
booby 198
borzois 202
bower bird 106, 117, 188, 241, 242
  spotted 106, 107
box fish 32
bristletail 59
brittle star 94, 217, 225
buffalo 189
bug 30, 34, 89, 91, 224, 237, 239
bullfinch 109
bunting, yellow 35, 152
Burmese jungle fowl 145, 195
bush cricket 133, 134, 139, 188
bush-shrike 134, 139, 232
bush snake 94
bustard, houbara 39, 101, 102, 212, 253, 255
butterfish 198
butterflies 28, 32, 35, 38, 57, 124, 144, 158, 218, 260
  cabbage white 58
  Camberwell beauty 209
  comma 173
  hairstreak 257, 260
  peacock 154
  small tortoiseshell 37, 40, 108, 111, 121, 128, 152, 154, 239, 252
  speckled wood 211
  swallowtail 260
buzzard 94, 100, 105, 232

cabbage white butterfly 58
Camaroun toad 110
Camberwell beauty butterfly 209
Cape horned viper 114
Cape hunting dog 27, 43, 46, 72, 145, 193, 201
carpet moth 221
carpet viper 114
carrion crow 51, 105, 208, 225, 242
cat 66, 98, 104, 109, 145, 195
caterpillar 218, 219, 234, 248
  cinnabar 133
  puss 125
  stick 35, 157, 216
cattle 243
chaffinch 35, 100, 170
chameleon 94, 124

cheetah 26, 46, 189, 200, 201
chough 51, 92, 105
cicada 130, 246, 248
cichlid fish 52, 147
cinnabar moth caterpillar 133
clam, butterfly 225
clawed toad 112
cleaner fish 257
coccinellid beetle 208
comma butterfly 173
common snipe 48, 49, 50, 116
convolvulus hawk moth 218
copepod 213
cormorant, 198, 199, 200
coyote 98
crab, shore 71, 95, 227, 261
cricket 27, 133, 134, 139, 188
crossbill 245
crows, 48, 170, 194, 255
  carrion 51, 105, 208, 225, 242
  chough 51, 92, 105
  hooded 51
  jackdaw 51, 65
cuckoo 232
cuttlefish 34, 36, 67, 68, 83, 113, 118, 128, 129, 147, 152

Dalmatian pelican 198
darter, Indian 198
deer 41, 65, 71, 180
deerhound 202
deermouse 62, 182, 237
dhole 27
dikdik 45, 46
dogs 41, 45, 90, 98, 122, 132, 147, 189, 202, 272
  Afghan 202
  borzois 202
  Cape hunting 27, 43, 46, 72, 145, 193, 201
  deerhound 202
  dhole 27
  eskimo 98
  greyhound 202
  lurcher 202, 203
  saluki 202
  sheepdog 202
  whippet 202
dolphin (mammal) 197
dolphin fish 86, 124
dolphin gull 105
double-crested cormorant 198
dowitcher 51, 81
drongo, racquet-tailed 259
ducks 48, 76, 77, 87, 95, 116, 139, 164–9, 175, 176, 184, 204, 212, 230
  eider 87, 88, 97, 169, 234
  mallard 65, 109, 169
  old-squaw (long-tailed) 63, 98, 99, 132
  scoter 51
  shelduck 88
  tufted 87
  wigeon 65
  wood 96
dragonfly 27, 34
ducklings 67, 116, 165, 168
  eider 6, 56, 86, 88, 106, 179, 208
  red-breasted merganser 52, 132
dunlin 79, 80, 81, 97, 176
dwarf mongoose 101
eagles 39, 66, 102, 106, 253, 255
  imperial 101
  short-toed 106
  tawny 101
earwig 28
eel 199
eider duck 87, 88, 97, 169, 234
elephant 65
emu 106
eskimo dog 98

falcons 26, 176
  gyr 229, 230
  merlin 170, 229
  peregrine 26, 45, 48, 76, 77, 78, 170, 229, 230, 232
field sparrow 111
fieldfare 100
finches 35, 100, 109, 170
fishes 54, 61, 67, 69, 70, 74, 82, 83, 88, 116, 132, 139, 175, 204, 213, 214
  albacore 123
  bonito 123
  box fish 32
  butterfish 198
  cichlid 52, 147
  cleaner fish 257
  dolphin fish 86, 124
  eel 199
  flounder 69, 199
  flying fish 123, 124
  garfish 133
  globe fish 124
  goby, spotted 52
  goldfish 154
  grouper 69, 129, 162
  guppy 52
  herring 200, 208
  loach, sucking 110
  mackerel 200

marling 123, 200
minnow 52
mullet, grey 3, 52, 74, 75, 76, 199
perch 52, 83
pike 83, 200
pilchard 133
pollack 199
porcupine fish 32, 124, 133
reef fish 130, 133
rockling 198
rudd 52, 224
sand eel 3, 199
sardine 129, 153
shark, thresher 89, 200
snapper, grey 223
sprat 133, 199
squirrel fish 70, 129
stickleback 52, 53, 54, 83, 132, 154, 181, 199, 208, 216, 234
swordfish 123, 200
trout 52, 200
tuna 123, 200
weever fish 31
whitebait 3, 198
wrasse 52
zebra fish 85
flatfly 39
flattid bug 89, 91
fleas 39, 59, 120, 123, 178
flies 27, 108
flatfly 39
gnat 55
hover 208
jewel 85, 126, 127
mosquito 55, 56, 116, 217
psychodid 57
syrphid 208
whitefly 57
flycatcher, tyrant 258
flying fish 123, 124
fowl 74
burmese jungle 145, 195
chicks 152, 216
fox 4, 62, 63, 72, 73, 74, 98, 105, 110, 114, 118, 184, 185, 186, 187, 192, 205
arctic 63, 98
Simien 63
frog 34, 94, 130, 137, 232
fulgorid bug 237, 239

gannet 42, 198
garden spider 235
garden tiger moth 135
garfish 133
gazelle 27, 43, 45, 46, 71, 91, 201, 202
Thomson's 181, 189, 254
gecko, flying 93
gerenuk 46
glaucous gull 232
glow-worm 222
gnat 55
goat 65
goby, spotted 52
godwit, bar-tailed 48, 81
godwit, black-tailed 49, 51, 52, 81
goldfish 154
goose, white-fronted 51
goshawk 229
grasshopper 27, 28, 30, 67, 130, 188, 194
green African 125
grass snake 110
great northern diver (common loon) 80, 199
great snipe 48
great tit 153, 210, 241
greater black-backed gull 106, 155, 231, 245
grebe, great crested 96
greenshank 81
grey mullet 3, 52, 74, 75, 76, 199
grey phalarope 81
greyhound 202
grouper 69, 129, 162
grouse 87
Guiana quail 134
guillemot 80, 231
gulls 4, 6, 88, 105, 117, 143, 194, 235, 255
black-headed 48, 65, 77, 100, 104, 136, 155, 170
dolphin 105
glaucous 232
greater black-backed 106, 155, 231, 245
herring 65, 105, 106, 155, 170, 184, 210
guppy 52
gyrfalcon 229, 230

hairstreak butterfly 257, 260
hamster 88, 178
hare 45, 116, 120, 122, 181
black-tailed jackrabbit 45, 181
Snowshoe 124
white-sided jackrabbit 71
white-tailed jackrabbit 182
harrier 229
harvestman 221

hawk 6, 26, 104, 177, 232
goshawk 229
red-shouldered hawk 96, 98
sparrowhawk 87, 229
hedgehog 32
herons 199, 232
herring 200, 208
herring gull 65, 105, 106, 155, 170, 184, 210
hog-nosed snake 110
honey-eater 49, 51
hooded crow 51
hornbill 101, 103, 117, 173, 174, 204, 253, 255
horse 65, 190
humming bird, marvellous spatuletail 258, 259
hover fly 208
hyena 27, 41, 46, 77, 100, 189, 192, 193

impala 4, 5, 38, 40, 46, 72, 133, 155, 159
Indian antelope 181
Indian mongoose 101, 103, 104

jackal 45, 46
jackdaw 51, 65
jackrabbit
black-tailed 45, 181
white-sided 71;
white-tailed 182
jack snipe 48, 50
jaguar 111
jay 35, 216
jellyfish 31
jerboa 270
jewel fly 85, 126, 127

kangaroo 63, 64, 114
kestrel 170
kittiwake 87
knot 81
krill 195, 197

lacewing 58
ladybird 208
lapwing (green plover) 48, 49, 51, 74, 76, 79, 104, 158, 159, 188
leaf-footed bug 257, 258, 260
leaf-hopper 218, 248
leaf insect 210, 248
lemming 182
leopard 46
lesser yellowlegs 51, 77, 81
limpet 207
lion 26, 27, 38, 46, 90, 109, 122, 133, 156, 178, 179, 188, 189, 202
little stint 81
lizard 8, 93, 94, 95, 111, 112, 131, 219, 232, (whiptail) 240
loach, sucking 110
lobster 257, 261
loons 153
black-throated 96
common 80, 199
red-throated 98
lurcher 202, 203
lynx 124

mackerel 200
magpie 242
mallard duck 65, 109, 169
mantis 26, 125, 127, 130, 158, 191, 211
marlin 123, 200
martin 194
purple 158
marvellous spatuletail 258, 259
mealworm 129, 144, 151, 152
membracid bug 30, 237
merganser, red-breasted 52, 132
merlin 170, 229
mice 8, 62, 88, 109, 178, 219, 232, 244
millipede 32
minnow 52
mongoose 124, 255
monkey
patas 95, 212
vervet 253
moon moth 260
moorhen 169
mosquito 55, 56, 116, 217
moths 8, 9, 29, 32, 33, 35, 38, 39, 117, 123, 124, 128, 144, 149, 150, 151, 152, 160, 179, 180, 196, 204, 205, 216, 217, 227
carpet 221
catocalid 233, 234
convolvulus hawk 218
eucosmid 210
eyed hawk 125
garden tiger 135
hawk 209, 218
moon 260
mottled umber 218
peppered 209
saturniid 258
sphinx 28

New World 219
yellow underwing 111, 121
motmots 258
racquet-tailed 257
turquoise-browed 259
mottled umber moth 218
mullet, grey 3, 52, 74, 75, 76, 199

nandu 44
nighthawk 97, 194, 258
nightjar, standard-winged 258, 259, 260
nutcracker, 244
Clarke's 212
thick-billed 212

octopuses 67, 69, 113, 137
old-squaw duck 63, 98, 99, 132
opossum 110
Virginia 191, 192
oribi 45
ostrich 169
otter 95, 98
owl 4, 232
barn 232
short-eared 170, 229
tawny 100
oystercatcher 81, 153, 207

pangolin 32
parrot 258
partridge 87, 96
patas monkey 95, 212
peacock butterfly 154
pelican 113, 198
penguin 105
peppered moth 209
perch 52, 83
peregrine falcon 230
periwinkle 223
petrel 232
phalaropes, grey, red-necked, and Wilson's 81
pheasant 30, 109
pigeon 188
pike 83, 200
plover 30
killdeer 97
ringed 81, 97, 98, 108, 121, 169
pollack 199
polynoid worm 94, 131
porcupine 32
porcupine fish 32, 124, 133
porpoise 123, 198
prawn 61, 69, 261
psychodid fly 57
ptarmigan 44
puffin 199, 231
purple sandpiper 164, 225
puss moth caterpillar 125

quail 94
bobwhite 86
Guiana 134

rabbit 2, 3, 6, 47, 62, 65, 70, 71, 90, 109, 120, 122, 132, 181, 202, 203, 225
racoon dog 191
racquet-tailed drongo 259
rat 16, 162, 178, 205, 264, 265, 270, 272
raven 51, 105
razor shell (jack-knife clam) 94
red deer 65
reef fish 130, 133
red-necked phalarope 81
redshank 48, 79, 153
red-shouldered hawk 96, 98
red-throated diver (loon) 98
rhinoceros 32
robin 62
rockling 198
roe deer 65
rook 51, 65
rudd 52, 224
ruff 81, 228, 229, 230, 248

saluki 202
sanderling 81
sandhopper 59, 61, 90, 178
sandpiper 48, 81
purple 164, 225
sandwich tern 234
sardine 129, 153
saturniid moth 258
sawfly 218
scale worm 94, 216
scallop 222
scorpion 31
scoter 51
sea anemone 31
sea cucumber 94
sea pansy 112, 118, 128, 133, 147, 152, 162
sea pen 112
sea slug 94

sea urchin 31
secretary bird 94
seedsnipe 48
shag 198
shark, thresher 89, 200
sheep 65, 106, 180
sheepdog 202
shelduck 88
shorehopper 59, 74, 92, 123, 178, 195
short-eared owl 170, 229
short-toed eagle 106
shrews, 88, 195
  American short-tailed 31
  pigmy 62
shrimp 61, 69
Siamese tree snake 128
silverfish 59, 60, 122
Simien fox 63
skink 94
skipjack beetle 125, 126
skua 87
  great 105, 106
skunk 20, 32
small tortoiseshell butterfly 37, 40, 108, 111, 121, 128, 152, 154, 239, 252
snail 207, 219, 221, 222
  African land 220, 227
snakes 5, 25, 31, 32, 94, 101, 104, 112, 114, 117, 173, 232, 255
  adder 94
    night 114
  asp 114
  African egg-eating 114
  bush 94
  grass 110
  hog-nosed 110
  Siamese tree 128
  viper
    cape horned 114
    carpet 114
    saw-tailed 114
snipe 48, 49, 50, 116
  common 48, 50
  great 50
  jack 48, 50
sparrows. 30, 66, 194
  field 111
  song 111
sparrowhawk 87, 100, 106, 229
speckled wood butterfly 211
sphinx moth 28, 219
spiders 30, 31, 34, 89, 191, 194, 209, 235, 236
  garden 235
spittlebug 136, 219, 248
sprat 133, 199
springbok 45, 65, 71, 72, 133, 158, 181, 245
spring-hare 46, 47
springtail 59
squid 66, 69, 83, 89, 124, 129
squirrel
  grey 47, 48, 88, 244, 270
  African ground 191
squirrel-fish 70, 129
starfish 71, 112
starling 70, 76, 77, 78, 79, 106, 107, 133, 159, 175, 176, 188, 194, 242
stick caterpillar 35, 157, 216
stick insect 210, 239
stilt, black-winged 96, 97
stints 81
stoat 62, 63, 98, 114, 118, 169, 184
sugar birds 258
swallow 4, 105, 194
  bank 100, 155
swallow-tailed butterfly 260
swift 194
swordfish 123, 200

tawny eagle 101
tawny owl 100
Temminck's stint 81
terns 105, 106, 117
  arctic 235
  Sandwich 234
thrushes 30, 207, 225
  fieldfare 100
  mistle thrush 105
  song 153
tinamou 110, 111
tits 80, 215
  blue 210
  great 153, 210, 241
toad 124
  Camaroun 110
  clawed 112
tortoise 94
touraco 130, 131
trout 52, 200
tufted duck 87
tuna 123, 206
turnstone 79, 81, 92
turtle 246

vampire bat 70, 257
vervet monkey 253
vipers 114
voles 88, 225, 244, 270
wasps 34, 219

chalcid 210
ichneumon 210
parasitic 153
water boatman 224
water flea 83, 153, 181
weasel 62, 63, 64, 184
weaver bird 210
weever fish 31
whale 195, 197, 198, 200
whippet 202
whitebait 3, 198
whitefly 57
white-fronted goose 51
white pelican 198
wigeon 65
wildebeest 27, 72, 77, 85, 133, 178, 189
Wilson's phalarope 81
wolf 5, 27, 41, 43, 63, 184
wood duck 96
woodlice 32
woodpigeon 45
worm
annelid 94
polynoid 94, 131
scale 94, 216
wrasse 52
wren, central American 134

yellowlegs, lesser 51, 77, 81, 85, 236

zebra 27, 193
zebra fish 85

Species index (Latin)

Acantholyda 217
Accipiter gentilis 229
Accipiter nisus 229
Accipitridae 26, 232
Acrididae 28
Aepyceros melampus 38
Aequipecten irradians 222
Aglais urticae 37, 152
Aix sponsa 96
Alcedo atthis 127
Alchisme grossa 238
Aleyrodidae 57
Alopex lagopus 63, 98
Alopias vulpinus 89
Amblyornis 241
Ammodytidae 199
Anas penelope 65
Anas platyrhynchos 65, 109
Anatidae 87, 175
Anatinae 166, 167
Anhinga anhinga 198
Anisoscelis foliaceae 258
Anolis carolinensis 111
Anolis lineatopus 219
Anopheles 55
Anser albifrons 51
Antidorcas marsupialis 45
Antidorcas melampus 46
Antilope cervicapra 181
Apidae 194
Apodemus sylvaticus 88
Aquila heliacus 101
Aquila rapax 101
Araneus alsine 235
Araneus diadematus 89, 235
Archaeopteryx 195
Arctia caja 135, 136
Ardea herodias 199
Ardeidae 232
Arenaria interpres 79
Arge fuscipes 218
Arge ustulata 218
Argiope 32
Argiope flavipalpis 236
Argiope pulchella 235
Argiopidae 235
Argonauta 113
Argyra diaphana 127
Argyrodes 235
Armina 112
Ascophyllum nodosum 223, 224
Asio flammeus 229
Astropecten 112
Athous haemorrhoidalis 125, 126
Automalis 152
Aythya fuligula 87

Balistes 31
Biston 35
Bitis caudalis 114
Blarina brevicauda 31
Bocydium rufiglabrum 237, 238
Brachydanio rerio 85
Bucorvus cafer 101
Bufo superciliaris 110
Buteo 232
Buteo buteo 100
Buteo lineatus 96
Buteo rufofuscus 94

Caenurgina erechtea 151, 179
Calanus 61, 213
Calidris alpina 79, 97, 176
Calidris maritima 164
Canidae 184
Canis azarae 110
Canis latrans 98
Canis lupus 63
Canis simiensis 63
Canis vulpes 72, 192
Capella gallinago 48
Capitonidae 134
Capreolus capreolus 65
Caprimulgidae 194, 258
Carcinus maenas 71, 227
Cardium 71
Cassiteridum 88
Catocala 227, 232, 234
Causus 114
Centronotus gunnellus 199
Cepaea 220, 221, 222, 223, 225
Cepaea hortensis 219, 221, 226
Cepaea nemoralis 219, 220, 221, 226
Cephalapoda 114
Cepphus grylle 80
Ceratophyllus gallinae 59

Ceratophyllus garei 59, 60, 132
Centropus sinensis 94
Cervus elephas 65
Charadriidae 30
Charadrius hiaticula 97, 121
Charadrius vociferus 97
Chiroptera 27
Chlamydera maculata 106, 241
Chlamydotis undulata 101
Chlorophoneus multicolor 232
Chordeiles minor 97
Cicinnurus regius 259
Ciliata mustela 199
Circaetus gallicus 106
Circus 229
Cissura 152
Clamator jacobinus 232
Clethrionomys 62
Clethrionomys glareolus 63
Clupea harengus 199
Cnemidophorus 240
Cnemidophorous tigris 240
Colinus virginianus 86
Columba palumbus 45
Conger conger 199
Connochaetes gnou 72
Corvus corax 51
Corvus corone 51, 208, 225
Corvus frugilegus 51
Corvus monedula 51
Corylus avellana 212
Crangon 61
Crenimugil labrosus 199
Crocidura suaveolens 88
Crustacea 32, 94, 261
Ctenophthalamus nobilis 59
Cuculidae 232
Cuculus canorus 232
Cuculus clamosus 232
Culex 55
Cuon alpinus 27
Cyclops 61
Cyclosa 236
Cyclosa mulmeinensis 236
Cyclosa tremula 236
Cyphonia clavata 237, 238
Cyphonia hirta 237, 238
Cypridina 69

Daphnia 61, 154, 155
Daphnia magna 83, 153
Dasypeltis 114
Dasypeltis scabra 114
Dermaptera 28
Diadema 31
Dicruridae 258
Dicrurus paradiseus 259
Didelphis azarae 110
Didelphis virginiana 191
Didontidae 32
Diplopoda 32
Dipsadomorphus dendrophilus 128
Diptera 27, 56
Discodoris fragilis 94
Dolichopodidae 85, 126, 127
Donax variabilis 225

Echinoidea 31
Echis carinata 114
Echiurus echiurus 94
Edentata 32
Elateridae 125
Emberiza citrinella 35, 152
Emphusis bakeri 238
Enarmonia conicolana 210
Ennomos 35
Ennya chrysura 238
Epinephalus striatus 69, 129
Eptesicus 150
Erannis defolaria 218
Erinaceus europaeus 32
Erithacus rubecula 62
Errinyis ello 219
Erythrocebus patas 95
Esox lucius 83, 200
Eudaemonia brachyura 258
Eumomota superciliosa 259
Euphausia 195
Euplectes 211
Euptychia 209
Euptychia hermes 209
Euptychia penelope 209
Euptychia renata 209

Fagus sylvatica 244
Falco columbarius 229
Falco peregrinus 26, 45
Falco rusticola 229
Falconidae 26, 232
Felis onca 111
Floronia bucculenta 235
Fratercula arctica 199
Fringilla coelebs 100, 170
Fucus 223, 224
Fulgora 239

Gallus gallus spadiceus 145, 195
Garrulus glandarius 216

Gasterosteus aculeatus 52, 83, 153
Gavia arctica 96
Gavia immer 80, 199
Gavia stellata 98
Gaviidae 153
Gazella cuvieri 45
Gecarcinus 95
Gekkonidae 93
Gerhosauridae 93
Glareola lactea 97
Glareolidae 96
Gobius ruthensparri 53
Gyrinocheilus aymonieri 110

Haematopus ostralegus 153, 207
Haemulon flavo-lineatus 129
Haliaster sphenurus 107
Halysidota tessellaris 128, 151
Harengula thrissina 84
Harmathoë tuta 94
Helicarion 71
Helogale undulata 101
Hemiptera 224, 236, 237
Herpestes mungo 101
Herse convolvuli 218
Hestiasula sarawaka 127
Heterocarpus 69
Heterodon platyrhinos 110
Heteronotus abbreviatus 237, 238
Heteronotus tridens 237, 238
Heteroteuthis 67, 69
Himantopus himantopus 96
Hippoboscidae 39
Hirundinidae 194
Holocentrus ascensionis 70, 129
Homo sapiens 269, 270, 271, 279
Hyena hyena 192
Hyloicus pinastri 216
Hymenoptera 218

Insectivora 32
Ipomaea aquatica 218
Isopoda 32

Labrus bergylta 53
Lacertidae 93
Lagomorpha 181
Lagopus lagopus 44
Laminaria 199
Lampyris noctiluca 222
Laniidae 134
Laniurus 134
Laothoë populi 209
Laridae 105
Larus argentatus 88, 210
Larus marinus 155, 245
Larus ridibundus 48, 100, 136, 155, 170
Larus scoresbii 105
Laternaria servillei 237
Leander 61
Lebistes reticulatus 53
Lepidoptera 34, 35, 57, 107, 114, 218
Lepisma saccharina 59, 122
Lepus alleni 71
Lepus americanus 65, 124
Lepus californicus 45, 181
Lepus callotis 71
Lepus europaeus 44
Lepus townsendi 182
Leuciscus erophthalmus 224
Leucorhampha 28
Limicolaria 227
Limicolaria martensiana 220
Limnodromus griseus 51
Limosa lapponica 48
Limosa limosa 49
Litocranius walleri 46
Littorina littoralis 223, 224, 226
Littorina obtusata 224
Loddigesia mirabilis 259
Loligo vulgaris 83
Loxia 245
Lutianus griseus 223
Lutra annectens 95
Lutra lutra 98
Lybius 134
Lycaenidae 260
Lycaon pictus 27
Lycophidion capense 94
Lymnocryptes minimus 48
Lynx canadensis 124

Macrodipterx longipennis 258, 259
Madoqua 45
Magicicada cassini 246
Magicicada septendecim 246
Makaira indica 200
Malaconotinae 232
Marginella 219
Megaptera novae-angliae 197
Melanitta fusca 51
Melanitta nigra 51
Melanitta perspicillata 51
Meles 152
Meles meles 65
Meliphagidae 49, 258
Melospiza melodia 66
Mergini 87

Mergus serrator 53
Metridia 69
Microtus 62
Motmotidae 258
Mugil chelo 53
Mus musculus 88, 219
Muscicapidae 30
Musophagidae 130
Mustela erminea 62, 98
Mustela nivalis 63
Mustelidae 32, 62
Myriapoda 32
Myotis 150
Myotis lucifugus 128, 129, 151

Nassarius 71
Natrix natrix 110
Nemeritis canescens 153
Neuroptera 58
Nothura maculosa 110
Notonecta glauca 34
Nucifraga caryocatactes 212, 244
Nucifraga columbiana 212
Nyctereutes procyonoides 191
Nycterybiidae 39
Nymphalis io 35

Ochrodota 152
Octopus 113
Octopus vulgaris 69
Odonata 27
Odontophorus gujanensis 134
Ophiocantha eurythra 94
Ophiopholis aculeata 225
Orchestia gammarella 59, 90, 92
Orcinus orca 198
Oryctolagus cuniculus 47, 132
Ostraciontidae 32
Otidae 39
Ourebia ourebi 45
Oxynoë 94

Panolis 217
Papilio dardanus 28
Papilio lysithous 28
Papilio polytes 28
Papilionidae 260
Paradisaeidae 258
Paralichthys dentatus 69
Pararge aegeria 211
Paroctopus bimaculatus 67
Parus 215
Parus atricapillus 80
Parus bicolor 80
Parus caeruleus 210
Parus major 153, 210
Passeriformes 88
Passerinae 30
Patella 207
Pedetes capensis 46
Pelecaniformes 198, 199
Pelecanus crispus 198
Pelecanus onocrotalus 198
Pelmatochromis kribensis 147
Pennatula 112
Perca fluviatilis 53, 83
Peromyscus maniculatus 36, 44, 62, 128, 136, 142, 145, 182, 183, 205, 237
Phalacrocorax aristotelis 198
Phalacrocorax auritus 198
Phasianidae 30, 87
Phasianus colchicus 109
Phasmida 210
Pheugopedius hyperythrus 134
Philaenus 226
Philaenus spumarius 136, 137, 219
Philanthus triangulum 34
Philomachus pugnax 228
Philothamnus semivariegatus 94
Pholidoptera aptera 134
Pholidoptera griseoaptera 134
Pholidota 32
Phonognatha wagneri 235
Photoblepharon 69
Phoxinus phoxinus 52
Phrynosoma cornutum 111
Phymateus viripides 125
Pinus cembra 212
Pisces 32
Plasmodium 243, 244
Plasmodium falciparum 243
Platichthys flesus 199
Ploceidae 30
Plotosus anguillaris 133
Podiceps cristatus 96
Poinsettia pulcherrima 219
Polistes crinitus 219
Pollachius pollachius 199
Polygonia c-album 173
Pomatocarcinus 95
Procellariidae 232
Progne subis 158
Promerops cafer 49
Promerops gurneyi 49
Pseudocreobotra wahlbergi 125
Psittacidae 258
Psocus venosus 89
Psychoda phalaenoides 57, 108, 120
Psychodidae 56, 57

Pterophylla camellifolia 134
Ptilinorhynchidae 106
Ptilonorhynchus 241
Ptychozoon kuhli 93
Pulex irritans 59
Pygoscelis papua 105
Pyrrharctia isabella 128, 151
Pyrrhocorax pyrrhocorax 51
Pyrrhula pyrrhula 109

Rattus norvegicus 65, 270
Renilla 112, 113
Rhea americana 44
Riparia riparia 100, 155
Rissa tridactyla 87
Rodentia 32

Sagittarius serpentarius 94
Salmo trutta 53, 200
Salticidae 195
Saturniidae 112, 260
Scardinius erythrophthalmus 53
Scincidae 93
Sciurus carolinensis 47
Scolopacidae 48, 225
Scorpiones 31
Sepia officinalis 36, 67, 68, 83, 142
Sepioteuthis 67, 89
Sigara distincta 224
Smerinthus ocellatus 125
Solen 94
Somateria mollissima 56, 234
Somateriini 87
Sorex araneus 88
Sorex minutus 62, 88
Sphingidae 29, 218
Spinachia vulgaris 199
Spisula 71
Spizella pusilla 111
Sponogophorus hoffmanni 238
Sprattus sprattus 199
Staphylinidae 28
Stercorariidae 105
Stercorarius skua 105
Sterna paradisaea 235
Sterna sandvicensis 234
Strigidae 232
Strix aluco 100
Struthiolaria 71
Struthionidae 169
Sturnus vulgaris 70, 242
Sula bassana 41
Sulidae 198
Syphonaptera 39
Systellaspis 69

Tadorna tadorna 88
Talitrus saltator 59, 90
Teiidae 93
Tenebrio 129
Tetragnatha 235
Tetraonidae 87
Thecla 260
Thecla togarna 261
Theridion 235
Thinochoridae 48
Thyrophilus modestus 134
Thysanura 59
Trachinidae 31
Trigonia 70
Tringa flavipes 51
Tringa solitaria 48
Tringa totanus 48, 153
Triphaena pronuba 111
Trochilidae 258
Trypanosoma brucei gambiense 243, 244
Trypanosoma brucei rhodesiense 243
Turdinae 30
Turdus merula 109, 111, 153, 207, 208
Turdus philomelos 153, 207
Turdus pilaris 100
Tyrannidae 258
Tytonidae 232

Uloboridae 235
Uloborus 236
Uria aalge albionis 231

Vanellus vanellus 48
Vanessa antiopa 209
Vanessa io 154
Vanessa urticae 154
Vimba vimba 210
Vulpes vulpes 62, 63

Xenopus 142
Xenopus laevis 112
Xerus erythropus 191
Xiphias gladius 200

Subject index

abnormal behaviour 183
acceleration 126
acoustic displays 128–9, 133–5, 145, 188
  alternation 134, 135, 188
  ultrasonic 151–2
action patterns 165–8, 191, 192, 254
acupuncture 190
adaptedness 18
adaptive relationship 15
aerial dances 55, 74
aerial plankton feeders 56, 194, 205
aerodynamics 176–7
aggregation 155
aggressive attacks 44
agonistic behaviour 143, 178
AIDS 244, 248
Akha people 278
alarm calls 4, 38, 75, 253–4, 257
alternation behaviour 143
altruism 99, 109, 169, 174, 253–6
  reciprocal 229, 254, 255
ambivalent behaviour 143, 145
ambush 20, 25
American football 'plays' 66, 267
anarchization 172, 204, 247–52, 280
  altruism 255
  anomaly 233
  biochemical 242–4
  coloration 220, 247, 250
  dispersion 250–2
  display diversification 193, 234, 241, 250
  erratic movement 193, 201
  group display 173–4
  life history 246, 248
  speciation 234
  structure 236–7, 244
anomaly 125, 233, 248
antennae of Daphniidae 155
anteroposterior confusion devices 257, 260–1
antibodies 243, 248
anticipation 145–7, 180, 181, 205, 206
antigenic drift 243–4, 248
antiparasite activities 120, 178
antiphonal singing 134–5
antiprotean devices 40, 193–203
antisubmarine warfare 213
aposematism
  see also warning coloration 151, 209, 217, 234
apostatic selection 222–5, 232
aquatic medium
  locomotion 53, 54
  detection of prey 213–14
area-concentrated searching 207–12, 214, 217, 226, 247, 250, 251
  habitat patches 210
  prey dispersion changes 208–10
armour 20, 28, 32
arms race
  biological 1, 2, 20, 21, 35
  human 1, 20, 21, 279
aspect diversity 233–6, 248, 250
  tropical chromatic polyphasy 226
asymmetry 218
attention distraction 5, 7, 39, 71, 93–107, 117, 155, 164, 255
attention-drawing properties 131, 158, 181, 183
audiogenic seizures 182–3, 205, 237
Australian aborigines 63, 64, 278
autoanalgesia 189–91, 205
autonomic activity 186
autotomy 93–5, 162
  antigen shedding 244

banding of snails 220–3
barbed wire 20
baseball 267
berserking 112, 128, 133, 147
biome maturation 193
bird flocks 76–82
bizarre shapes 236–8
Blitzkrieg 267
blocked escape 182
bluff 28
booby-traps 20
bouncing, *see* leaping, pronking, stotting
boxing 267
breeding displays 48–51, 175–6, 258
bubble net 197
bunching of prey 76, 78, 83

camouflage, *see* crypsis

catalepsy 44, 108–11, 121, 136, 183, 191, 192
channel capacity 142
  human 143, 263
chaos 14
charging by predators 200
chemical defence 35
chemical warfare 20
chess 267
choice 148
chorus line hypothesis 176
chorusing of crickets 133–4
chromatophores 129, 162
clicking 128
close packing of prey 76–8, 83
colour changes
  against searching image 216, 247
  cephalopods 36, 67, 89, 114
  fish 69, 70, 129
  luminescence 112, 114
  mammals 71
  orientation 129
  relation to immobility 108
  unexpected onset 132
  wing and thigh patches 130
  zigzagging 37, 67
combinatorial selection 23
comedy 267–68
commensalism 24
communication channel 142, 149, 159
competition 24, 25, 193, 210
complexity 263
compromise activities 143
concealed spin 26
concealment 30, 86, 199, 235
conflict 142–59, 183, 233, 251, 263
  intensity 147
  in prey 164
conformity 278
confusion 83, 140–59
  colour change 85
  immobility displays 192
  information overload 265
  multiple erratic 75, 90
  searching image 231
  tail streamers 258
  untied erratic 77, 82–4
  vocal mimicry 241
control systems 33
convulsoid behaviour 8, 111–15, 117, 235, 250
  audiogenic seizure 36, 188
  begging displays 113
  causation 182–8
  human 113
  inheritance 136
  loss of discrimination 186
  novelty 147
  protean elements of 128, 138
  reptiles 114
  single erratic 64
  situation 33, 182–3
cooperation
  hunting 198–9, 214
  mobbing 174
  Prisoner's Dilemma 256
coordination
  bird flocks 79–82, 175–6
  fish schools 85, 175
crash dive 6, 86, 87, 88
crazy-dances 62-4, 115
  model 184–7, 205
crazy-flight 49
crime 265
crypsis 25, 26
  after flash displays 37, 130–1
  anarchization 250–2
  association with immobility 37, 108, 122, 124, 182
  clumped dispersion 212
  colour change 70
  convergent evolution 15
  denial of stimuli 34, 107, 141
  detection distance 209
  disruptive coloration 250–2
  flattening 61
  interspecific diversity 233–4
  on leaf edges 218
  orientation of fish 129–130
  overcome by predators 35, 157
  polymorphism 221, 223, 228
  scattered dispersion 207–10
  searching image 216
  special resemblance 29–30, 89–91, 238
  stripes 85
  terrestrial habitats 214, 227
  territoriality 211
cryptic fragility 217, 248, 250
cultural evolution 10
culture shock 264
cybernetics 13

death feigning 33, 110, 117, 191–2
deception 20, 110
defecation 106
defence integrator 165, 166
defensive reaction 147
deflection 257–61
deimatic effect 68, 70
delinquency 265

denial of key stimuli 107, 108, 114, 117
deployment inconsistency 250, 252
derived erratic displays 72–117
descent in birds 51, 52
detection
  factors determining 213–14, 252
  frequency 207, 216
deterrence
  convulsoid displays 8, 28
  mobbing 100
  predatory resemblance 125
  united erratic 76, 133
  vocal mimicry 242
Dinka herders 278
direct detection distance 209, 214
directing responses 144, 188
directing stimuli 177–8, 180, 183, 204, 205
disinhibition 165, 186, 205
disorder 12, 13
dispersion 86, 206–9, 240
  changed by predation 210–11
  clumped 212–14, 247, 250, 252
  covert 249, 250
  crypticity and scatter 116, 209, 250
  egg-laying patterns 209–10
  missiles 267
  wind scattering 209
displacement activities 63, 143, 163, 183, 184
display diversification 240–2, 248–9, 250–2
  moths 233
  vocal mimicry 241
disruptive colouration 172, 250–2
distastefulness 27, 133, 217, 218
distraction displays 35, 92–107, 117, 250
  altruism 253
  autotomy 93–5, 159, 217
  bustards 39
  colour flashes 71
  definition 93
  model 164–9, 203–4
  primates 95
  tail extensions 260
  value of grouping 212
distress calls
  see screaming 125
diversity
  interspecific 206
  intraspecific 193, 206
  serial 250
  simultaneous 250
diurnal movements 59
dodging 2, 26
dribbling 66
drive conflict
  in injury feigning 163–9, 203–4
  in mobbing 170–1
duetting 134

early experience 272–3
echo-location 150–2
egg
  aspect diversity 234–5
  laying 209–10
  polymorphism 231–2
electrocution 20
electroencephalogram 191
elements of unpredictability 115, 119–39
emotional withdrawal 265
endoparasites 242
endorphins 190
endurance 26
enkephalins 190
entropy 12, 13
epigamic displays 48–51, 175–6, 258
epileptic fits 8, 113
erection of hair 124
ergonomics 261
escape drive 177–83, 186, 204–5
ethology 16, 17
euphoria of gadgets 18
evolution 15, 161–2, 169–74, 192–4, 203–5
  behavioural diversity 239–40
  capture mechanisms 158, 194
  convergent 85, 140, 169, 181
  emancipation 163–4
  erratic flight 188
  grouping behaviour 82, 212
  human 269–71, 273, 276–80
  immobility displays 188–92
  moths' ear 150
  of protean categories 115
  psychosocial 280
  reciprocal altruism 255–56
  scattered dispersion 208
  social organisation 173–4, 176
evolutionarily stable strategy 229, 230
evolutionary machine 274
expectation 109, 136, 164, 233
experimental reasoning 18
eye spots
  deterrence 26, 28, 34, 107
  experimental evidence 144
  falsification of information 141
  overcome by learning 35, 157

fainting 192

falling randomly 222
fear 145
feint 267
flares 258, 260
flash coloration
  enhancing crypsis 130
  experimental studies 152
  intraspecific functions 158
  motion changes 67, 84
  overcome by learning 157
  skipping lepidopterans 37, 111, 128
  startle and mood effects 125, 145
  transformational aspect 137
fleeing 37
flight streamers 257–9
football 66, 122, 267
form changes 124–5, 138
  illusion 128, 131
  immobility 108
  searching image 216, 247
fossorial habits 20
frequency-dependent selection 109, 220, 224–6
frequency of deployment 239

gambling 267, 278
game theory 15, 157, 256
gape 194–5
gene flow 13
generalization 147
genetic fitness 253, 257
Goons 268, 276
grooming 184
group-decision mechanism 82
group selection 245
guerrilla warfare 9
gull colonies 105–6

habitat selection 38
habituation 35
handling time 217
heat-seeking missiles 258
heterosematism 220–2, 248
hierarchical organization
  behaviour mechanisms 165–9, 184–6
  life 16
hijacks 203
HIV 244
hissing 128
homeostatic mechanisms 16
hopping 61, 108
human
  brain 273, 274–7
  decision making 261–3
  evolution 269–71, 273, 276–80
  versatility 270, 272, 279, 280
humour 276, 279
hunter–gatherers 277–9
hunting in groups
  benefit of experience 212
  countering jinking and scatter 193, 197–202
  increased visual range 214
  predatory success 26, 27, 46
hydrodynamic efficiency 83
hysterical fits 8, 113

immobility 33, 107–11
  denial of key stimuli 34
  in predators 25, 26
  in prey 107–11
  interrupting motion 57, 120
  mechanism and evolution 188–92
  overcome by learning 35
  protean aspects 37
  special postures 121
  sudden action 108–9, 188, 205
immunity 31, 242–4
inclusive fitness 255
incongruity 148, 156
inconsistency 263
inertia 122
inflorescence mimicry 89–90
information 149, 155, 262–3
  entropy 13
  noise 13, 157
  overload 263, 265
  theory 142, 146
  transactions 16
injury feigning 7, 8, 95–9, 117, 120
  action on predator 159
  changes in speed 120
  definition 95
  evolution 162–3, 169–70
  lure and prize 93
  mechanism 99, 163–9
ink clouds 31, 67, 89, 94
insect nervous system 178–80
insectivorous birds 178, 188, 219
integration 273, 275–7, 279
integrators 184–5
intelligence
  biological 274–80
  factors 275
  IQ tests 273
  multiple intelligences 277
intelligence system 9
interfemoral membrance 194, 196
interprey distance 208–9, 213, 225, 251

interrupted motion 120–2
Inuit hunters 278
irregularity-generating mechanisms 204–5
irridescence 85, 126–7
irruptions 245

jinking 2
  avoidance of fast predators 193, 202
  sport 66, 67
  structural modifications 47

Karate 5
Key stimuli in predation 34
Kin selection 217
Krill 195, 197
Kung of the Kalahari 277

language 271–3, 279
lateral-line of fish 85
leaping
  causation 180–1
  click beetle 126
  colour polymorphism 219
  disturbed arthropods 59, 112, 235
  mammalian herds 39, 45, 75
  single erratic 45
learning
  aposematism 217
  blocking by flash display 152
  cryptic fragility 217
  experience predators 146
  interception by bats 58, 152
  intraspecific diversity of colour 225, 232
  limited by protean effects 156–8
  overcoming systematic defences 35
  polyethism 237
  skills 264
  unlearning 276–7, 279
  varied defences in spider webs 236
learned countermeasures 117, 157
lek 228–30
levels of biological organization 23
lizard's tail 93, 94
locatory calls 100
locomotory mode changes 131–2, 138
  in fleas 120–1
  in injury feigning 98–9
luminescence
  anteroposterior differences 131
  attention-drawing effect 159
  berserking 112, 133
  cephalopod ink 31, 67, 94
  crypticity 131, 217
  directional changes 69
  scales 94, 131
lurchers 202
lures
  acoustic 133–4
  associated with leaping 71
  autotomy 71, 93–5
  compact ink clouds 31
  conspicuity 93, 131
  effect of multiple lures 145–6
  illusion of transformation 131
  injury feigning 97
  luminescent 94
  mobbing 99, 117
  primates 95
  racquets and tail streamers 258–60

malaria 243, 248
manoevrability 26, 47, 122
Marilyn Monroe 268
martial arts 267
mathematical models 17, 18
metamorphosis 29
microtymbal organs of moths 128, 151
migration
  aspect diversity 234
  irruptions 245
  mobbing during 106
military strategy 203
mimicry
  Batesian 27
  behavioural 28, 117, 164
  calls of predators 107, 117, 242
  Müllerian 27, 234
  non-vocal sounds 106
  vocal 106–7, 188, 241–2
  vocal diversity effects 242, 248
mobbing 4–5, 100–6
  anarchization 204
  by a single individual 101, 104
  causation and evolution 170–4, 212, 256–7
  class demonstration 5
  compared to multiple erratic 73
  effect on mood 145
  evidence of effectiveness 100, 155
  intraspecific responsiveness 100, 176
  killing snakes 101
  mock attacks 100
  protean nature 36, 100–1, 117, 133
  reciprocal altruism 254, 256–7
  turbulence 104

Mood-induced changes 37, 145
moth-bat interactions
  arctiid clicks 151–2
  causation of erratic flight 179–80, 204
  survival value 149–150
  variation in single erratics 58–59, 123
motivational change in predators 145–7, 250–1
motorway matrix signals 268
multiple erratic display 3–5, 72–6, 116, 250
  aerial 74, 76
  effect on gregariousness 73, 155
  fish in pools 3–4, 75
  in herds 75
  irregularity in sandhoppers 178
  space restriction 39, 74
mutations 12, 271
mutualism 24

natural selection
  adaptedness 18
  autiprotean devices 194
  compared with intelligence 274, 276
  conflict induction 159
  Darwinian approach 15–9
  death feigning 191
  erratic behaviour 181
  hornbills killing snakes 173
  human evolution 273, 280
  injury-feigning intensity 97
  linguistic communication 273
  moth's ear 150
  parallel solutions 10, 15, 21
  producing protean effects 14, 41–2, 205
  survival value 159, 180
  tail streamers 258
  unaddressed variation 252
  warning calls and mobbing 252
  whole-organism focus 23, 271
nervous breakdowns 264
nest dispersion 211
nets 20
niche 23–4, 279
  hindwing diversity 234
nidicolous broods 88
nidifugous broods 88
nocturnal habits 20
noise in offices 266
North American Indians 277
novelty
  attention-drawing properties 131
  effect on insectivorous birds 144
  increased reaction time 144, 149
  inducing conflict 146–7, 149, 159
  naïve opponents 146–7
  overcome by learning 157
  transformation 147
number changes 138–9
number of stimuli
  conflict induction 144, 159
    protection by schools and swarms 153–4
nuptial flights 194
nut crop variation 244–5

oligophagy 233
optimal foraging theory 153, 207, 231
orientation mechanism 37, 145–7, 250–1
orientation reaction
  arrest of actions 147, 184
  causation 144, 147
  characteristics 147
  induced by convulsoid displays 183–4

pain 188–91
paired-associate learning 156
parachutes 20
parasite–host interactions 24
  blood parasites and HIV 242–4
  insect ectoparasites 39–41
  parasitic cuckoos 232
parasitoids 210
partnership 24
perception 141, 262
  confusion of 148, 233–4, 250–1, 258
  species differences 137, 274
  speed of recognition 148, 216
periodic cicadas 246
philosophical difficulties 12
pilomotor activity 186
play-fleeing 65–6, 180
play-pronking 65, 181
playing dead 108–9, 191
playing possum 191
poison 20, 31–2
poisonous spines 31
poker 267
polyethism 37
  aposematic displays of *Arctia* 135–6
  death throes of grass snake 110
  display diversification 240–2, 248, 250–2
  diversity of response 239
  element of unpredictability 138
  immobility 108

overcoming learning 237–9
rodent convulsions 135–6, 237
polymorphism
anarchization 172
apostatic 219–33, 248, 250–2
behavioural freedom 227
biochemical 242
crypsis 218–19, 224–8
cuckoos and predators 232
display diversification 225, 240–1
effection on dispersion 31
females 136–7, 219, 232
genetics 220, 226
individual recognition 229
land snails 220, 222–3, 225
mimetic 28
overcoming learning 136, 206
population density 220
ruffs 228–30
searching image 218–20
tropical 224, 226
polyphagy
prey's scattered distribution 210
producing interspecific display diversification 239
variations in immobility displays 192
population control 22, 25
population genetics 15
power to mass ratio 122
predation 24
energy costs 201–2
predator-specific defences 136, 170
predatory strategies 25–32
predictability
fundamental limit 158
importance in modern society 11
number of possibilities 157–8
placement of prey 211
preparatory responses 145
vulnerability to countermeasures 2
predictive tracking 145, 158
preparatory responses 145
prey density
changes 133
predatory success 153–4, 156
prey preference 216
primary protean displays 41, 76, 116
Prisoner's Dilemma 174, 256
prize 93, 117
probability 11–3, 142
production booms 244–6, 248
pronking
altruism 254, 257
anticipatory arousal 180–1, 205
confusion of predators 45–6
dorsal signal line 45, 182
evolution 181, 205, 254
intraspecific display 46, 158
protean behaviour
basic survival value 149–53
basis of validation 140–1
classification of displays 41, 116–7
definition 35–6, 40
intraspecific signals 158
predators' countermeasures 40, 89, 193–203
protean dispersion 86
see dispersion, covert
protean insurance
birds losing height 51–2
breeding displays 49–51, 107, 188
crustacean locomotion 60
diurnal habit 58
evolution 188
insects erratic flight 56–8
leaping when undisturbed 45
side-slipping flight 45
vocal mimicry 107, 188
protean phenomena
summary 249–51
Proteus 36
punishment and reward 264–5
pursuit predators 26–7
pyramid of numbers 24

randomness
contrast to clumped covert dispersion 214
entropy 12–3
relationship to 'protean' 36, 137, 214
scatter hoarding 212
racquets 258
rare behaviours 70–1
reaction diversity 250
see repertoire of escape displays
reaction times
alternative stimuli 230, 263
anarchization and display diversification 251
avian visual 82, 176
elements of unpredictability 119, 138, 181
experimental evidence 144–9, 262–3
flock coordination 80–2
human 263–4
neurological conflict 146, 159
in prey 26, 82
relationship to errors 262
recognition

individuals 241, 255–6
nest parasites and predators 232
prey 29–30, 233, 251
redirection
aggression 265
caused by conflict 184–5
'pounce' of crazy-dance 184–5
reflection of light 214
reflexive selection 223, 225, 248
regeneration 93–4
relative position changes 133, 138
inducing conflict 145, 154
releasers 172
releasing mechanisms 33–4, 165–7, 185
evolutionary aspects 34
releasing stimuli 166, 168, 177, 185
repertoire of escape displays
causes of deployment 135
display diversification 206, 250–2
element of unpredictability 135–6
predator-specific displays 136
uncertainty and conflict 146
variation in deployment 146, 239
web-building spiders 236
resemblance 30
resilin pads 121
reversal of body 35, 257, 260–1
reward for effort 266
Rikki-Tikki-Tavi 101
risk to prey 215, 233, 257
ritualization 15, 172, 204
rodent-run displays 131, 163
role changes 139, 145
rugby football 66, 267
rump patch and back displays 71

scatter 6, 85–92, 250–2
avian and mammalian broods 87–8
changes in number and density 132–3
closer to predator 178–9
clumped dispersion 212–4
conflict in directing responses 144
crash dive 87–8
fish 75, 86, 88, 132
inflorescence mimicry 90
innate 88
invertebrates 89
naval warfare 267
predator very close 38, 85
predators' countermeasures 89, 197
random dispersion 116
sandhoppers 90, 92, 178
scatter hoarding 212
scent polymorphism 219, 227
scent trails 72
schooling fishes 82–5, 132–3, 153, 213
scientific method
awareness of anomaly 247
diversity of approaches 14
ill-attuned to irregularity 2, 206
screaming
captured passerines 70, 108, 125
rabbits and bats 70
startle effect 108
variation and sporadic use 125, 136
searching image 29, 207, 215
biome maturation 193
bizarre shapes 236–7
central perceptual changes 216
cryptic fragility 217
dispersion 212, 216–7
ecologically defined 215
flash colouration 130
heterosematism 220–3
interspecific diversity 233–5, 240
oligophagy 233
polymorphism 218–20, 223–8, 230–2
protean countermeasures 247–51
secondary protean displays 42, 116
selective predation
colour polymorphisms 219, 223–8
birds' eggs 231
origin of bounding displays 181
origin of erratic flight 188
reanalysis of Popham's data 223–4
selfish gene 22
selfish herd 82
semantic openness 273
sensory attunement 26, 34
Serengeti ecosystem 46
sexual dimorphism 54
sexual recognition 80
sexual selection 231
shape changes
see form changes
compared with leaping in herds 74–5
shuffling in groups
fish 83–4
mammals and birds 39, 76–7
side-slipping 45, 51
sideways movement 71
silvery coloration 84, 129–30, 154
single erratic display 2–3, 44–72, 116
against host grooming 120–1
anarchization 250
avian 45, 48–52
component of other displays 115

conflict induction 145–6
crustacea 60–2, 197
elements of unpredictability 108, 120, 122, 132
evolution 181, 188, 193, 205
fish 52–4, 199
individual differences 57
insects 55–60, 149–51, 179–81, 196
mammals 2, 45–7, 132, 201–2
mechanisms 179–80, 213
on plains and tundra 44–7
predators 62–4
small tortoiseshell 37
skill 145, 148, 230
buffer stores and reaction time 263–4
skipping
butterflies 37–8, 121, 152, 252
moths 8, 33, 111–2, 128, 147
sky-pointing 41–2
sleeping sickness 243, 248
smoke-screens 20, 31
snake bites 31–2
snow 124
social behaviour 22
see cooperation
social insects 253
social signals in fish 129
sonar of bats 149
sound production
fish 70, 129
insects 127–9, 134–5
multiple erratic of impala 75
passerine distress calls 125
rugby 67
spacing-out against predaton 208, 210–2
see dispersion
spatial restriction 64, 74
spears 200
speciation 234
speculum of ducks 80, 175–6
speed 20, 26
changes 120–2, 132, 138
predators 26, 124, 193, 201–2
prey 54, 108, 154
spines 20, 31
spinning 57, 61–2
splitting of flocks and herds 77, 79
startle effect
anomaly 233
berserking 112
complex transformations 127–8, 251
convulsoid displays 113
cuttlefish colour changes 68, 113
eye spots 125
flash display 125
origin of bounding displays 182
screaming 70, 108, 125
stealth 20, 25, 193
stimulus complexity 128, 131, 146
stinging cells 94
stings 31
stink glands 32
stotting 45, 181, 254, 257
strategy 21
stridulation 134
stripes 85
struggling 108–9
suboptimal stimuli 184
subversive activities 9
see terrorism
surprise 20, 26, 146, 199
swim bladder 162
swirling displays 114
systematic phenomena 11

tactics 21
tail in turning 201
teleonomy 15
territorial displays
behavioural conflict 143
duetting in birds 135
protean insurance in birds 48–51, 74
vocal mimicry 241–2
territorial disputes
erratic fleeing 54, 65
territoriality and protean dispersion 211
terrorism 203, 267
thanatosis 107
see immobility
thermodynamics 12, 14
Three Mile Island 262
timing
changes inducing conflict 144–6, 159
element of unpredictability 138
tit for tat strategy 174, 256–7
transformation
of apparent species 131, 133
effects 147, 250–2, 263
relationship to randomness 137–8
transparency
crypsis in open water 214
crypsis in pools 61
traps 20
tumbling in pigeons 188
turacin 131
turacoverdin 131
'twitch' in horses 190
tympanic organ of moths 149–51, 179, 205

ultrasonics

acoustic jamming 152
arctiid clicks and palatability 151–2
bats' efficient prey-interception 150–1, 195
moths erratic-flight mechanism 179–80, 204–5
reactions of moths 58–9, 123
survival and moth's ear 149–50
Umwelt 137
unaddressed variations 252
uncertainty
anticipation of alternative stimuli 146
culture shock 264
entropy 13
experienced predators 146
military strategy 266
novelty 147
stimulus irregularity and number 146
tolerance 276
unsystematic response-deployment 146, 156, 206
undulating flight 176–7
uniformity within groups 154–5
united erratic display 76–85, 116
anarchization 176
birds 76–82, 175–6
chorus line' hypothesis 176
conflict induction 145–6, 250
fish 74–5, 82–4
flock tightening 176
original reasons for grouping 212
predator specificity 170
risk reduction not division 82–3
shuffling 83–4
synchronization 77, 79–80, 85
visual coordination 174–6, 204
wing bar and speculum 80–2, 175
unpalatability 151–2
unpredictability 2, 11
anarchization 172–3
class demonstration 5
communication with people 268
conflict induction 146
elements 84, 119–39
faced by contemporary man 11, 264, 269, 278–9
game theory 157
immobility 108
injury feigning 8, 99
interest in gambling 278–9
mobbing attacks 101, 256
organisation of neurological conflict 170–1
polyethism and display deployment 239
relationship with randomness 11, 36
scatter 86–7
subversive activities 9
within structural system 54, 157
unprofitable prey hypothesis 231
upper and lower surfaces 222

vacuum activities 184, 186
visual reflexes 80, 176
visual search 213–4
vocabulary 272
vocalizing unexpectedly 67

waltzing behaviour 188
warfare 9, 266–7, 279
Waorani of Ecuador 278
warning coloration
chemical defences 27, 32, 135
dense packing 133
mimicry 27
warning signals
armament race 20
altruism and alarm calls 253–4
pronking 46
pursuit predators 26
water surface 123–4
webs 32
which-hand game 157, 222
whiffling 51–2, 72
whirling 123, 145, 178
wind-dispersed insects 56
wing bars of waders 80–2, 154, 175
wing scales 32

zigzagging
absence of directing stimuli 178
aerodynamics 176–7
birds 48–52
colours and sounds 67–71
conflict induction 145
crustaceans 61
distraction display 93, 168
endogenous 132
fish 3, 53–4
immobility 37, 108
insects 55–60
limits of predictability 157
mammals 3, 44–7, 201
natural selection 186
plains, steppe and tundra 44
play 65–7
protean nature 37, 119
united erratic 76, 83–4